Methods in Neurosciences

Volume 10

Computers and Computations in the Neurosciences

Methods in Neurosciences

Edited by

P. Michael Conn

Department of Pharmacology
The University of Iowa
College of Medicine
Iowa City, Iowa

Volume 10
Computers and Computations
in the Neurosciences

ACADEMIC PRESS, INC.
Harcourt Brace Jovanovich, Publishers
San Diego   New York   Boston   London   Sydney   Tokyo   Toronto

Front cover photograph (paperback edition only): Nucleus basilis of Meynert.
The coordinates of the neurons have been plotted with an image analyzer (Biocom
200). Colors were attributed to each polygon according to its area. Photo courtesy
of Professor Charles Duyckaerts, Hôpital de la Salpêtrière, Paris, France.
For more detail see the legend for the lower panel of Figure 6 of Chapter [32].

Academic Press, Inc.
1250 Sixth Avenue, San Diego, California 92101-4311

United Kingdom Edition published by
Academic Press Limited
24–28 Oval Road, London NW1 7DX

International Standard Serial Number: 1043-9471

International Standard Book Number: 0-12-185269-5 (Hardcover)

International Standard Book Number: 0-12-185270-9 (Paperback)

PRINTED IN THE UNITED STATES OF AMERICA
92 93 94 95 96 97   EB   9 8 7 6 5 4 3 2 1

Table of Contents

Section I Data Collection

Section III Data Modeling and Simulations

Contributors to Volume 10

Article numbers are in parentheses following the names of contributors. Affiliations listed are current.

PETER ÅRHEM (17), Nobel Institute for Neurophysiology, Karolinska Institute, S-104 01 Stockholm, Sweden

U. ARVIDSSON (7), Department of Anatomy, Karolinska Institute, S-104 01 Stockholm, Sweden

PETER H. BARRY (29), School of Physiology and Pharmacology, University of New South Wales, Sidney, New South Wales 2033, Australia

PASCALE BOULENGUEZ (8), Laboratoire de Neurobiologie, Centre National de la Recherche Scientifique, 13402 Marseille, France

MICHAEL BRUNNER (3), Fachbereich Biologie, Universität Kaiserslautern, D-675 Kaiserslautern, Germany

KEVIN BUSH (30), Department of Molecular Biology, School of Osteopathic Medicine, University of Medicine and Dentistry of New Jersey, Stratford, New Jersey 08084

JAMES P. BUTLER (16), Department of Environmental Science and Physiology, Harvard School of Public Health, Boston, Massachusetts 02115

K. CARLSSON (7), Department of Physics, The Royal School of Technology, S-100 44 Stockholm, Sweden

L. CHARRON (23), Departments of Medicine and Physiology, University of Montreal, Montreal, Quebec H2W 1T8, Canada

BRENDA J. CLAIBORNE (19), Division of Life Sciences, University of Texas at San Antonio, San Antonio, Texas 78249

J. H. COCATRE-ZILGIEN (14, 15), Department of Entomology, University of Illinois, Urbana, Illinois 61801

G. COLLIOT (5), INSERM, U314, Hôpital Maison Blanche, 51092 Reims, France

R. JOHN CORK (25), Department of Biological Sciences, Purdue University, West Lafayette, Indiana 47907

G. CORKIDI (5), Centro de Instrumentos, Universidad Nacional Autonoma de Mexico, Cd. Universitaria, C. P. 045510, Mexico D.F., Mexico

FRANS CORNELISSEN (18), Department of Physiology, Janssen Research Foundation, B-2340 Beerse, Belgium

C. COSTA (32), Department of Neurology, Santa Maria Hospital, 1600 Lisbon, Portugal

P. DELAÈRE (32), Laboratoire de Neuropathologie Escourolle, Hôpital de la Salpêtrière, 75651 Paris, France

F. DELCOMYN (14, 15), Department of Entomology, University of Illinois, Urbana, Illinois 61801

VICTOR H. DENENBERG (9, 10), Biobehavioral Sciences Graduate Degree Program, University of Connecticut, Storrs, Connecticut 06269

C. DUYCKAERTS (32), Laboratoire de Neuropathologie Escourolle, Hôpital de la Salpêtrière, 75651 Paris, France

SVERKER ENESTRÖM (6), Department of Pathology, Linköping University, S-581 85 Linköping, Sweden

WILLIAM S. EVANS (16), Department of Medicine, University of Virginia Health Sciences Center, National Science Foundation, Center for Biological Timing, Charlottesville, Virginia 22908

M. FREIRE (20), Instituto Cajal, E-28002 Madrid, Spain

HUGO GEERTS (18), Department of Physiology, Janssen Research Foundation, B-2340 Beerse, Belgium

GEORGE L. GERSTEIN (4), Department of Neuroscience, University of Pennsylvania School of Medicine, Philadelphia, Pennsylvania 19104

JAMES L. HARGROVE (11), Department of Foods and Nutrition, University of Georgia, Athens, Georgia 30602

DIANE K. HARTLE (11), Department of Pharmacology and Toxicology, College of Pharmacy, University of Georgia, Athens, Georgia 30602

J.-J. HAUW (32), Laboratoire de Neuropathologie Escourolle, Hôpital de la Salpêtrière, 75651 Paris, France

E. C. HIRSCH (5), Laboratoire de Médecine Expérimentale, INSERM, Hôpital de la Salpêtrière, 75651 Paris, France

WILLIAM R. HOLMES (31), Department of Biological Sciences, Ohio University, Athens, Ohio 45701

CHRISTIAN HOOCK (3), Fachbereich Biologie, Universität Kaiserslautern, D-675 Kaiserslautern, Germany

DALE HUFF (30), Department of Pathology, Magee Women's Hospital, University of Pittsburgh School of Medicine, Pittsburgh, Pennsylvania 15213

CLAIRE E. HULSEBOSCH (21), Department of Anatomy and Neurosciences, Marine Biomedical Institute, University of Texas Medical Branch, Galveston, Texas 77555

MARTIN G. HULSEY (11), Department of Foods and Nutrition, University of Georgia, Athens, Georgia 30602

STAFFAN JOHANSSON (17), Nobel Institute for Neurophysiology, Karolinska Institute, S-104 01 Stockholm, Sweden

MICHAEL L. JOHNSON (16), Department of Pharmacology, University of Virginia Health Sciences Center, National Science Foundation, Center for Biological Timing, Charlottesville, Virginia 22908

STEPHEN W. JONES (28), Department of Physiology and Biophysics, Case Western Reserve University, Cleveland, Ohio 44106

GERRY H. KENNER (9, 10), Thermoluminescence Laboratory, Division of Radiobiology, University of Utah, Salt Lake City, Utah 84112

UWE T. KOCH (3), Fachbereich Biologie, Universität Kaiserslautern, D-675 Kaiserslautern, Germany

DIMITRI M. KULLMANN (22), Department of Pharmacology, School of Medicine, University of California, San Francisco, San Francisco, California 94143

HSIN-YI LEE (30), Department of Biology, Rutgers University, Camden, New Jersey 08102

O. LEJEUNE (5), INSERM, U289, Hôpital de la Salpêtrière, 75651 Paris, France

ERICH LIETH (24), Department of Neurobiology, Harvard Medical School, Cambridge, Massachusetts 02138

J. P. MESSIER (23), Departments of Medicine and Physiology, University of Montreal, Montreal, Quebec H2W 1T8, Canada

K. MOSSBERG (7), Department of Physics, The Royal School of Technology, S-100 44 Stockholm, Sweden

PETER J. MUNSON (12), Analytical Biostatistics Section, Division of Computer Research and Technology, National Institutes of Health, Bethesda, Maryland 20892

ROBERT G. NAGELE (30), Department of Molecular Biology, School of Osteopathic Medicine, University of Medicine and Dentistry of New Jersey, Stratford, New Jersey 08084

RONY NUYDENS (18), Department of Physiology, Janssen Research Foundation, B-2340 Beerse, Belgium

ROGER NUYENS (18), Department of Physiology, Janssen Research Foundation, B-2340 Beerse, Belgium

J. M. PEYRONNARD (23), Departments of Medicine and Physiology, University of Montreal, Montreal, Quebec H2W 1T8, Canada

ROBERT PINARD (8), Laboratoire de Neurobiologie, Centre National de la Recherche Scientifique, 13402 Marseille, France

PIERRE RAGE (8), Laboratoire de Neurobiologie, Centre National de la Recherche Scientifique, 13402 Marseille, France

G. ENRICO RAVATI (12), Laboratory of Molecular Pharmacology, Institute of Pharmacological Sciences, University of Milan, 20133 Milan, Italy

JOHN ROSS (33), Department of Chemistry, Stanford University, Stanford, California 94305

JEAN P. ROYET (26), Laboratoire de Physiologie Neurosensorielle, Université Claude-Bernard, 69622 Villeurbanne, France

LOUIS SEGU (8), Laboratoire de Neurobiologie, Centre National de la Recherche Scientifique, 13402 Marseille, France

W. TERRELL STAMPS (21), Department of Entomology, University of Missouri—Columbia, Columbia, Missouri 65221

BRIAN R. STROMQUIST (1), Research and Development, Unimax Systems Corporation, Bloomington, Minnesota 55431

STÉPHANE SWILLENS (27), Institut de Recherche Interdisciplinaire, Université Libré de Bruxelles, B-1070 Brussels, Belgium

M. TAJANI (5), INSERM, U289, Hôpital de la Salpêtrière, 75651 Paris, France

MARIO TIBERI (13), Department of Cell Biology, Duke University Medical Center, Durham, North Carolina 27710

WARREN G. TOURTELLOTTE (2), Department of Pathology, Washington University, St. Louis, Missouri 63110

B. ULFHAKE (7), Department of Anatomy, Karolinska Institute, S-104 01 Stockholm, Sweden

JOHANNES D. VELDHUIS (16), Department of Medicine, University of Virginia Health Sciences Center, National Science Foundation, Center for Biological Timing, Charlottesville, Virginia 22908

LIPO WANG (33), Neural Systems, National Institutes of Health, Bethesda, Maryland 20892

Preface

The speed with which data must be collected and the complexity of its analysis have made computers an absolute necessity in the neurosciences. Computers, in addition to being used for collecting and analyzing data, have been very useful for constructing models of function at all levels—from gene expression to behavior.

This volume is divided into three sections describing, in a pragmatic manner, data collection, analysis, and modeling. The chapters are written in a way that will allow readers to adapt the technology described for the study of other systems. The use of personal computers for these purposes is emphasized.

Methods presented range from the collection of neurophysiological data, through the control of laboratory equipment, the recording and identification of pulses, spikes, and bursts, to cartographic methods and analysis of immunocytochemical and confocal microscopic data. Analysis of the kinetics of gene expression, receptor binding, and behavioral data is described, as is the analysis of graded potential and fast axonal transport, and physical characterization. Modeling and simulations are included for calcium oscillations and the IP_3-sensitive calcium channel, ion gating, membrane potentials, individual neurons, and neuron networks.

The goal of this volume, as well as others in this series, is to provide in one source a view of the contemporary techniques significant to a particular branch of neurosciences, information which will prove invaluable not only to the experienced researcher but to the student as well. Of necessity some archival material will be included, but the authors have been encouraged to present information that has not yet been published, to compare (in a way not found in other publications) different approaches to similar problems, and to provide tables that direct the reader, in a systematic fashion, to earlier literature as an efficient means to summarize data. Flow diagrams and summary charts will guide the reader through the processes described.

The nature of this series permits the presentation of methods in fine detail, revealing "tricks" and short cuts that frequently do not appear in the literature owing to space limitations. Lengthy operating instructions for common equipment will not be included except in cases of unusual application. The contributors have been given wide latitude in nomenclature and usage since they are best able to make judgments consistent with current changes.

I wish to express my appreciation to Mrs. Sue Birely for assisting in the organization and maintenance of records and especially to the staff of Academic Press for their energetic enthusiasm and efficient coordination of production. Appreciation is also expressed to the contributors, particularly for meeting their deadlines for the prompt and timely publication of this volume.

P. Michael Conn

Methods in Neurosciences

Edited by P. Michael Conn

Section I

Data Collection

[1] Neurophysiological Data Acquisition Based on Personal Microcomputer System

Brian R. Stromquist

Introduction

The personal computer (PC) has quickly become a commonplace sight in the laboratory. PCs are used in every facet of the research process, from controlling instruments and gathering data to preparing manuscripts and graphics for publication. Part of the appeal of these machines, which hastened their prevalence, is that they brought to the desktop of the researcher much of the abilities that traditionally were provided by outside centralized service organizations. PCs have brought increased autonomy to those who have put them to use. But as these resources become transformed and distributed as PC hardware and software, integrating them so information flows smoothly from one to another becomes a significant challenge.

The system described here attempts to address this challenge as well as two more: given that a custom application must be written to accomplish a certain task, one should make it run on all future hardware as it becomes available and make it flexible so as to accommodate as many different user requirements as possible. Meeting these challenges is desirable in any line of research. This system attempts to do so for two widely used techniques in electrophysiological research: recording of evoked synaptic potentials and analysis of poststimulus histograms.

Over the years, collection and analysis of such data have advanced from visual comparison and manual calculations of oscilloscope traces to acquisition and analysis by computers. One particular limitation of most of these systems, however, has been the software. Although in many cases the software may be very powerful in its ability to analyze a particular set of waveforms, its flexibility in the final analysis and presentation of data is usually very limited.

The system software operates within a Microsoft (MS) Windows (Microsoft, Inc., Redmond, WA) environment and employs custom software for the acquisition and review of data in conjunction with commercially available software that can be directly linked to the Acquire and Review modules. This combination provides a powerful package not only for automated on-line data acquisition and review but also for further analysis and presentation of electrophysiological data.

The main advantages of the system are threefold. First, it allows for automated, on-line collection and analysis of evoked potentials and poststimulus histograms with a minimum of effort by the experimenter. Second, the extracted data are automatically linked to a commercially available spreadsheet/graphics program (Microsoft Excel). This program allows the data to be further analyzed and graphed. Finally, all results can be output to either a laser printer or plotter. It should be emphasized that data collection, analysis, and all graphical presentations are accomplished on-line. A hard copy of all intermediate and final results can be obtained immediately following the data collection.

System Overview

The system has two types of input channels: analog and histogram. There are four analog and two histogram channels. Each can be triggered independently, or all can share a common trigger. Both pretrigger as well as posttrigger data can be collected. The analog input channels are used for recording field potentials that have been amplified to ± 1 and ± 10 V. The sampling rate is individually selectable and can be one of the following values: 20, 50, 100, 500 μsec or 1, 2, 5 msec. A maximum of 2000 sample points per channel can be collected. The histogram channels are digital inputs and are intended for collecting transistor–transistor logic (TTL) pulses from the output of window discriminators (although pulses of up to 80 V are allowed). The bin width for each histogram channel can be set from 2 to 50 msec with a resolution of 2 msec. Histograms have a maximum duration of 2 sec. All channels can be used simultaneously at a sample rate of 100 μsec or greater. At 50 μsec two analog and two histogram channels can be used. At 20 μsec, one analog and one histogram channel can be used simultaneously.

Hardware

The system hardware, as described, typically centers around a 80386 IBM-AT compatible computer with 4 megabytes of memory, a 40-megabyte hard disk, EGA video adapter, and a color monitor. The only hard requirements for the computer are that it be capable of running Windows and be fast enough to keep up with the data acquisition. A standard IBM-AT has proved to meet these requirements.

Experimental data are acquired through a Burr–Brown PCI-20000 modular subsystem. The PCI-20000 consists of a carrier board which fits into one slot within the computer and up to three module boards that plug into the carrier. The entire subsystem occupies two PC slots. The carrier board provides 32 bits of digital input or output. In the system described, two modules are used. The analog-to-digital converter (ADC) module has 12-bit resolution, 8 multiplexed input channels, and a conversion time of 11.25 μsec. The timer module provides a time bases for the analog channels and digital counters for the histogram channels.

A locally designed interface module provides a front panel for the PC-20000. The interface module is rack mountable and provides BNC connectors for easy connection of analog, histogram, and trigger inputs. For the analog inputs a switch-selectable gain amplifier provides $\times 1$, $\times 2$, $\times 5$, and $\times 10$ settings which can be used to match the input signal range to the ADC inputs on the PCI-20000. The setting of the switch is ready by Acquire and the gain is taken into account for the acquisition and analysis of the data. The interface box also buffers histogram and trigger inputs.

Software

The goal of the system is that it be intuitive to use, relieving the researcher of much of the repetitive aspects of the experiment and presenting results in a variety of meaningful forms. Ideally, the experiment should be configured and run with a minimum of operation on the part of the experimenter.

To achieve this, Microsoft Windows was chosen as the most suitable operating environment. MS Windows is a graphical user interface that handles much of the details of the computer hardware (i.e., display, mouse, printer, etc.). This is an advantage because it allows the system some independence of specific hardware and it is better able to take advantage of newer devices as they become available. Windows also features the ability to have several programs running and sharing the display at the same time. By using Windows Dynamic Data Exchange (DDE) protocol or its clipboard, concurrently running applications can share data on-line. This makes possible a flexible system comprised of several applications that can be used separately or in combination. Since Windows provides standard mechanisms for interprogram communication, commercially available as well as custom applications can easily be integrated into the system. Of the three applications that constitute the current system, two, Acquire and Review, were developed locally, using the Windows System Development Kit Version 3.0. Microsoft Excel was purchased commercially.

Acquire

The Acquire application is responsible for collecting and averaging data according to settings selected by the user. On issuing the run command, data are collected and displayed as graphs. Other commands are then available to

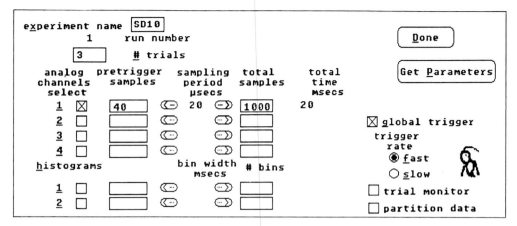

FIG. 1   Parameter Window. This window functions as the "front panel" of the system. With the settings shown, a single analog channel will be recorded. A run will consist of the average of three trials. Each trial will consist of 1000 samples taken every 20 μsec. Forty samples will be recorded before the trigger occurs.

display the data numerically, print them as graphs or tables, save them on the disk, and send them to the other applications (i.e., Review, discussed below).

Configuration of experimental parameters is easily accomplished through the parameter panel, a set of buttons and text boxes that resemble a digital oscilloscope front panel (Fig. 1). Each parameter in the panel can be easily modified by selecting it (either with the mouse or by typing a command sequence) and inputting or selecting the desired parameter. All user input is checked for consistency, and error messages warn of missing or incorrectly specified parameters. After all the parameters are set as desired, the configuration is saved to a disk file. Saved configurations can be retrieved later through the Get-Parameters option.

Following configuration, the Acquire and Review window can be seen. At this point, the user can collect data. Choosing the run command from the menu initiates acquisition (see Fig. 2). In this mode the program waits for the trigger(s), and once detected posttrigger samples are collected, averaged with previous trials, and displayed. This trial cycle (i.e., collect–average data) is repeated for the number of times specified in the configuration setting. By choosing the save-data command the averaged results can now be saved on the disk. The collect–save cycle is fully interlocked; the program will not proceed to the next run until the current data have been saved or discarded. Also, it will not allow the saving of the same data more than once. If, in the

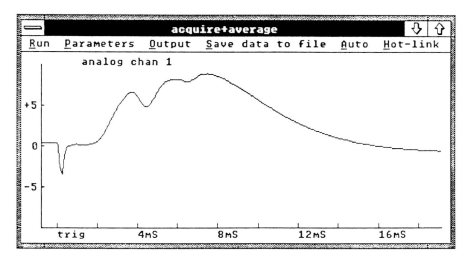

FIG. 2 Acquire Window displaying a hippocampal field potential on analog channel 1. Acquire collects digitized signals from the four analog and two histogram channels and displays the running average with each new trial. The response shown is the average of three trials. All six channels can be displayed simultaneously. Acquire can be run independently or can be hot-linked to Review for analysis of the response.

experiment, data collection occurs at regular time intervals, the collect–save cycle can be automated. By choosing the auto command and specifying the interval and number of runs desired, the experiment can proceed automatically. This has eliminated much drudgery from experiments involving upward of 300 runs. Another feature, the trial monitor, does somewhat the opposite; it allows for the sorting and discarding of trials within a run. The trial monitor, when enabled on the parameter panel, displays each trial as it is collected along with the current average of previous trials. Each trial can be either included in the average or discarded. A further enhancement of this feature, the data partition, allows for each trial to be placed into one of two running averages to be discarded.

A plot may be printed by choosing the print–plot command from the output menu. Additional commands are available for printing and displaying the raw data (ample values) for each plot.

The last feature of the Acquire program is the hot-link. This is the method by which data in Acquire can be shared on-line with the Review program. By selecting a channel name in the hot-link menu any data collected on that channel will automatically be sent to Review in addition to being displayed in Acquire.

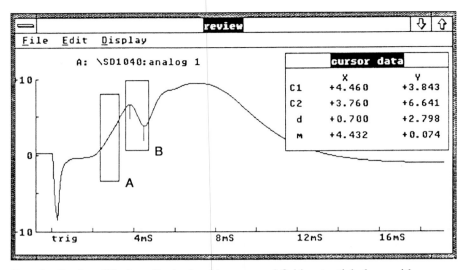

FIG. 3   Review Window displaying hippocampal field potential along with cursors and cursor data window. The two rectangles are positioned and sized by the user. The cursor data window automatically displays an analysis of the regions outlined by the rectangles. C1 and C2 are the coordinate points of the peak and trough within rectangle B. Line d is the absolute difference in volts and milliseconds between peak and trough. The slope of the data covered by rectangle A (determined by linear regression) and its goodness of fit are displayed in line m.

Review

The Review program is used to display and analyze data collected with the Acquire program. This can be done either on-line through a hot-link, or for files saved on disk. Review can operate in either a single-channel or a dual-channel mode. In single-channel mode the chosen data are displayed using the entire window area (see Fig. 3). Cursors can be added and adjusted to provide time, voltage, and optionally slope measurements. In dual mode, runs are selected for display, and the window is subdivided into four graphs: two for the selected data (channel A and channel B) and one each for the sum and difference of the first two. The four graphs displayed in this mode are adjusted so their scales are the same. This facilitates visual comparisons as shown in Fig. 4.

Analysis pertaining to the EPSPs is accomplished by selecting the desired portions of the waveform to be analyzed (using two rectangles designated A and B, shown in Fig. 3). Within these rectangles various calculations per-

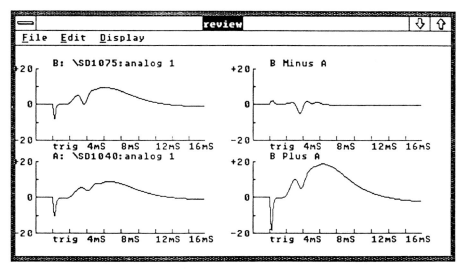

FIG. 4   Review Window displaying two field potential channels, A and B, along with the sum and difference plots B + A and B − A. All plots are automatically given uniform scales. Data cursors (shown in Fig. 3) can be added to any of the four plots. Channel A displays a user-selected saved response to which the on-line responses in channel B are referenced. Off-line comparisons could also be made between any two saved responses.

taining to the waveform are performed. Within rectangle A the slope of the waveform is calculated via linear regression. The results are given in V/msec. A goodness of fit statistic (chi-square, χ^2) is also calculated and displayed in the cursor data window. This is taken into account in accepting or rejecting a particular response. Within rectangle B the maximum (peak) and minimum (trough) values are obtained along with their difference. [This is performed in order to calculate the population spike (see below).] The cursor data window also displays the absolute position and difference of the cursors. The bounding rectangles can be adjusted using the mouse to outline the region of interest. Whenever this happens, the cursors will automatically be repositioned if necessary.

The min and max cursors are convenient because they gravitate toward the most significant features of the region. However, they can also function more generally. Simply by clicking on one and dragging it horizontally, the cursor will ride along the waveform continuously updating its position in the cursor data window. This can be used to determine the timing of a particular deflection within a waveform. The portion of the waveform within either bounding rectangle may be examined in closer detail by double clicking the mouse inside the rectangle. Doing this expands the boxed region to the full plot size. If greater magnification is needed, the process can be repeated. Another double click restores the original view.

Review uses the clipboard to share data with other applications. From within Review, cursor data, file name, run number, the time the run was collected, or an entire waveform can be copied to the clipboard. All or any of these parameters can subsequently be picked by a spreadsheet, drawing or graphing program, or word processor.

Excel

The electronic spreadsheet has become an invaluable laboratory tool. Because of its general purpose numerical analysis and graphic features, the spreadsheet has replaced many previously custom-written scientific applications. These programs have also won favor among researchers owing to the flexibility they allow in entering, editing, and organizing data. It is, therefore, extremely advantageous to develop software which takes advantage of a full-featured spreadsheet.

Powerful as they have become, however, spreadsheets cannot efficiently manipulate data graphically. For example, the interactive cursor features of the Review program are not possible in any current spreadsheet application. Also, data acquisition is rarely performed directly by a spreadsheet. In the development of this system, however, it became apparent that flexibility of data collection and analysis had to be a key feature. The data acquisition, analysis, and output formats preferred by one researcher were completely unsatisfactory to another. It was therefore decided that these capabilities could best be met by incorporating MS Excel into the system.

Excel is a full-featured spreadsheet for Windows with graphic, database, macrolanguage, and interprogram communication capabilities. A set of macrocommands has been developed to facilitate the sharing of data between Review and Excel and to allow Acquire and Review to be controlled through Excel.

Using the System

The system can be used for the acquisition and analysis of a wide variety of evoked potential and poststimulus histogram data. A typical session begins by first activating Acquire and setting the desired data parameters for the experiment. Acquire is hot-linked to Review and run in order to obtain and

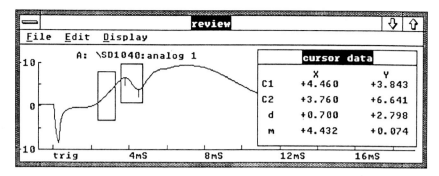

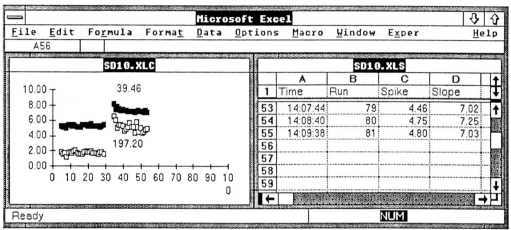

FIG. 5   Run time arrangement of the Acquire, Review, and Excel windows. Acquire (depicted by the icon in the upper left corner) is minimized to avoid cluttering of the screen, but it is still active. Review displays the averaged waveform and feature analysis. Excel shows the tabulated data from Review and a graph of the spike and slope values so far obtained as well as the percent change, calculated in the spreadsheet. All operations concerning Acquire and Review are executed from within the Excel menu.

save a sample response. Review is then activated to recall the response. This is performed in order to set the cursors at the desired locations in the field response. Finally Excel is loaded with a template and initialized for data collection (see (Fig. 5).

The template contains specified, prelabeled cells or groups of cells for the input of various experimental observations (i.e., current level required to induce minimum response, maximum response, intensities to be used for input–output curves, baseline observations, etc.) In addition, certain cells contain user-defined formulas (many of these are built into Excel) for various calculations. These can range from simple descriptive statistics, such as averages and standard deviations, to more complex user-defined computations. The templates are easy to create and can be altered as necessary.

The windows of the three programs are then modified in size and arranged in a configuration that allows for maximum viewing of the field potentials, cursor data, spreadsheet, and graphs (Fig. 5). With the Acquire, Review, and Excel programs ready, data collection is initiated by a trigger switch connected to the trigger input of analog channel 1. This last action begins data collection. The averaged response is displayed by the Review module. Having reviewed the response along with the cursor data, the response can be saved to disk (for later review if required) or ignored. The cursor data are also transferred to Excel, which automatically updates the spreadsheet and the graph. This process is repeated until all data are collected for a particular experiment. At the end of a session, the spreadsheet and graphs are saved to disk. If data reduction is required for specific experimental trials, the data can be accessed through Review. The off-line analysis is identical to the on-line analysis. Automated off-line analysis of the entire experiment or segments of it is also possible.

Conclusions

A microcomputer-based system has been described for the acquisition, reduction, display, and storage of electrophysiological data. The system consists primarily of commercially available hardware and software along with custom-built interface hardware and acquisition and review programs. The present system offers a considerable advantage over other systems since almost all of the required hardware and software packages are sold commercially and may already exist in most laboratories. The Acquire and Review modules as well as the computer interface can be obtained from The Rockefeller University Electronics Laboratory, New York, NY. This combination of software and hardware presents a very cost-effective data acquisition system for neurophysiologists. The main advantages of the system lie in its ease of use (including automation) and flexibility of data manipulation, analysis, and display. It can handle a wide variety of electrophysiological data without any software modification because of its modular design and broad flexibility of its individual components. By fully exploiting the features of the MS Windows environment, these custom modules will run on any PC that is Windows capable and are able to share data with any application that supports the Windows clipboard and DDE, thus giving them added life and value.

[2] Personal Computer-Based Motion Control: A Tutorial and Application Using Simple Motor Controller to Drive Light Microscope Stage

Warren G. Tourtellotte

Introduction

Personal microcomputers (PCs) are an essential component of most modern research laboratories. They have become especially valuable because they are affordable and excellent high-speed data processors, and there are a variety of "ready-to-use" interface devices available from many commercial sources for high-speed data acquisition and analysis. Consequently, one can affordably implement a versatile computer system capable of performing routine laboratory administrative needs as well as data acquisition and analysis with only minimal knowledge of the technical aspects of computer hardware (electronics) or software (computer program) operation.

The PC can also interact with the laboratory environment in an active manner, for example, by controlling motorized devices for the accurate positioning of electrodes, stereotaxic devices, or other laboratory instrumentation. Motor control interfaces designed for PCs are available from a variety of commercial sources, but their application within the laboratory requires skillful connection of a motor together with the proper electromechanical coupling to an appropriate instrument. Moreover, the software required to control precisely the electronic circuitry is often not available in a ready-to-use format that can suitably control the motors for the defined purpose. These difficulties add an undesirable order of complexity to implementing motor-controlled devices, especially for investigators who have a primary interest in making their applications work and not developing it. Nevertheless, situations often arise when commercial systems are not available to meet the needs of a specific application and must be designed either *de novo* or modified from existing technologies.

This chapter is intended to demonstrate some basic principles for PC-based motor control. A detailed description of an electronic circuit designed in our laboratory and used to control the motion of a light microscope stage demonstrates the interaction between its electronic circuitry and controlling software (1). The circuit is not intended to replace the myriad of motor control circuits available commercially, but rather is a workable alternative that has a well-documented circuit diagram and programming characteristics. The latter two features are all too often obscured by manufacturers for proprietary reasons. Finally, the implementation of a motor-controlled light microscope stage is specifically described, although the circuit can be readily implemented for other purposes once its operation and programming characteristics are adequately understood.

Light Microscope Charting System

Analyzing experimental data generated by neural tracing studies, immunocytochemistry, or *in situ* hybridization often requires an initial form of reduction that localizes the marker in neuropil or other tissues. One traditional method for plotting the distribution of such cellular markers in tissue has been to connect modular transducing devices to a microscope stage in order to detect and measure its movement in terms of a voltage that can be interpreted by a standard $X-Y$ recorder (2–4). The motion of the microscope stage is recorded by the $X-Y$ plotter to obtain a macroscopic view of tissue contours and the spatial distribution of labeled cells representing more than a single microscopic field. In the last decade, efforts have been made to digitize the macroscopic plots so that the data are available for quantitative analysis and manipulation by a computer. Several research groups have developed various schemes for encoding the movement of the microscope stage thereby allowing the computer to access the digitized data (5–11). However, this technology has proved difficult to disseminate since many of the implementations are based on outdated and nonstandard computer hardware or difficult to duplicate hand-wired circuitry.

A modern computerized charting system (CCS) based on a standard IBM-compatible microcomputer and a motorized stage will demonstrate a sophisticated application of the motor controller circuit (12, 13). The electronics have been designed to fit onto a standard IBM PC-compatible circuit board that plugs directly into the microcomputer card socket (Fig. 1A). The stepping motors are configured with a threaded screw and mounted to the microscope stage to generate a linear movement as the motor is advanced by discrete rotary steps (Fig. 1B).

Methods

Hardware Design

Overview

The stepping motor circuit is based on the CY512 intelligent positioning stepper motor controller (Cybernetic Micro Systems, San Gregorio, CA). The microprocessors simplify the design and provide stand-alone motor

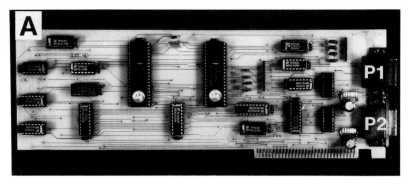

FIG. 1   (A) The motor controller circuit has been placed onto a custom IBM PC-
compatible circuit board. Two connections, P1 and P2, provide access to the joystick
and motor control signals, respectively. (B) The circuit has been used to control the
movement of a light microscope stage. The stage position is digitized by sampling the
discrete motor movements and recorded by the computer to trace tissue section
contours and mark the location of appropriate tissue landmarks and chemical markers.
[Reproduced with permission from W. G. Tourtellotte, *J. Neurosci. Methods* **35**, 157
(1990).]

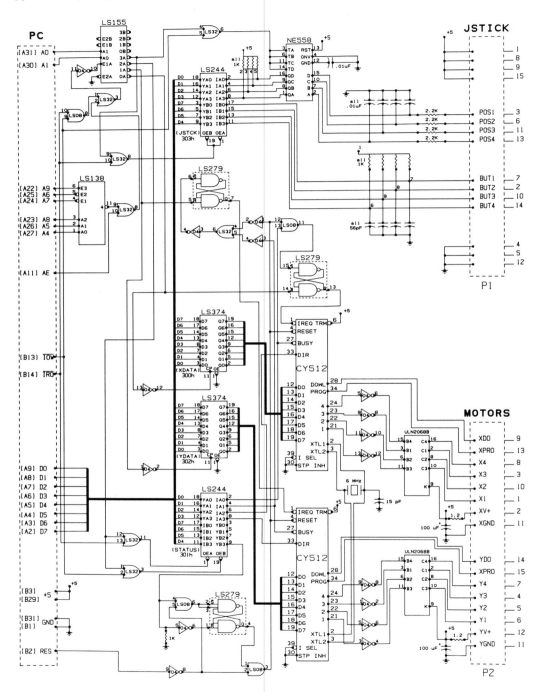

control. Once programmed, each CY512 generates the required timing signals for the prescribed stepping sequence. A list of additional components required to build the circuit can be derived from the wiring diagram (Fig. 2). The components are extremely common low-power Schottky-clamped transitor–transistor logic devices (LS TTL) with the exception of two linear components: the NE558 quad timer and the ULN2068B Darlington power transistor (Spraque Electric Co, Hudson, NH).

The circuit is designed to drive four-phase stepping motors (5-V DC motors, 1 A/phase maximum) since they are very common and affordable. They can be configured for linear tracking (Fig. 1B) or, more conventionally, for rotary motion. In one form or another, four-phase motors can be obtained from a variety of motor manufacturers (e.g., Superior Electric Co., Bristol, CT; HSI Inc., Waterbury, CT; Sigma Instruments Inc., Braintree, MA).

To control the stepping motors interactively, a joystick interface has been provided within the circuit. The wiring conforms to standard joystick connector wiring (e.g., Mach III joystick; CH Products, San Marcos, CA), and the joystick can be plugged directly into the circuit without modification (Fig. 2; connector P1, type DB15, JSTICK).

Theory of Operation

The motor control circuit has two functionally distinct sections accessed by the PC via 4 input/output (I/O) ports. One section is responsible for the joystick analog-to-digital (A/D) conversion logic, and the other is responsible for controlling the stepping motors. A single I/O port is assigned to the joystick interface, and the remaining three are used to submit commands to each microprocessor and to monitor the appropriate status signals.

Joystick Control Circuitry

The joystick can be accessed by read and write operations to I/O port 303 hexidecimal (Fig. 3; 303h, where h means hexidecimal, base 16 representation). The circuit is an adaptation of the IBM standard game control interface (International Business Machines, Boca Raton, FL), although the sensitivity has been substantially increased. The circuit converts the analog joystick position to a discrete numerical format representable by the PC. The deflec-

---

FIG. 2   Wiring diagram for the motor controller circuit. The input signals on the left (PC) are derived directly from the personal computer. The output signals on the right provide connections with the joystick (connector P1) and the motors (connector P2). [Reproduced with permission from W. G. Tourtellotte, *J. Neurosci. Methods* **35,** 157 (1990).]

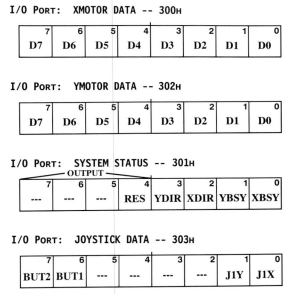

I/O PORT: XMOTOR DATA -- 300H

7	6	5	4	3	2	1	0
D7	D6	D5	D4	D3	D2	D1	D0

I/O PORT: YMOTOR DATA -- 302H

7	6	5	4	3	2	1	0
D7	D6	D5	D4	D3	D2	D1	D0

I/O PORT: SYSTEM STATUS -- 301H

OUTPUT

7	6	5	4	3	2	1	0
---	---	---	RES	YDIR	XDIR	YBSY	XBSY

I/O PORT: JOYSTICK DATA -- 303H

7	6	5	4	3	2	1	0
BUT2	BUT1	---	---	---	---	J1Y	J1X

FIG. 3 Four 8-bit input/output (I/O) ports are used by the host computer software to program the motor controller circuitry. Each small box represents a single bit location and the assigned function. Write operation to I/O ports 300h and 302h send instructions to the X and Y motor microprocessors, respectively. Read and write operations to I/O port 301h either reset the entire controller circuitry (RES) or read the current motor direction (XDIR, YDIR) and the status of the microprocessors (XBSY, YBSY). Read instructions from I/O port 303h provide information regarding joystick position (J1X, J1Y) and button depression (BUT1, BUT2). [Reproduced with permission from W. G. Tourtellotte, *J. Neurosci. Methods* **35,** 157 (1990).]

tion of the joystick is linearly represented by variable resistors coupled to the movement along each axis. When the resistors are placed in the circuit (i.e., the joystick is plugged into the socket), and changed accordingly, the time constants of the resistor–capacitor networks connected to the NE558 are altered (Fig. 2; NE558 input C and D). Once each timer is started, the time duration is characteized by the time constant, and hence the absolute joystick position. Converting the analog joystick position to a number is then a matter of starting the NE558 timer for both X and Y axes and counting the time it takes for each timer to "time-out." Two numbers representing the current X–Y position of the joystick are subsequently obtained. With the appropriate software, successive operations are performed at a rate sufficient to query the joystick position almost continuously.

Motor Control Circuitry

The motor controllers are programmed via I/O ports 300h and 302h, while status signals are read from I/O port 301h (Fig. 3). When the data are written to the appropriate motor controller by the host computer software, they are latched by the octal D-type flip-flops (Fig. 2; 74LS374) and read into the controllers by switching the SR latches (Fig. 2; 74LS279). Initially, while the command is being processed, the controller will reset the BUSY line to logic 0 (Fig. 2; CY512, pin 27). The BUSY line can be monitored by the host software via I/O port 301h, bits 0 and 1 (Fig. 3), and must return to logic 1 prior to submission of the next instruction (Fig. 4, see WRITEPORT procedure).

A combined software and hardware reset circuit has been incorporated into the circuit. If either the PC sends a reset signal (typically occurring during power-on conditions) or the host software writes a logic 1 to I/O port 301h bit 4 (Fig. 3), the controllers and their data latches will be reset. This ensures that the controllers are always initialized to the proper state prior to receiving any programming commands.

Finally, although the CY512 generates the appropriate four-phase timing signals (Fig. 2; CY512, pins 21, 22, 23, 24), they are not capable of supplying the electrical current required to energize and drive the motors. Thus, the circuit includes an extremely simple and inexpensive unipolar power-transistor Darlington driver. Such a design requires a four-phase motor with six leads (a motor with two windings tapped at each end and at the center). The positive voltage is attached to the center of each winding (Fig. 2; P2, pin 2, 12), and the end of each winding is pulled to ground through power transistors controlled by one of the phase output lines from the CY512 (Fig. 2; ULN2068B, pins 3, 6, 11, 15). A current-limiting resistor has been placed on the positive voltage line (Fig. 2; ULN2068B, pin 9) in order to decrease the field decay time constant of the motors and provide a faster step response. When the motors are mounted to a microscope stage (see Results), stepping rates in excess of 3000 steps/second can be achieved with this driving circuit.

Software Design

Overview

The CY512 simplifies the software design because it supports many high-level motion control commands (13). For example, the ATHOME command instructs the microprocessor to mark the current motor position as a coordinate system origin. As the motors are stepped in any direction, a 16-bit position register is automatically updated to reflect the number of steps

```pascal
PROGRAM OBCCS_DEMO;
USES
  Crt;

TYPE
  Button_Range = $01..$02;
  PortType = (XMotor, YMotor);
  CommandType = String[8];

CONST  {Define OBCCS data ports as global constants}
  XData = $300; YData = $302; Status = $301; JStick = $303;

VAR
  Button_Number : Button_Range;
  IOPort : PortType;
  Ch : Char;

FUNCTION JoyStick (Axis:Integer) : Integer;
  {Returns the position value (0..255) of the joystick.  Assembly language
  is used to optimize execution speed}
BEGIN
  inline ($B8/$00/$00/  {MOV  AX, 00h -- Initial output value}
    $BA/$03/$03/  {MOV  DX, 303h -- Joystick control port 303h}
    $B9/$00/$00/  {MOV  CX, 00h  -- Reset A/D counter}
    $FA/  {CLI -- disallow program interuption}
    $EE/  {OUT  DX, AX -- Output to port 303h}
{loop}  $EC/  {IN  AX, DX -- Read port 303h value}
    $41/  {INC  CX -- Increment A/D counter}
    $23/$86/Axis/  {AND  AX, AXIS[BP] -- Test if AXIS has reset}
    $75/$F8/  {JNZ  loop -- IF not then GOTO LOOP}
    $FB/  {STI -- Restore system interrupts}
    $89/$8E/JoyStick)  {MOV  JOYSTICK[BP], CX -- Assign value}
  Delay (2);  {For fast machines allow I/O to settle}
END;  {Function JoyStick}

FUNCTION Button (Button_Number:Button_Range) : Boolean;
  {Returns the status of the joystick button referenced by the parameter
  Button_Number. The function returns a boolean value of TRUE if the queried
  button is depressed.}
BEGIN
  Button := Boolean(Port[JStick] AND (Button_Number SHL 4));
END;  {Function Button}

PROCEDURE System_Reset;
  {Writes a string of 30 reset pulses to ensure that the board circuitry is reset
  appropriately}
VAR
  I : Integer;
BEGIN
  For I := 1 to 30 do  {write 30 pulses for reset}
    Port[Status] := $FF;
END;  {Procedure System_Reset}

PROCEDURE WritePort (Data:Char; IOPort:PortType);
  {Data are sent to either the XMotor or the YMotor controller data ports.  The
  procedure checks the status and waits a limited time for controller status ready}
VAR
  Wait : Byte;
BEGIN
  Wait := 0;
  Case IOPort of
    XMotor : Begin
      While (Port[Status] AND $01) = 0 do
        Begin
          Delay (1000); {wait 1 second}
          Inc (Wait, 1);
          IF (Wait=20) THEN
            Begin
              Writeln ('XMotor timeout ...');
              Halt;
            End;
        End;
      Port[XData] := Ord(Data);
      Delay (1); {give time to settle}
    End;
    YMotor : Begin
```

FIG. 4 A Turbo Pascal program to demonstrate motor and joystick control by the host personal computer. The program interacts with the circuitry through the four designated I/O ports. The motors will move in the direction that the joystick is deflected. The demonstration program can be terminated by pressing any key on the keyboard. A detailed description of specific subroutines is addressed in the text. [Reproduced with permission from W. G. Tourtellotte, *J. Neurosci. Methods* **35,** 157 (1990).]

```
                 While (Port[Status] AND $02) = 0 do
                 Begin
                    Delay (1000); {wait 1 second}
                    Inc (Wait, 1);
                    IF (Wait = 20) THEN
                    Begin
                       Writeln ('YMotor timeout ...');
                       Halt;
                    End;
                 End;
                 Port[YData] := Ord(Data);
                 Delay (1); {give time to settle}
              End;
     End; {Case IOPort}
  END; {Procedure WritePort}

  PROCEDURE Command (St:CommandType; IOPort:PortType);
     {Each command is submitted to the controllers one character at a time via
     WRITEPORT. This procedure parses the entire command referenced by the ST
     parameter into single characters and submits them to the procedure WRITEPORT}
  VAR
     I : Integer;
  BEGIN
     For I := 1 to Length(St) do WritePort (St[I], IOPort);
     WritePort (#$0D, IOPort);
  END; {Procedure Command}

  PROCEDURE Initialize_Motor (IOPort:PortType);
     {A typical initialization sequence for each motor controller.  See text for
     details}
  BEGIN
     Command ('A', IOPort);
     Command ('F 1', IOPort); {Set coordinate origin}
     Command ('R 250', IOPort);
     Command ('S 50', IOPort); {Set acceleration/deceleration step count}
     Command ('H', IOPort); {Set half step mode}
     Command ('N 1000', IOPort); {SET MOTION RESOLUTION}
  END; {Procedure Initialize_Motor}

  PROCEDURE JoyDrive;
     {A primitive demonstration of the joystick motion control logic.  The joystick
     positions are read via JOYSTICK and tested whether the position is outside the
     set threshold (140..180).  If a change is detected and the value is less than
     threshold, the motors are set to move counterclockwise.  For values greater
     than 180, the motors are set to move clockwise.  Finally, the movements are
     executed accordingly.}
  VAR
     JoyX, JoyY : Integer;
     XChange, YChange : Boolean;
  BEGIN
     XChange := FALSE; YChange := FALSE;
     JoyX := JoyStick(1); JoyY := JoyStick(2); {Read JSTICK position immediately}
     IF (NOT (JoyX IN [140 .. 180]) AND ((Port[Status] AND $01) < > 0) THEN
     Begin
        IF JoyX < 140 THEN Command ('-', XMotor) {Change direction}
           ELSE Command ('+', XMotor); {Change direction}
        XChange := TRUE;
     End;
     IF (NOT (JoyY IN [140 .. 180])) AND ((Port[Status] AND $02) < > 0) THEN
     Begin
        IF JoyY < 140 THEN Command ('-', YMotor); {Change direction}
           ELSE Command ('+', YMotor); {Change direction}
        YChange := TRUE;
     End;
     IF XChange THEN Command ('G', XMotor); {Execute motor movements}
     IF YChange THEN Command ('G', YMotor);
  END; {Procedure JoyDrive}


  BEGIN {**** MAIN PROGRAM ****}
     System_Reset;
     Initialize_Motor(XMotor); Initialize_Motor(YMotor);          {Initialize both motors}
     REPEAT
        IF Button(1) Then Writeln('Button 1 pressed ...');
        IF Button(2) Then Writeln('Button 2 pressed ...');
        JoyDrive;
     UNTIL KeyPressed; {Touch keyboard to stop program}
     System_Reset;
  END. {OBCCS_Demo}
```

FIG. 4   (*continued*)

moved from the coordinate origin. Thus, the POSITION command can be used to step the motors automatically to a prescribed absolute coordinate (relative to the ATHOME position), without any additional "record keeping" by the host computer (e.g., POSITION 0 would automatically position the motor to the origin from any arbitrary location). Of course, a coordinate system would have an important physical meaning if, for example, the motors were configured to track linearly and they were calibrated such that the distance of each step were precisely known.

In addition, relative positioning is also painlessly achieved with built-in instructions provided by the CY512. The NUMBER command can be used to set the number of relative steps the motor will make from the current location in a single operation. There are also commands for adjusting the motor direction (CLOCKWISE, COUNTERCLOCKWISE) and the stepping speed (FACTOR, RATE), an acceleration/deceleration command (SLOPE), and a variety of other instructions related to program sequencing.

The CY512 can operate in program mode or command mode. When in program mode, the submitted commands are stored in a program buffer. From the command mode, a stored program can either be executed (DOITNOW) or parameter commands (e.g., NUMBER, CLOCKWISE, COUNTER-CLOCKWISE) executed immediately and the stepping sequences initiated according to the set parameters (GO).

The stepping motor circuit is easy to program from the host computer. A simple program has been written to control two stepping motors with a joystick in order to demonstrate software control of the motor controller circuitry (Fig. 4). It does not use the program mode of the CY512, but rather adjusts certain parameter commands in the command mode and initiates the stepping sequences directly (GO). The demonstration program (Fig. 4; written in Turbo Pascal, Borland International, Scotts Valley, CA) provides the basic motor/joystick control logic used for controlling the stepping motors in a functional computerized light microscope charting system (12, 13).

Joystick Programming

Two functions have been written to demonstrate reading the joystick position (Fig. 4; Function JOYSTICK) and the status of its buttons (Fig. 4; Function BUTTON). The JOYSTICK function has been written in machine language to increase execution speed. It requires one parameter (Axis) which specifies the axis of the joystick position to calculate. An output instruction to I/O port 303h starts the quad timer (Fig. 2; NE558). The bit corresponding to the requested axis (Fig. 3; I/O port 303h bit 0 for X axis and bit 1 for Y axis) is read continuously while incrementing a counter until a reset of the timer is detected. The value of the counter when the time-out is detected represents a numerical index of the absolute joystick position for the requested axis.

The joystick button status can be read easily by using the BUTTON function. A simple read instruction from I/O port 303h bit 6 (button 1) and bit 7 (button 2) indicates whether either button is currently depressed.

Motor Programming

The CY512 has a relatively large command set (14). The commands are submitted to the motor controller data ports by the COMMAND and WRITEPORT procedures (Fig. 4). The INITIALIZE_MOTOR procedure demonstrates a typical programming sequence for setting the motor coordinate origin to the current position ('A'), a maximum step rate ('F 1', 'R 250'), an acceleration/deceleration parameter ('S 50'), a half-step mode to double the stepping resolution ('H'), and the number of quantal stepping units to take ('N 1000') when instructed to do so (GO; e.g., if the motor is mechanically configured for 1-μm linear stepping, each GO command would then step the motors 1 mm). Prior to receiving the GO command ('G') the motors must at least receive parameters prescribing the stepping rate and the number of steps to execute.

Interactive Motor Control

The procedure JOYDRIVE demonstrates the simple manner by which the controller circuit can be programmed to control two stepping motors interactively with the joystick (Fig. 4). First, the current X (JoyX) and Y (JoyY) position is read by calling the procedure JOYSTICK. For each axis in succession, the value is then tested to determine whether the joystick has been moved (i.e., if the current position value of an axis is outside an arbitrary range, 140–180) and whether the CY512 is ready to accept commands (i.e., assuring that the BUSY line is reset for each controller on I/O port 301h, bits 0 and 1). If the joystick value is less than 140, the motor direction is set to COUNTERCLOCKWISE ('−'), but if the value is greater than 180, the direction is set to CLOCKWISE ('+'). Finally, if the joystick has been moved and after the new direction for each motor is set, the command to move each motor is submitted (GO, 'G'). Since no new NUMBER commands are sent to the motor controllers after the INITIALIZE_MOTOR procedure, the motors will step in 1000-step increments according to the joystick position.

The sample program demonstrates rather primitive joystick/motor control logic. For example, the motor movements are set by the program at a fixed 1000-step increment. In a more functional version of the joystick control program, it would be desirable for the user to alter the step resolution manually as more precise motor movements are required. Similarly when the joystick is deflected beyond a certain threshold (e.g., 140–180), the motors begin to move in repetitive 1000-step increments at a constant rate regardless of a change in the magnitude of joystick deflection. A more sophisticated approach would vary the rate of the motor movements according to the magnitude of the joystick deflection. Such a scheme has been implemented and previously described in detail (12, 13).

Results

The open-loop motor control circuit described in this chapter is both accurate and reliable. It was originally designed to control stepping motors mounted to the stage of a microscope. For this application, the motors have been configured to track linearly by moving a threated motor shaft through a tapped bushing mounted to a microscope stage (Fig. 1B).

Motor Accuracy

The motors used (Warner Electric, South Beloit, IL; Model SM-400-006-BU) were purchased for $100 each and were engineered to provide a 1.0-μm linear step displacement for each single half-step rotary movement. However, calibration with a microscope stage micrometer revealed that they actually produced a 1.1-μm displacement with approximately 10-μm accumulative error over a 1-cm linear travel (0.1% linear accumulation). For applications requiring well-defined stepping criteria, it is essential to calibrate the motors and adjust the software accordingly. For example, to obtain accurate 1-mm movements from these particular motors, they should be instructed to make 909 quantal steps (1.1 μm each; 'N 909') rather than 1000 ('N 1000').

Reliability Testing

While open-loop motor control is simple and can be performed rapidly, overall reliability is an important consideration since it is vulnerable to stepping errors (see Discussion). To test the potential for error by the control circuit and motors in the absence of feedback circuitry, discrepancies between the number of step commands and the physical movement of the microscope stage are directly measured with a microscope stage micrometer. For each axis independently, the origin is established (COMMAND 'A') and the starting position manually recorded on the stage micrometer. The motor controllers are programmed to make four 1000-step movements either forward (COMMAND ' + ') or backward (COMMAND ' − ') in random sequence and then to move back to the origin (COMMAND 'P 0'). The entire process is repeated twice for each trial. After each trial (a total of 16,000 steps) the final stage position is again read on the micrometer in order to detect potential

Charting System Software

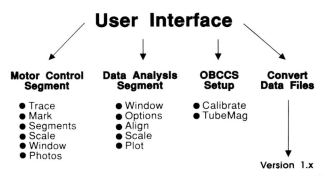

FIG. 5   A substantially more sophisticated motor control program provides a variety of features for controlling a microscope stage and plotting tissue section contours and the location of labeled cells in tissue.

stepping errors between the number of steps executed by the motor controller circuitry and the number that actually occurred. The test results for 100 such trials, applied to each motor, revealed very reliable stepping movements (e.g., a single step error per trial represents a 0.0063% error) with a single step error occurring in a maximum of 5% of the trials.

Programmable Light Microscope Stage

When the stepping motors are mounted to the stage of a light microscope, the circuit can be used to digitize the stage position interactively with joystick control. The CCS software, an extensive modification of the simple program in Fig. 4, has been designed to provide data acquisition (Fig. 5; MOTOR CONTROL SEGMENT) with stage control similar to that demonstrated in Fig. 4. For example, the stage movements can be recorded as the operator traces tissue section contours and the location of appropriate cellular or chemical entities (12, 13). Another component of the charting system software (Fig. 5; DATA ANALYSIS SEGMENT) is used to manipulate, plot, or scale, enumerate labeled groups of cells, and do area measurement calculations on the charted sections (Fig. 6A).

Unlike traditional pantographic charting systems (2–4, 8), digitizing the microscope stage movement in quantifiable increments (e.g., at 1.0-μm resolution) generates data in a format that can be readily analyzed quantitatively by a microcomputer. The areas of traced tissue sections can be calculated and the marked locations enumerated. Of course with the appropriate software, it

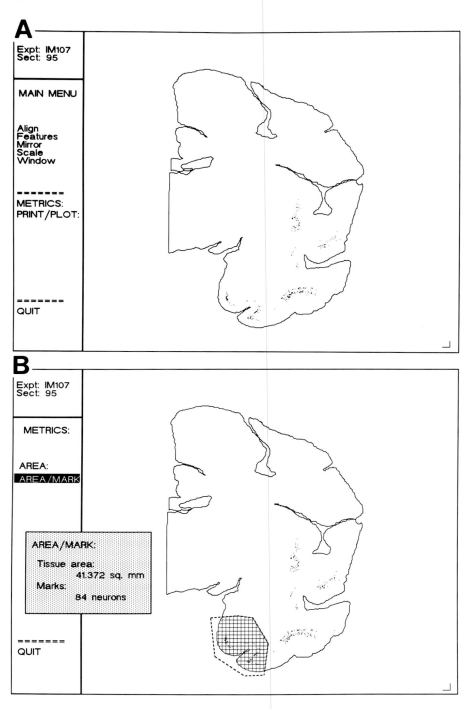

A

Expt: IM107
Sect: 95

MAIN MENU

Align
Features
Mirror
Scale
Window

METRICS:
PRINT/PLOT:

QUIT

B

Expt: IM107
Sect: 95

METRICS:

AREA:
AREA/MARK

AREA/MARK:

Tissue area:
41.372 sq. mm
Marks:
84 neurons

QUIT

is also possible to calculate the area of only a portion of a charted tissue section and count the number of marked locations within the specified region. For example, the use of appropriate test area contours can provide an accurate method for calculating cellular densities in specified tissue section regions (Fig. 6B).

Discussion

Our motor control circuit provides a versatile and uncomplicated approach to remote stepping motor control. Unlike some motor controllers which must receive the motor timing signals directly from the host computer (15), the circuit described here relieves the host computer of time-consuming control tasks. The host processor can issue and monitor the progress of motor control operations while the bulk of its processing capabilities are used for more complex graphical and numerical operations. Since the motor control microprocessors accept rather high-level instructions, the simplified programming should prove to be a desirable feature for less motivated programmers.

Open-Loop versus Closed-Loop Control

Open-loop controllers do not require verification that the programmed stepping sequences have been performed. In fact, using this scheme it is tacitly assumed that the stepping sequences have been performed properly and that the physical motor movement is accurately represented by the host computer. While this approach is simple to design it can be risky, particularly when physical loads are placed on the motors (e.g., when they are connected to a microscope stage) since they are required to surmount the static inertia of the load. Stepping errors can occur when the motors are instructed to step and the load is accelerated immediately to move at the maximum step rate (slew). Thus, discrepancies between the number of programmed steps and

---

Fig. 6 The motor control circuit can control stepping motors mounted to a microscope stage. The precise movements of the stage can be registered by the host computer and used to trace tissue section contours and mark specific cell locations. (A) A computer charting system which uses the circuit to control the stepping motors can archive, scale, plot, and perform a variety of metric calculations. (B) Under the METRICS menu, the AREA/MARK command can calculate areas circumscribed by arbitrary contours (with $1.0\text{-}\mu m^2$ resolution) and can count the number of locations marked within a specified area. Bar: 1 mm. [Reproduced with permission from W. G. Tourtellotte, *J. Neurosci. Methods* **35,** 157 (1990).]

the actual physical displacement can occur if the load does not respond rapidly to the accelerating forces provided by the motors. The problem can be solved by accelerating the motors more gradually to their slew rates and decelerating the motors to the appropriate target position. Such operations are handled automatically by the motor controllers through the implementation of ramping instructions that control the graded acceleration and deceleration of the stepping motors (e.g., COMMAND 'S 50', acceleration/deceleration control).

To maintain maximal control over stepping operations, and although we have used the circuit quite successfully in open-loop mode, the circuit can be configured for closed-loop operation. One method by which this might be achieved would be to control the DOWHILE line (Fig. 2; connector P2, pins 9, 14) with feedback pulses generated by optical encoders directly coupled to the microscope stage (e.g., absolute or incremental optical encoder; Parker Hannifin Corp., Compumotor Division, Petaluma, CA). To verify the stepping operations, the CY512 could be programmed to test the line (COMMAND 'T') each time a step is executed. This could, however, produce inadequate performance since the added verification operations may take a considerable amount of time to perform.

For motion control of the microscope stage, I selected an open-loop system because of the simplicity involved in mechanically coupling the microscope stage, its speed, and practical reliability.

Limit Sensing

Currently, the motor-controlled microscope stage has no limit sensing. Reaching the physical motor limits during stepping operations can generate discrepancies between the physical stage movement and the stage movement assumed by the computer. Simple limit-sensing circuitry would be an appropriate solution to ensure that limits are not exceeded. Optical microsensors (e.g., Digi-Key Corp., Thief River Falls, MN; about $3.50 each) could be coupled to the stage and calibrated so that metal tabs mounted to its moving portion would trigger the switches when certain physical limits are exceeded. The status of the switches could then either be monitored by the host computer or used to control the stepping motor controllers directly (e.g., DOWHILE line).

Choice of Encoder

Instead of stepping motors, potentiometric devices, either linear (4) or nonlinear (2, 3), could potentially be used as digitizing devices with the addition of an intermediary analog-to-digital converter to convert the analog voltage delivered by the potentiometers to a digital numeric value readable by the computer. In general, however, these devices are not sufficiently sensitive to reliably detect stage displacement in the micron range. Alternatively, shaft encoders (either absolute positioning or incremental positioning) could be directly attached to the microscope stage to monitor its motion. Such passive digitizing methods have been reported for both electron (11) and light microscopic systems (16). I have selected a more active digitizing method using stepping motors for two fundamental reasons. First, a motorized stage is versatile and provides an added element of computerized control over the microscope. Second, although shaft encoders can provide high resolution, in practice their accuracy and reproducibility are questionable. This is not due to the methodology per se (the encoders themselves are extremely accurate), but more so to implementation. Typically, one encoder is used for each axis of the stage (16). The stage manipulation is awkward and most easily accomplished by manipulating one encoder at a time in succession while following along a contour. The result is often a jagged appearing tracing not accurately reflecting the smooth contour. In our experience stepping motors (in conjunction with manual data sampling) provide better control over the digitizing process although they may often be slower.

Method of Tracking

Certain methods for tracking tissue section contours and labeled cells are obligatory, depending on the encoding method used. The stage can be controlled remotely (e.g., with a joystick or digitizing tablet) or directly. Directly manipulating the microscope stage, which has been used for common pantographic systems (2–4) and, for example, by Mize (11, 16), is necessary with passive encoding devices such as shaft encoders. However, systems that use active encoding devices, such as stepping motors, will necessarily require remote control. Glaser et al. (17) use a digitizing tablet or mouse to control the motorized stage. Their method relies heavily on a combined computer-generated and video camera image to trace neuronal processes and perform morphometric analyses. By contrast, our CCS was designed to display the entire tissue section on the video monitor independent of the microscope objective selected to view it. Moreover, using the joystick to track along tissue section contours and to locate specified regions on the section is relatively simple to master.

Method of Data Display

The CCS displays the digitized data in real time on the microcomputer graphics display. It uses data structures for internal data manipulation and high-speed graphics routines to optimize the performance of the graphics interface. To redisplay even the most complicated tracing requires only a few seconds. The real-time display is intended only as an aid for data acquisition and preliminary analysis; hence, its low resolution is adequate for this purpose. The full resolution is exploited when high-quality plotting devices or other analytical methods are used to plot the data.

An alternative method of data display superimposes a camera image of the microscopic field onto the computer-generated tracings (17, 18) and has also been used in conjunction with a motor-controlled stage. This sophisticated method was not selected since we desired a system that could display the entire digitized section on the screen, orienting the operator during the charting process.

Data Sampling Interval and Resolution

With the CCS, the digitizing resolution is based on both the sampling interval and the charting speed selected during data acquisition. The maximal 1.0-μm resolution is a hardware constraint imposed by the stepping motors. In practice, the sampled resolution may be somewhat less than optimal (e.g., points separated by 50 μm connected by line segments). For the purposes of mapping the spatial distribution of labeled cells on charted contours (our present application), this resolution yields excellent quality. For more quantitative methods, this resolution may not be adequate. The problem can be solved either by samping more data (using either a slower charting speed or greater data collection rate) or with mathematical approximation (e.g., B-spline interpolation; Ref. 19) generating additional points based on an approximated curve passing through the sampled points. We often use the latter method since tissue sections typically have smooth contours and hence curvatures that are highly predictable mathematically. Decreasing the sampling interval does substantially increase the time spent tracing.

In my experience, manual data sampling is appropriate when remote control of the microscope stage is used. For example, when points are sampled automatically as the operator traces with a joystick, sampling errors will frequently occur when the cross-hair reticle is not aligned properly with the section contour. When manual methods are used, however, the operator samples the point only when the contour is aligned, thus assuring that each sampled point lies precisely on the section contour. This added precaution ensures particularly accurate mathematical interpolation, since such contours will then best approximate the actual tissue section contour.

Summary

Microscope stages coupled to X–Y recording devices are used routinely by many laboratories to trace tissue sections and mark the locations of labeled cells. The extraordinary expense and difficulty in implementing a functional computerized version of the basic charting methodology have been impediments to many laboratories. With the described motor control circuitry and the appropriate control software, such a device can be implemented with relative ease, and to such an affordable degree that microscope computer charting systems can now become routine instrumentation in each laboratory desiring their sophisticated capabilities.

Acknowledgments

The author is grateful to Dr. Gary W. Van Hoesen for his support and contributions to this project. Supported in part by U.S. Public Health Service Grants 5T32 GM07337 (W.G.T.) and NS14944 (G.W.V.H.).

References

1. W. G. Tourtellotte, *J. Neurosci. Methods* **35,** 157 (1990).
2. J. Boivie, G. Grant, and H. Ulfendahl, *Acta Physiol. Scand.* **74,** 1A (1968).
3. G. Grant and J. Boivie, *Brain Res.* **21,** 439 (1970).
4. E. Eidelberg and F. Davis, *J. Histochem. Cytochem.* **25,** 1016 (1977).
5. D. F. Wann, T. A. Woolsey, M. L. Dierker, and W. M. Cowan, *IEEE Trans. Biomed. Eng.* **20,** 233 (1973).
6. T. A. Woolsey and M. L. Dierker, *in* "Neuroanatomical Research Techniques" (R. T. Robertson, ed.), p. 47. Academic Press, New York, 1978.
7. D. J. Forbes and R. W. Petry, *J. Neurosci. Methods* **1,** 77 (1979).
8. D. J. Reed, R. Gold, and D. R. Humphrey, *Neurosci. Lett.* **20,** 233 (1980).
9. J. J. Capowski and M. J. Sedivec, *Comput. Biomed. Res.* **14,** 518 (1981).
10. T. J. DeVoogd, F.-L. F. Chang, M. K. Floeter, M. J. Jencius, and W. T. Greenough, *J. Neurosci. Methods* **3,** 285 (1981).
11. R. R. Mize, *J. Neurosci. Methods* **8,** 183 (1983).
12. W. G. Tourtellotte, D. T. Lawrence, P. A. Getting, and G. W. Van Hoesen, *J. Neurosci. Methods* **29,** 43 (1989).
13. W. G. Tourtellotte and G. W. Van Hoesen, *J. Neurosci. Methods* **41,** 401 (1992).
14. Cybernetics Micro Systems, *CY512 Technical Data Manual,* San Gregorio, California, 1983.
15. J. J. Capowski and S. A. Schneider, *J. Neurosci. Methods* **13,** 97 (1985).
16. R. R. Mize, *in* "The Microcomputer in Cell and Neurobiology Research" (R. R. Mize, ed.), p. 111. Elsevier, Amsterdam, 1985.
17. E. M. Glaser, M. Tagamets, N. T. McMullen, and H. Van der Loos, *J. Neurosci. Methods* **8,** 17 (1983).
18. A. Alvarez-Buylla and D. S. Vicario, *J. Neurosci. Methods* **25,** 165 (1988).
19. T. Pavlidis, "Algorithms for Graphics and Image Processing," p. 247. Computer Science Press, Rockville, Maryland, 1982.

[3] Single-Unit Isolation from Multiunit Nerves by Dedicated Recording and Computation Methods

Uwe T. Koch, Christian Hoock, and Michael Brunner

Introduction

In neuroethological studies, it is often desirable to have an overview of the activity of a large nerve. The intracellular method cannot yield such an overview because it often depends on chance which units are hit, and, in each experiment, only a few units can be recorded. Thus, extracellular whole-nerve recordings combined with a system to decompose complex spike patterns has gained considerable attention. For simple cases involving a few units, there are computer programs available providing automatic on-line analysis of the data. In more complex situations, multichannel recording electrodes have been used. The data analysis of all these systems is based on the concept of templates, that is, sample waveforms of the different units active in the nerve. Once the templates have been formed in a "learning" mode by observing single-unit events, even complex waveforms can, in principle, be analyzed. The analysis of a summated signal of overlapping individual spikes (overlapping complex), however, presents difficulties, since it cannot be made unambiguous. In addition, there may be cases where templates cannot be set up, because some units may always tend to fire close to each other in response to a stimulus.

We have developed a system for recording and analyzing complex spike signals which does not require the concept of templates. It uses a high-precision multichannel cuff electrode and a dedicated analog computing system for data analysis.* This chapter describes the cuff electrode, the evaluation system, and a selection of measurements obtained from recordings in an insect connective.

---

* The initial version of this "velocity filter" was presented by Koch and Brunner (1).

Methods in Neurosciences, Volume 10

Methods

Construction of Chamber Electrode

Because the evaluation method requires that the signals from the individual electrodes be very similar, great care was taken to develop a multichannel recording electrode with very stable recording properties. The factors influencing the reproducibility of extracellular recordings had been investigated by simulation and experimental methods.

To describe the results, a number of definitions are useful.

> *Electrode length* L_E: In an extracellular electrode, a certain (active) length of the nerve is capacitively coupled to the recording electrode (e.g., a hook). This length is limited by the insulating medium (e.g., petroleum jelly) which usually surrounds the electrode and the nerve to prevent shunting of the recording signal to ground potentials as encountered in the surrounding body fluids.
>
> *Insulator leakage resistance* R_L: The quality of the insulating medium and its way of application determine the leakage resistance which creates an unwanted pathway for the nerve signal to ground.
>
> *Source resistance* R_S: The recorded signals as generated in the nerve can be adequately described as produced by a signal generator with a defined output resistance, the source resistance.
>
> *Differential recordings:* There are two "single-ended" electrodes placed on the same nerve close to each other. The distance between the electrodes is called the recording point distance D.
>
> *Time interval* Δt: As a measure of the "sharpness" of a spike, we use the time between the maximum and the minimum voltage excusion of a (basically) biphasic spike, the peak–peak interval Δt.
>
> *Length constant* λ: The length constant λ is the length that an electric signal can passively travel in an axon until it is attenuated by a factor of e. For practical considerations $\lambda \approx v\ \delta t$, where v is the spike propagation velocity and δt the duration of the spike as it would appear in an intracellular recording.

The results of our simulation and experiments were the following: (1) In single-ended recording electrodes, the amplitude is kept at a maximum value, and the peak–peak interval is kept to a minimum value, if the electrode length L_E fulfills the condition $L_E < 0.3\lambda$. Because λ decreases with decreasing axon diameter, this condition is most critical for very small axons. (2) The amplitude of the recorded spikes depends critically on the values of R_L and R_S. To achieve stable and high recording amplitudes, the condition

$R_L > 10 R_S$ should be fulfilled. Because R_S is in the range of 50–100 kΩ, this means that quite high value of R_L are needed. (3) In differential recordings, the form and amplitude of the recorded spike depend strongly on the recording point distance D. The amplitude decreases monotonically when D decreases below a characteristic value D_c which is of the order of 0.5λ. For decreasing D, the peak–peak interval Δt drops sharply near D_c and then almost remains at a constant minimal value if D is reduced further. Thus, for optimal differential recordings, a D value should be chosen that is small enough to obtain the minimum Δt value, yet still large enough to have a reasonable spike amplitude. The optimal D value is in the range of 0.3D_c, where the spike amplitude is still 76% of the value at D_c, and Δt is of the order of 0.1 msec, close to the minimum. For a propagation velocity of 2 m/sec, this means a D value of 0.2 mm. (4) As for single electrodes, the leakage resistance R_L is of prime importance to the quality of a differential recording. In addition, the spike quality is influenced by the leakage between the two recording points. At the optimal value of D (0.2 mm), it becomes extremely difficult to provide sufficiently good insulation between the two recording points.

These results led to the development of a novel multichannel cuff electrode (Fig. 1). It uses a rigid insulating capsule milled from plexiglass surrounding the nerve (N) to be recorded from. The actual contact defining the electrode length L_E is provided by a small volume of electrolyte [stick insect saline (2)] which is also completely enclosed in the capsule. The connection to the amplifier is then made by a wire which is inserted into the liquid-filled chamber without touching the nerve.

The electrode capsule is made from two parts of plexiglass, the electrode body (B) and the lid (L). The body contains a groove (G) accommodating the nerve (N). The recording chambers (C) are cut as small slits perpendicular to the groove. The recording wire (R) is inserted through a hole from the side. It is spiral shaped to increase the wire surface in contact with the ringer. The lid is a plexiglass plate carrying a silicon rubber pad (P) which fits onto the electrode body. At the position of the groove, the silicon pad has a wedge (W) protruding into the groove when the lid is mounted. The rubber pad seals off the recording chambers when the lid is pressed onto the electrode body via a set of miniature screws. By varying the lid pressure, the wedge is pushed down into the groove in a controlled manner. When it touches the nerve, the nerve is sealed off at the entry and exit of each recording chamber, thus providing the necessary seal between the external hemolymph and between adjacent chambers. When a recording is set up, the lid pressure is increased slowly while watching the spike amplitudes on an oscilloscope. The slowly improving seal creates increasing spike amplitudes. At the point where the spike amplitude is not further increased by increasing the lid pressure, the seal is sufficiently good, and lid pressure is not further increased to avoid

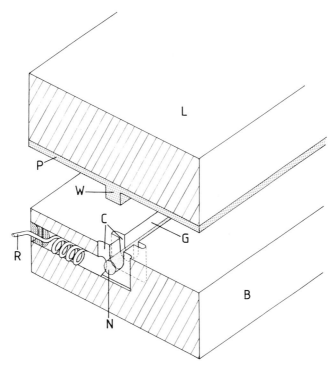

FIG. 1   Perspective drawing of the chamber electrode. The electrode body B and lid L have been cut at the site of one of the 12 chambers C to show its inside form. The nerve N runs along the groove G. The chambers provide a liquid conducting bridge between nerve and wire R leading to the amplifier. The lid L carries a silicone rubber pad P having a wedge W which protrudes into the groove providing a seal of the nerve and the chambers. [Adapted with permission from Brunner and Koch (4).]

crushing the nerve. A total of 12 electrode chambers are provided on one electrode body, arranged in pairs to form 6 differential electrodes with a fixed recording point distance (D) of 190 μm and a distance of 1.5 mm between the differential electrodes. The overall size of the electrode including lid was 12 × 2.5 × 2.5 mm.

This electrode provides very stable and reproducible differential recordings. The sizes of individual spikes usually do not vary by more than 20% between the different recording sites. The chamber electrode is of course reusable, if it is thoroughly cleaned after each recording session.*

* A detailed description of the chamber electrode and its construction methods is given in Refs. 3 and 4.

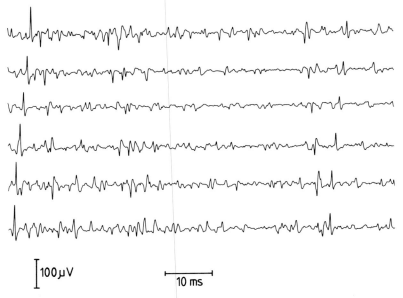

$$\Big[100\,\mu V \qquad \underset{\text{10 ms}}{\vdash\!\!\!-\!\!\!-\!\!\!\dashv}$$

FIG. 2   Sample of a recording with the six-channel differential chamber electrode. The spike sizes are very similar in all channels. Time shifts of spikes between the individual channels can be easily recognized. [Reprinted with permission from Brunner and Koch (4).]

Properties of Chamber Electrode

In a series of experiments, the properties of the signals recorded using the 6-fold differential electrode were investigated in a preparation involving the promesothoracic connective of the large stick insect *Acrophylla wulfingii*. Figure 2 shows a sample of the recordings used in these experiments. The amplitudes of the spikes reached values up to 100 μV.

The electrode has six sections of adjacent differential recording sites. The distance W between the six recording sites was 1500 ± 10 μm. When measuring the time shifts τ_m of the signals over each section, we found that they deviated by up to 10% from the expected value, τ_{ex}, of W/v. These deviations varied from section to section and from preparation to preparation. They were clearly affected by the manner in which the nerve was manipulated during insertion into the groove. However, the relative values of the deviation $(\tau_m - \tau_{ex})/\tau_{ex}$ were the same within measurement accuracy for all units analyzed, regardless of their propagation velocities. We concluded that the deviations arise from unintentional lengthening or shortening of the nerve in the section. In addition, these deviations did not change as long as the nerve remained untouched in the electrode.

Data Evaluation

To extract the relevant features, namely, velocity and spike amplitude of each unit from the six differential recordings, we used an analog computing system. It consists of of a set of delay units, each providing a voltage-controlled time shift for one channel and a series of adding circuits. The concept is to apply an appropriate time shift to each channel, compensating the traveling times of a given spike along the axon. Thus, the recording is transformed in such a way that a given unit will appear simultaneously in each channel, and, when added linearly, the signals superimpose to form an enhanced spike, while the signals from other units do not superimpose favorably and are not enhanced.

To use such a system efficiently, it is convenient to adjust all time shifts by one control. Then, one control knob is sufficient to set the analyzing system to a selected velocity value. Figure 3 shows the construction of the delay circuits. These are built around a bucket brigade analog delay line (Reticon R5106, EG&G Reticon, Sunnyvale, Ca.). They feature a linear relationship between input voltage and time shift. This is achieved by measuring the period of the voltage-controlled oscillator (VCO) (74LS629) using a constant current-driven integrator and a sample and hold circuit. At the output of the sample and hold, a voltage proportional to the VCO period is available. A controller with an integrative characteristic compares this voltage to the command voltage and adjusts the VCO input such that the deviations disappear.* At its front end, each delay unit carries a precision attenuator, permitting the sensitivity of the unit to be set according to the position of the recording site associated with it.

In principle, for a six-channel recording, five delay units would be sufficient to generate the overlap of all signals. However, this would imply time shifts to be generated which are below 1 msec, which is the shortest time shift the bucket brigade circuit can generate. To circumvent this limitation, all six channels are passed through a delay unit, which produces a general time shift of 1 msec for all six signals. The individual time shifts are then added onto this general offset value.

A substantial improvement of the velocity resolution of the system can be

---

* Detailed circuits are found in Ref. (3).

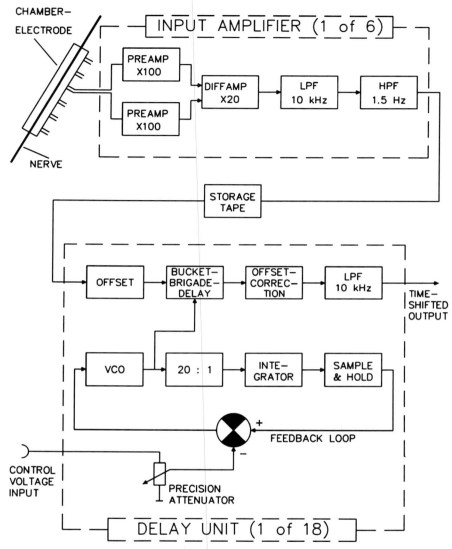

FIG. 3   Semischematic drawing of the main analog hardware components. The input amplifier consist of two battery-operated preamplifiers, a differential amplifier, and lowpass and highpass filters (12 db/octave). The analog delay unit is built around a bucket brigade analog delay line as used in acoustic equipment. A clock frequency generator with a feedback loop is used to ensure a linear relationship between control voltage and amount of delay. The attenuator in the delay input stage is used to scale the delay value such that the signals from the same unit coincide temporally at the output of all delays.

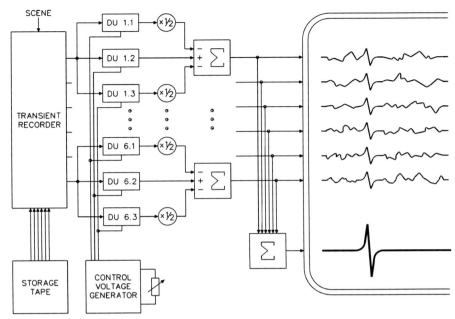

FIG. 4   Schematic drawing of the setup used for the precision evaluation of spike velocity, amplitude, and latency. A part of the six-channel recording is stored in a transient recorder, and a 20-msec section ("Scene") is selected and repetitively sent to the delay units. There are three chains of delay units (DU). DU 6.1 means delay unit for channel six, first chain. Delay chain 2 is the main "straightforward" chain. Chains 1 and 3 are "side chains" which are slightly detuned. Slow background signals pass the side chains in similar manner as chain 2, and, when the signals reunite, most of the slow background signals and spikes with other velocities are canceled. The delay outputs are displayed on the oscilloscope screen. The lowest trace shows the summated signal, where the spike is enhanced and much of the background canceled out.

achieved by using three delay units in parallel for each channel. In this "lateral inhibition" scheme (3, 5), the two additional delay units are slightly "detuned" as if they were filtering out a velocity $v^+ = v_0 + \Delta v$ or $v^- = v_0 - \Delta v$. These detuned delays treat slow potential variations almost in the same way as the "center" delay tuned to the velocity v_0, but the sharp spikes will be much reduced in amplitude by the detuned units. Thus in the signal $v' = v_0 - 1/2 (v^+ + v^-)$, the slow potentials are strongly suppressed, while the spike amplitude is much less affected. Figure 4 shows the setup incorporating all 18 delay units, as well as the arrangement used for the high-resolution data evaluation.

In an initial step, a section of the six-channel recording is read into a six-channel transient recorder (Physirec*), which has a sampling rate of 20 kHz and a depth of 512,000 samples for each channel. A 20-msec section of the recording is selected and is repetitively displayed on the screen of an eight-channel Tectronix oscilloscope. The Physirec settings are chosen such that the data are played back at the original "real time" speed. Before reaching the oscilloscope, the signals of the six channels are passed through the array of delay units as shown in Fig. 4. The linear sum of all time-shifted signals is also generated and displayed on the seventh trace of the oscilloscope. When the control voltage for the delays is varied, the spikes in the different channels are shifted back and forth on the oscilloscope screen. The size of the spike in the adder output varies, depending on the delay adjustment as shown in Fig. 5. In Fig. 5A, the shapes of a spike in the adder output are shown for different settings of the delay control, indicated as filter setting; in Fig. 5B, the amplitude of the spike is plotted versus the filter setting. This plot shows a clear maximum near $v = 1.41$ m/sec. The sharpness of this maximum permits the determination of the velocity with a precision of better than 0.5%.

When the filter setting is optimized for one spike, the amplitude of the spike can be read and used for further characterization of the unit. Owing to the linear superposition of the signals, the amplitude of the spike is measured with much better reproducibility than from a single-recording channel, since the effects from other potentials are canceled to a large extent.

Sample Experiments

To illustrate the capabilities of the system, we report the results of an experiment on the stick insect *Acrophylla wulfingii*. The group of Bässler has investigated the processing of information from the chordotonal organ measuring the angle between femur and tibia in each leg.† Hofmann and Koch (7) and Hofmann *et al.* (8) had shown in the stick insect *Cuniculina impigra* that this sensory system encodes different modalities of angular movement: position, velocity, and acceleration are represented in pure form in the activity of dedicated units and also in certain combinations. Büschges (9, 10)

---

* Designed and manufactured by L. Neumann, Riedfeld 7, D-8031 Seefeld, Germany.

† For a review on this work, see Ref. (6).

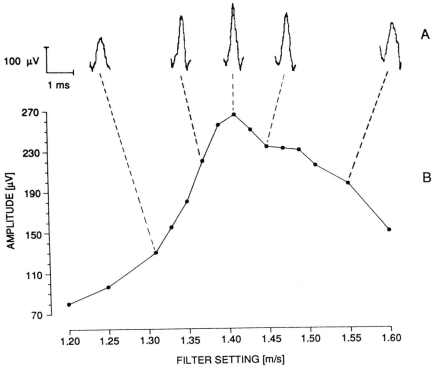

FIG. 5   (A) Replicas of the lowest oscilloscope trace in Fig. 4 showing a spike of one unit. As the filter setting (i.e., the delay values of the delay chains) are gradually changed, the amplitude and shape of the spike varies. (B) Amplitude of the observed spike plotted versus the setting of the delay lines ("velocity filter"), calibrated in m/sec. The curve has a clear maximum at 1.41 m/sec. The resolution of the velocity scale and the shape of the curve make it plausible that the velocity value of the maximum can be located with 0.5% precision. [Adapted with permission from Hoock (11).]

has begun to investigate the interneurons processing information from this sensory system. Using intracellular recording and staining techniques, he described ascending and descending intersegmental interneurons being activated by signals from the femoral chordotonal organ. Among the descending interneurons, he was able to distinguish 17 different morphological and physiological types (9). As our method should enable us to get a more complete knowledge of the number of units in the connectives, we set out to check how many descending interneurons we would be able to detect.*

Procedure

The animal is fixated ventral side up on a foam plastic plate. The left frontal leg is fixated and incised on the femur near the femur–tibia joint. The apodeme transmitting position information from the joint to the chordotonal organ is fixed into a miniature clamp, and the apodeme is cut distal from the clamp.

The clamp is mounted on the center of a loudspeaker with position feedback regulation, to transmit stimuli to the chordotonal organ mimicking tibia movements. The stimulus waveforms, ramp and hold stimuli with controlled curvature at the edges, are generated in a specially build ramp generator (U. T. Koch, 1990, unpublished) which permits control of the parameters of position, velocity, and acceleration independently.

To record intersegmental interneurons, the chamber electrode is inserted on the left connective between pro- and mesothorax. To this aim, the thorax is opened ventrally between the pro- and mesothoracic ganglion, and the left connective is freed from surrounding tissue. The chamber electrode is filled with *Carausius* saline (2), and the connective is flushed with the saline before inserting it gently into the groove of the electrode. The lid is mounted and tightened until the spike amplitudes are optimized. The amplified signals from the six differential electrodes are stored on a seven-channel instrumentation recorder (Racal Store 7DS) together with a replica of the stimulus waveform.

The stimuli always have an amplitude of 200 μm, mimicking a change of femur–tibia angle of 20°. The velocity of the stimulus is varied from 0.1 to 27 mm/sec, and the acceleration from 300 to 4×10^4 mm/sec^2. For each setting, the stimulus is presented 10 times, with pauses of 10 sec between stimuli. One recording session can last up to 2 h without any noticeable degradation of the recorded signal.

The spike pattern recorded as reaction to the ramp and hold stimuli consists mainly of two bursts of spikes, marking the onset and the end of the ramp. The spikes are largely overlapping. The data are analyzed using the high-resolution technique envolving lateral inhibition and the Physirec transient recorder. Over a section of 100-msec duration following each of the ten stimulus presentations, the spike velocities are measured as well as their latency with respect to the stimulus onset. The amplitudes of the spikes are also measured. The data are stored in a file on an IBM AT computer for statistical analysis. We use a specially developed plotting routine to display

---

* A detailed account of this investigation is found in Ref. 11. A small section of the results was published in Ref. 12.

the occurrence of each spike in a plane with the velocity of the spike as the
x axis, and the latency as the y axis. The amplitudes of the spike are represented by the different sizes of the plotter symbols.

As an example, Fig. 6 shows the results obtained from a ramp with 2.23
mm/sec velocity and 200 mm/sec^2 acceleration. One can clearly distinguish
a large number of small clusters around a latency value of 2.5 msec. These
units marking the onset of the stimulus are apparently acceleration-dependent
units. Other units respond over the duration of the ramp, which characterizes
them as velocity-encoding units. To distinguish the different units, it is also
helpful to consider the velocity histograms plotted on the same velocity scale
for each amplitude class, in the lower part of Fig. 6.

Many of the units characterized in Fig. 6 can be reidentified in the plots
constructed from the responses to the other ramp stimuli. Because each unit
has a certain response range and threshold, it is necessary to evaluate the
responses to all the different stimuli.

As a final result of such an analysis, we were able to distinguish a total of
13 units responding to acceleration, 22 responding to velocity, 9 responding
to position, and 5 responding to both velocity and acceleration. Thus, a total
of 49 units could be detected and characterized by the evaluation of a single
recording session. Comparing this number to the 17 units described by
Büschges (9, 10) illustrates well the capabilities offered by this method.

Discussion

The problem of extraction of single-unit activity from multiunit extracellular
recordings has been studied by numerous authors. Papers published before
1983 on hardware and software systems will in general not be quoted here
because they are covered in excellent reviews by Schmidt (13, 14). Since the
reviews by Schmidt, a number of papers have appeared documenting the
ongoing development in this field. Cocatre-Zilgien and Delcomyn (15) present
an elegant new hardware system for spike discrimination and characterization
using the slope of the spikes as an additional criterion. Other interesting
hardware was developed by Wörgötter et al. (16). They used a set of eight
parameters characterizing the spike form which are stored as the setting of
eight potentiometers. Using an analog correlation computing algorithm, this
system can sort out one spike form on-line from an extracellular recording.

There are numerous reports on computer software systems for spike classification. Okada and Maruyama (17), Ikeda and Hoshimiya (18), Vibert et al.
(19), Smith and Wheeler (19a), and Salganicoff et al. (20) have used fast
microprocessors or workstations providing on-line analyzing capabilities.
With the exception of Okada and Maruyama (17), they all use a dedicated

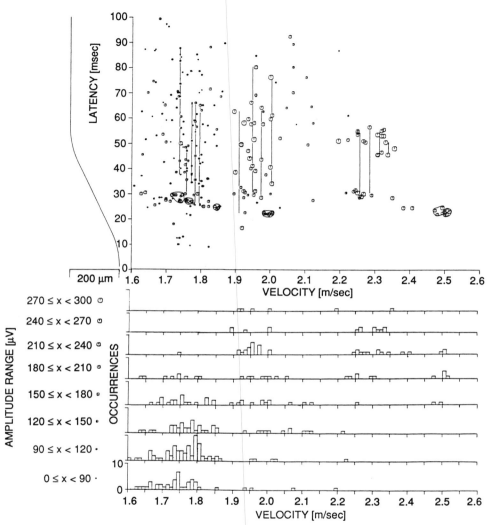

FIG. 6   (Top) Plot of individual spike events in the latency–velocity plane. The spikes recorded in the connective were elicited by ten presentations of a stretching stimulus (waveform displayed at left margin) applied to the chordotonal organ in the front leg of a large stick insect. A set of sharply defined clusters can be seen in the 20- to 30-msec latency range. Other units have activity distributed along a line. (Bottom) Velocity histograms for eight ranges of spike amplitudes. This diagram helps to distinguish different units with similar velocity but different amplitudes, as near the velocity of 1.8 m/sec. This plot shows only a section of the original velocity scale (1–2.8 m/sec) to permit resolution of the details. [Adapted with permission from Hoock (11).]

hardware system as a "preprocessor" to ease requirements of computing speed necessary for on-line operation. Edin *et al.* (21) and Forster and Handwerker (22) report partial on-line characterization capabilities, whereas Eberly and Pinsker (23), Mankin *et al.* (24), and Mitchell *et al.* (25) describe off-line systems.

Common features to all these software approaches are (1) the use of the concept of templates which are defined in a first step of the analysis (often called "learning phase") and used later to classify all spikes in the recordings. Vibert *et al.* (19) have designed their algorithm such that it establishes and modifies templates continuously during the classification process. (2) They are unable to deal with situations where several spikes superimpose and thus generate complex waveforms. Le Fever and De Luca (26) and Wheeler and Heetderks (27) tried to resolve these "overlapping complexes" by identifying individual templates within the complex, subtracting them, and searching for further templates to subtract until the complex is resolved. These schemes, however, only worked under the supervision of a skilled operator. An approach capable of unsupervised decomposition of overlapping complexes was described by Roberts (28) and Roberts and Hartline (29). Here, the signals from a multichannel electrode were analyzed by a special mathematical filter algorithm. The system, however, still required the establishment of templates. Fiore *et al.* (30) have described an approach using the algorithm of cross-correlation. This system is capable of establishing the number of individual units in a recording and their propagation velocity without the need for templates. But in this scheme, the information on the latency of a spike with respect to, for example, a stimulus is lost, as well as most features of the spike like amplitude or shape.

The requirement to establish templates may not seem difficult to meet. There are many situations, however, where the reaction to a stimulus consists of a densely packed burst of spikes, and these spikes may never occur spontaneously. Thus, they cannot be studied in isolation, and a template cannot be established.

The method described here does not require the establishment of any template. It is able to extract and characterize the spikes even if they overlap in large complexes. In addition, the waveforms of the spikes are reconstructed quite well, even when they occur within an overlapping complex.

The system will only fail if two spikes of the same velocity overlap almost completely. Then the system will describe the event as a new spike class of larger amplitude. At present, the system is hand operated, and the selection of spikes is quite time consuming. The operator, however, does not need to make special decisions as to the classification of the signal.

The method of analysis therefore is suitable to be emulated on a digital computer. In addition to the possibility of unsupervised analysis operation, the need for analog delay units and other dedicated hardware would also be abolished. We have established an algorithm mimicking the operation of the optimation of the delay chain setting for each spike. Trial runs have shown the feasibility of such a computer program. Presently we are working on the program parts for display and statistical analysis of the resulting data.

The velocity resolution of the system is sufficient insofar that the measurement uncertainty for a given spike (~0.5%) is somewhat smaller than the jitter in velocity that is observed between different spikes (~1%, see Fig. 6) from the same axon. In the present version, the system will tend to overlook small spikes within an overlapping complex of large spikes. The suppression of spikes with a "wrong" velocity is $1/n$, where n is the number of recording electrodes. As the current electrode spacing is 1.5 mm, and one differential electrode is only 0.27mm long, it seems feasible to use 18 electrodes on an electrode body with the same overall dimensions. Apart from improving the suppression effect on unwanted spikes, this arrangement would also improve the signal-to-noise ratio of the recorded spike waveforms.

In addition to the measurement of amplitude, latency, and propagation velocity, the spike waveforms appearing at the output of the linear adder as in Fig. 4 could be subjected to the usual waveform classification algorithms using templates. In this case, templates can be established even if the spike only appears in an overlapping complex. Such a postprocessing could enhance the discriminating capabilities of the system even further.

Acknowledgments

We gratefully acknowledge support by the Deutsche Forschungsgemeinschaft (KO 630/2). We wish to thank Winfried Galm for decisive help in some of the measurements, and Sybille Watt for the typing. We are particularly grateful to Professor U. Bässler for his constant support of this work.

References

1. U. T. Koch and M. Brunner, *J. Comp. Physiol. A* **157,** 823 (1988).
2. D. J. Weidler and F. P. J. Diecke, *Z. Vgl. Physiol.* **64,** 372 (1969).
3. M. Brunner, Ph.D Thesis, Universität Kaiserslautern, Germany, 1988.
4. M. Brunner and U. T. Koch, *J. Neurosci. Methods* **35,** 93 (1990).
5. M. Brunner, G. Karg, and U. T. Koch, *Neurosci. Methods* **33,** 1 (1990).
6. U. Bässler, "Neural Basis of Elementary Behaviour in Stick Insects." Springer-Verlag, Berlin, New York, and Heidelberg, 1983.
7. T. Hofmann and U. T. Koch, *J. Exp. Biol.* **114,** 225 (1985).

8. T. Hofmann, U. T. Koch, and U. Bässler, *J. Exp. Biol.* **114,** 207 (1985).
9. A. Büschges, Ph.D Thesis, Universität Kaiserslautern, Germany, 1989.
10. A. Büschges, *J. Exp. Biol.* **144,** 81 (1989).
11. Ch. Hoock, Diplomarbeit, Universität Kaiserslautern, Germany, 1990.
12. Ch. Hoock, W. Galm, and U. T. Koch, *in* "proceedings of the 18th Göttingen Neurobiology Conference" (N. Elsner and G. Roth, eds.), Thieme Verlag, Stuttgart and New York, 1990.
13. E. M. Schmidt, *J. Neurosci. Methods* **12,** 1 (1984).
14. E. M. Schmidt, *J. Neurosci. Methods* **12,** 95 (1984).
15. J. H. Cocatre-Zilgien and F. Delcomyn, *J. Neurosci. Methods* **33,** 241 (1990).
16. F. Wörgötter, W. J. Daunicht, and R. Eckmiller, *J. Neurosci. Methods* **17,** 141 (1986).
17. M. Okada and N. Maruyama, *Comput. Programs Biomed.* **14,** 157 (1981).
18. M. Ikeda and N. Hoshimiya, *Med. Biol. Eng. Comput.* **23,** 23 (1985).
19. J. F. Vibert, J. N. Albert, and J. Costa, *Med. Biol. Eng. Comput.* **25,** 366 (1987).
19a. S. R. Smith and B. L. Wheeler, *IEEE Trans. Biomed. Eng.* **35** (No. 10), 875 (1988).
20. M. Salganicoff, M. Sarna, L. Sax, and G. L. Gerstein, *J. Neurosci. Methods* **25,** 181 (1988).
21. B. B. Edin, P. A. Bäckström, and L. O. Bäckström, *J. Neurosci. Methods* **24,** 137 (1988).
22. C. Forster and H. O. Handwerker, *J. Neurosci. Methods* **31,** 109 (1990).
23. L. B. Eberly and H. M. Pinsker, *Behav. Neurosci.* **98,** 609 (1984).
24. R. W. Mankin, A. J. Grant, and M. S. Mayer, *J. Neurosci. Methods* **20,** 307 (1987).
25. B. K. Mitchell, J. J. B. Smith, B. M. Rolseth, and A. T. Whitehead, *Chem. Senses* **14,** 730 (1989).
26. R. S. Le Fever and C. J. De Luca, *IEEE Trans. Biomed. Eng.* **29,** 149 (1982).
27. B. C. Wheeler and W. J. Heetderks, *IEEE Trans. Biomed. Eng.* **29,** 752 (1982).
28. W. M. Roberts, *Biol. Cybernet.* **35,** 73 (1975).
29. W. M. Roberts and D. K. Hartline, *Brain Res.* **94,** 141 (1975).
30. L. Fiore, L. Geppetti, D. Ricci, and B. Di Vizio, *J. Neurosci. Methods* **27,** 109 (1989).

[4] Computers in Study of Simultaneously Recorded Spike Trains: Organization of Neuronal Assemblies

George L. Gerstein

Introduction

Neurobiologists readily accept the idea that neurons do not act alone, and that they must enter into organized assemblies in order to accomplish the complex computational processes that correspond to behavioral performance. However, the concept of "neuronal assembly" can be defined in a number of different ways, each with different, partly nonoverlapping properties and computational possibilities. We will limit the discussion here to assemblies that are determined through preferred relative timing in the several observed spike trains.

Only recently has a partially satisfactory technology for direct observation of such neuronal assemblies at work *in situ* been developed. Involved are special electrode structures, some special purpose electronic devices, and a large number of computer applications which are absolutely essential to this class of experimentation in order to obtain the data and subsequently analyze and interpret them. It is these computer applications which we will systematically examine below.

Stimulus or Behavioral Control

Problem

In most neurophysiological recording situations it is necessary to provide appropriate stimulation and/or to monitor behavioral responses. Both tasks are well handled by computers, although there is little that can be bought "off the shelf." Considerable thought should be given to the design of the arrangements both for operator convenience and for efficient use of hardware and programming. The requirements for control resolution and timing accuracy must be decided at an early stage of design. Flexibility in the arrange-

Methods in Neurosciences, Volume 10

ments of both hardware and software is highly desirable to allow meeting future, possibly as yet undefined needs.

Control Structure and Methods

Generally the hardware immediately attached to the computer should be able to produce logical, analog, and timing signals, and should also be able to accept such signals from the laboratory equipment. There are many commercially available boards that plug in to a standard personal computer (PC) bus and provide the necessary services. In turn these output signals should be used to control various special purpose devices (tone generators, light projectors, tactile stimulators, feeders, etc.) that are appropriate. Input signals would come from manipulanda or other sensors embedded in the experimental apparatus.

The program structures to operate the interface card should be as modular as possible. The lowest hardware control level should provide a module that defines a series of macrocommands for input and output (I/O). The rest of the controlling program can be written in higher level languages, and should call functions in the lowest module as needed. Should the hardware ever be changed, only this lowest module will need modification.

One of the most important aspects of this type of computer application is the handling of time and of interrupts. Some of the available I/O boards have on-board counters and oscillators for the timing function. Interrupt handling must take into account the vagaries and requirements of the operating system, but it should be able to provide service delay within the requirements of temporal resolution and accuracy required by the experiment.

Recording Multineuron Activity

Problem

To study the activities and properties of neuronal assemblies we must record simultaneously the activities of as many neurons as technically possible. At present extracellular recording is the only available approach and requires specialized electrode structures that sample more than one point in the tissue under study. It is usually necessary to process the signal from each pickup point in order to isolate the signals of individual neurons properly. Subsequently the identification and timing of all neural and laboratory signals must be put into a form that is convenient for analysis. Thus the final form of the data represents some number of spike trains, one from each of the N observed neurons. Current practical limits put N at about 60 neurons.

Spike Shape Sorting

It is quite common for electrode structures of the size used in such research to pick up signals (action potentials) from more than one neuron. The waveforms of the spikes will differ for each neuron because the detailed electric field will depend on the geometry of the neuron, relative to the position of the electrode. Thus it should be possible to distinguish the several spike trains (three or so) appearing on one electrode by shape, assuming that each spike waveform remains constant. The assumption of constant shape requires that the electrode position does not drift during the collection of data and that changes of spike shape during high-frequency burst discharges are minimal.

A large number of algorithms and devices have been developed over the years for both on- and off-line sorting of waveforms in this type of recording. Reviews of hardware and software methodologies are given by Schmidt (1, 2). Additional methods are constantly being devised (e.g., Refs. 3 and 4). Few of these approaches are commercially available [rare exceptions are the integrated data and analysis systems of Brainwave Inc. (5) and of Spectrum Scientific (6)].

Realizations of the various solutions to the waveform sorting problem differ in the extent to which they use hardware and software. We have had most success (7) with a hardware device based on principal components sorting (8) whose output is read and appropriately displayed by a computer. The problems inherent in such a computer connection are similar to those discussed above for stimulus control, and they are best solved with a general purpose I/O interface card. On the other hand, far better realizations of the peripheral sorting hardware device are now possible using programmable digital signal processor (DSP) chips. A number of such devices are under development in several laboratories and companies, but as yet there is nothing available commercially.

Event Timing and List Construction

Once the incoming data streams have been sorted out according to spike waveform, it is necessary to preserve the sequence and timing of all relevant neural and laboratory events. A time resolution must be chosen, usually in the range of 100–500 μsec. An identification code is assigned to each type of neural event (spikes from a particular neuron, e.g., as identified by a particular waveform). Similar codes are assigned to each type of laboratory event (e.g., a particular type of stimulus). A list is created in which each list element consists of an identification code (neural or laboratory event) and a time at which it occurred. The time can be specified either in absolute clock ticks since the beginning of the recording, or as an elapsed time (δt) since the previous event. Note that with most hardware it will be necessary to deal with clock register overflows in keeping event timing.

Although there is obviously an infinity of detailed arrangements which satisfy the general requirements, we find that it is advisable to end with a list that is in ASCII, consisting of one event code and one absolute time count per line. If necessary some event codes can indicate the presence of additional numbers on the line. Although this is not a compact way of storing the data, it has the great advantage that it is readily read by analysis programs without complex decoding of bits or bytes, and in addition is easily transferred between different laboratories. Considerable space saving can be obtained by storing such data files in "compressed" form, using standard ASCII text utilities for the purpose.

In general, the creation of appropriate real-time programs and integration with available hardware is a user problem. There is little commercially available that meets the requirements, especially at a reasonable price [but see the systems mentioned earlier (5, 6)].

Analysis of Neuronal Interactions

Problem

Once the data acquisition system has produced an event list, we can leave the restrictions of real time, and perform various appropriate calculations with the list. The goal of analysis is to identify "interesting" timing relationships among the several recorded spike trains and events (stimulus or behavioral) in the laboratory. Generally, favored timing relationships, namely, those that depart significantly from chance or expected values, are the signature of interactions among the observed neurons or of the effects of the laboratory events on these neurons. In seeking and interpreting favored timing relationships it is important to decide on an appropriate time scale, since different scales correspond to different physiological mechanisms. It is also essential in all such calculations to develop and use appropriate controls and statistical tests of significance. A general overview and many computational details of such analyses are in the books by Glaser and Ruchkin (9) and by Eggermont (10). Additional references to other recent work are given below. We note that none of these analyses are practical without computer technology, and in fact that the original developments in this area occurred in close temporal juxtaposition to availability of computers.

Traditional Pair Cross-Correlation

The usual method of examining relative timings of spike events in two trains is to use cross-correlation (11). Peaks and troughs in this measurement represent favored delayed coincidences and can be interpreted as signs of excitatory or inhibitory influences between or to the observed neurons. If the stimulus modulates the firing rates of either or both neurons, the rate of delayed coincidences is also modified. An appropriate data shuffle or shift with respect to stimulation instants can be used to destroy any detailed neuronal correlations while preserving the effects of stimulus modulation. The resulting shift predictor [or an alternate computation based on cross-correlation of the two peristimulus time (PST) histograms] is used to estimate the amount of the raw correlogram that represents stimulus-associated changes. The remainder represents neuronal interactions and is parsed into (1) direct interaction between the observed neurons and (2) shared input to both neurons. The distinction between these two possibilities is made on the basis of peak width and offset from zero delay. Final results of cross-correlation analysis can therefore be expressed as an "effective connectivity" among the observed neurons. This is the simplest "wiring" diagram that can explain the observed correlogram; we emphasize, however, that it is *not* the anatomical connectivity, but may be a subset of it. (The effective connectivity only shows synapses and influences that were active during the data acquisition, and it cannot specify any interneurons that are interposed between the observed neurons.)

Gravity Transformation

When more than two neurons are simultaneously recorded, analysis of relations in the entire group can of course be built up by analysis of all possible pairs by the traditional cross-correlation methods. Unfortunately, however, this is a combinatorial disaster: even 10 simultaneously observed neurons means 45 pairs to be calculated, interpreted, and put into a coherent overall scheme. Obviously a method that treats all observed spike trains in a single computation is desirable.

The gravity transformation (12, 13) represents each of the N observed neurons by a particle in a Cartesian N space. Each particle is given a "charge" which is a low pass filtered version of the spike train of the corresponding neuron: the charge is incremented by a fixed amount at each spike and decays between spikes with an appropriate time constant. Now we assume that there is a force between each two particles that is proportional to the product of their charges (but not to their distance) and that acts along the line between them. Thus the net force on a given particle is the vector sum of the individual pair forces. In a viscous medium, this will result in a proportional velocity.

At the beginning of the analysis we position all particles equidistant from each other, that is, at the vertices of a N-dimensional hypercube. In the subsequent movement, particles which represent neurons that have excessive near coincident firing will tend to aggregate. The final positions of the particles will therefore show a spatial clustering that is a representation of the timing relationships in the original data.

The clustering can be examined by standard methods in the full N-space. Alternatively, it is possible to project the particle movements from the N-space to a plane in order to provide a visualization of the neuronal relationships. (Projection loses information and can be quite misleading: caution and controls are essential.) The order and speed of aggregation of particles can be interpreted in terms of strengths and types of interactions among the observed neurons. Changes in the average (noisy) aggregation speed are signs of dynamic aspects of the neuronal interactions. As in the case of ordinary cross-correlations, it is possible to correct the gravity representation for the rate modulatory effects of stimulation in order to study the purely neuronal interaction effects.

The computational requirements of the gravity analysis are quite demanding, putting it out of the range of most PCs (possibly excluding the 486 variety) and suggesting the use of workstations. At the heart of the calculation is a three-dimensional matrix (the force between particles i and j has N components in the N space) which must be updated every time step. With some interesting variations of the analysis in which the force of particle i on particle j is not equal to the force of j on i, the entire matrix (rather than half of it) must be calculated (see Ref. 13). The calculation time thus increases as the third power of the number of recorded neurons. With typical workstations and recordings of 10–15 neurons, it is possible to make gravity analysis of the functional groupings in several minutes of raw recording with about 5 min of computation. This is not real time, but near enough to use during an experiment if necessary for guidance of parameters or other conditions.

Joint Peristimulus Time Histogram

Ordinary cross-correlation produces an estimate of neuronal interaction that is averaged over all the available data. It is entirely possible that the actual interactions are a function of time, varying in response to both uncontrolled and controlled (e.g., stimuli) factors. The joint peristimulus time (JPST) histogram uncovers the stimulus–time-locked aspects of interactions in a way similar to that of the PST histogram for the stimulus–time-locked firing of a single neuron (14–16).

Using spike train data obtained from two neurons during repeated presentation of physiological stimuli or during repeated behavioral events, we make the following construction on an x–y plot. Represent activity and time after each stimulus presentation for one neuron along the x axis and for the other neuron along the y axis. (Each successive time marker for presentation of a particular type of stimulus falls at the origin; obviously other fixed positions for the time marker along both x and y axes are more appropriate for the successive markers of particular behavioral events). Now, for each stimulus presentation (marker event) find the logical AND of the x and y neuron spike occurrences and plot them as dots on the x–y plane. After many stimulus presentations, the plane will likely show a nonuniform distribution of dot densities. Bands of increased (or decreased) dot density that lie parallel to the axes can be ascribed to stimulus-associated modulation of firing of either neuron alone and can be related to the PST histogram measurement of time-locked firing. A band of increased density near the 45° line represents the stimulus–time-locked course of near coincident firing of the two neurons, and hence can be associated with their interactions in our usual sense. Variations of density along this diagonal represent stimulus modulations of the interaction in the same average time-locked sense as does the PST histogram for the individual firing probabilities.

In the above "scatter diagram" it is difficult to make more than qualitative assessments, and in particular it is impossible to make corrections for the direct modulations of firing rates of the individual neurons. (Such modulations would affect the near coincident firing in a trivial and uninteresting way: we want to study only the neural components of the interactions and their stimulus-related time course.) The appropriate methodology (15) involves binning the scatter diagram to produce a JPST histogram. Subsequently we use the PST histograms and their standard deviations to make a bin by bin correction to the raw joint PST histogram.

The result is a complete elimination of "ridges" lying parallel to the axes; any remaining topography is due entirely to neuronal interactions rather than direct stimulus effects. In particular, the counts in any ridge lying near the principal diagonal can be exhibited as the corrected PST coincidence histogram, and they show the stimulus–time-locked course of the purely neuronal interaction.

Summation of bins along paradiagonal lines in the original JPST histogram produces the ordinary cross-correlation of the two spike trains. Similar summation on the corrected JPST histogram gives the cross-correlogram properly corrected for direct stimulus modulation of the individual firing rates. This is a far more accurate correction of the cross-correlation than can be obtained by the usual methods of shift or PST based predictors.

In addition it has been possible to develop significance testing for the topography (16), so that it is possible to establish confidence limits for the effects seen in the corrected JPST histogram. This is expressed in the form of "surprise" values and is most convincing when regions of the JPST histogram (i.e., groups of adjacent bins) show high significance.

Computational requirements of the JPST histogram are trivial, and times are limited largely by disk access to read the original spike occurrence time list. The memory requirements are less modest. The present program we use makes use of four arrays, each 400×400 of real numbers, that is, something like 3 megabytes (counting other smaller needed arrays) for the data alone. This is usually no problem on a workstation unless too many jobs are running simultaneously, but it may be awkward on a PC.

Repeating Patterns and Unitary Events

All the measurements above are averages, either over all available data or in a time-locked sense over the stimulus presentations or behavioral occurrences. However, we know that real computational functions in the nervous system do not require averaging over time or over many repetitions of identical external context. Thus we need analysis that detects events in the nervous system that are basically unitary (covering some time period up to a few seconds), but may repeat occasionally. We may associate such events with some aspect of behavior or with significant computational processes.

One example of event is a particular pattern of firing among the observed neurons. As usual, one must choose an appropriate time scale and duration in the defining criteria. If we write the activity of the several observed neurons in a musiclike notation (one neuron per staff line), we are looking for melodies. The pattern detector algorithms and programs (17) identify any such melodies in the data that repeat two or more times, and they also make a data-based prediction of how many such melodies of a given duration and repetition level would be encountered if the spike trains came from neurons that were completely independent. This means that we can detect the presence of an excess number and variety of melodies, but the programs do not identify which of this set of repeating patterns represent chance occurrences and which represent real, physiologically significant events. As a final step, we examine the times of occurrence of detected patterns relative to any external stimulus or behavioral events in an attempt to associate some of them with particular computational processes in the nervous system.

The straightforward brute force approach to repeating pattern detection among some number of simultaneously observed spike trains is impractical even with large workstations. Appropriately devious algorithms allow the analysis to be carried out in a few minutes for a typical data set. Memory requirements are not great, but computational power remains important.

Simulation Applications

Problem

Computer simulation of neural elements and networks can be used for many purposes and in many ways. It is extremely important to define the goals of the work in order to make an intelligent choice between available simulator programs and to define the ranges of whatever parameters are to be studied. The two most useful applications of simulation for us have been (1) validation and sensitivity testing for analytic computations and (2) modeling of the possible arrangements in the nervous system that could account for the observed real data or for various functional (possibly behavioral) computations. The first of these applications requires generating spike trains according to some set of underlying interactions of their sources and then using the analytic tools to see the extent to which they can recover these underlying relationships. The second application is much less restricted and requires considerably more choices.

Realism and Scale

The extent to which a simulation has biologically realistic elements and the scale or size of the simulation must be chosen with care, taking into account the goals of the process. Some of the available simulator programs allow a very large choice of detailed channel properties, ionic currents, spatial and temporal membrane parameters, dendritic and synaptic geometry, threshold and refractory properties, action potentials, etc. Other simulator programs simplify the individual neural elements, sometimes to being geometric point neurons with minimal ionic and membrane properties. Although detailed simulation incorporates large amounts of known biological data, it inevitably results in slower computation. Vast element detail is only practical if the scale, that is, the number of and connectional complexity between elements to be used, is relatively small. This inevitably poses a quandary: it is hard to establish whether the phenomena being simulated are the result of intrinsic element properties or of network properties or of a combination of both.

The optimal overall strategy would seem to be to include the maximum amount of detail and realism consistent with the required scale and reasonable computation time. At this maximum amount of detail, the simulation might be termed a "replica model." Subsequently, the model can be simplified in an effort to find the truly critical elements or properties that determine the overall performance under study. An alternate strategy, which is sometimes more practical, is to start near the minimum amount of detail and attempt to attain the necessary function. When this proves impossible, additional realism can gradually be added. Neither strategy necessarily guarantees understanding, even if the desired functions are adequately reproduced.

Programs

A number of different simulator programs are available, differing in capabilities for realism and scale, and with various levels of convenience in setup and output. At their least biological extreme, the elements are represented as adders (inhibition is either subtraction or division) with some nonlinear in–out function and a threshold. At the most biological extreme it is possible to model channels, dendritic geometries, and all sorts of other computationally demanding detail. Reviews of some of the various simulators can be found in the books edited by Koch and Segev (18) and by Eeckman (19). A traditional "connectionist" stimulator is available from Rochester University (20). Another large set of simulation programs (which we have used extensively) is described by MacGregor (21). Still other simulators are passed around privately without benefit of formally published descriptions.

The choice among these various programs is difficult without extensive evaluation of capabilities relative to the projected needs. This is by no means an easy problem because of the heavy investment of time required to make use of any of the systems. Often enough, none of the available systems will be satisfactory, and it is necessary to write a program. In a sense this is of course reinventing the wheel, but special requirements can in fact rarely be met by systems designed for great generality and flexibility.

Computers

The machine requirements for neural simulation projects can best be described as having no upperlimit. With typical workstations it is barely possible to do useful things (at the cost of long runs). The moment that the simulation must exceed networks of 1000 even relatively simple elements, it is necessary to think of large machines. Unfortunately most of the available simulation systems are not easily optimized for the current, somewhat parallel, larger machines. Massively parallel machines are available at some universities but would require extensive *de novo* programming efforts.

Interpretations

Goals

Although not central to the thrust of this chapter, it is appropriate to discuss briefly the physiological results, implications, and future development of multineuron spike train recordings. The principal purpose of all the rather complex and computer-based technologies we have been describing is to allow study of neuronal assembly properties, particularly with respect to their static and dynamic organization, but also with regard to their computational processes. Additional detail about these subjects can be found in Refs. (22) and (23).

Effective Connectivity

Analysis of multineuron separated spike train recordings with cross-correlation and the gravity representation allows easy detection of any functional assemblies among the observed neurons. This static organization is a description that represents an average over the observation period, and it is most easily expressed as the "effective connectivity" and its weights. We stress again that this is the simplest wiring diagram between and to the observed neurons that explains the amount and time course of favored near-coincidence firing. The actual anatomical connectivity may be far more complex, may contain elements or influences that are inactive or that modulate the spike trains too weakly for detection, and should be studied by other specifically anatomical methods. (Those methods, conversely, give little insight into activity or strength of connections.)

Application of static analyses to multineuron observations from a number of laboratories, each dealing with different preparations and brain regions, has shown that static assembly organization depends partly on context. The organization detected among the observed neurons may depend on which stimulus is being presented or what behavior is in progress. Thus we can speak of an organization that is an average over many presentations or occurrences of the same context.

Dynamics

Analysis of data from multineuron recordings with the JPST histogram shows that assembly organization is often dynamic on several time scales within a stimulus presentation or behavioral act. Membership of a given neuronal assembly may vary on time scales of 20 msec to 1 sec, and individual neurons may change their affiliation between different assemblies on the same time scales.

Analysis with the gravity representation sometimes shows short periods (perhaps 15–30 sec) where a particular aggregation process slows, halts, or even reverses. These are unitary dynamic events in assembly organization that are usually not tightly and repeatedly time locked to stimulus or behavioral events as are the modulations detected by the JPST histogram. Presumably they represent other events in the organism that are not under laboratory control.

The computational or behavioral significance of these widely observed assembly dynamics is presently not known. Additional analyses and experimental paradigms must be developed to progress along these lines.

Evaluation of Models

We have described above some of the considerations limiting neuronal simulation models that mimic particular brain regions or processes. Usually when such models are studied, the principal criterion for "appropriate" performance is the overall output for given inputs. The degree of physiological verisimilitude will vary with circumstances but is usually concerned with parameters that define element properties and specification of the connections between elements and groups of elements. We would strongly suggest that it is also appropriate to mimic the dynamic aspects of neuronal assembly organization that have been described above. At this stage we do not know how (if at all) such dynamics contribute to the possible performance, but surely working with models that incorporate such dynamics will be informative. It should be possible to examine the extent to which such new classes of model do mimic real neuronal assembly dynamics by conducting the same multineuron "recordings" and analyses on them.

Conclusion

In this chapter we have examined computer applications in the study of neuronal assembly organization and processes. Computers of various degrees of power are essential to all aspects of the enterprise, starting from acquisition of multineuron data in controlled situations, proceeding through analysis of the observed spike trains in terms of neuronal interactions and effective connectivities, and culminating in constructing models that can be used to test our understanding of the observations.

This is truly an area of neurobiology that could not have been studied before computers were widely available, and even now it is sufficiently demanding of both computer and other technology that relatively few laboratories are appropriately equipped for such work. As technical development continues and ever-increasing computer power becomes available in the laboratory, it should be possible to study larger numbers of neurons simultaneously than is currently possible. Given the number of neurons in even selected small parts of mammalian nervous systems, the observations will still represent very small samples out of very large populations. It remains to be seen if this resource-demanding path will lead to major gains in understanding assembly contributions to the computational processes of the brain.

References

1. E. M. Schmidt, *J. Neurosci. Methods* **12**, 1 (1984).
2. E. M. Schmidt, *J. Neurosci. Methods* **12**, 95 (1984).
3. M. Salganicoff, M. Sarna, L. Sax, and G. L. Gerstein, *J. Neurosci. Methods* **25**, 181 (1988).
4. W. E. Faller and M. W. Lutges, *J. Neurosci. Methods* **37**, 55 (1991).
5. Brainwave Inc., Broomfield, Colorado 80020.
6. Spectrum Scientific Co, Dallas, Texas 75206.
7. G. L. Gerstein, M. Bloom, I. Espinosa, S. Evanczuk, and M. Turner, *IEEE Trans. Syst. Man Cybernet. SMC-13*, 668 (1983).
8. M. Abeles and M. H. Goldstein, *Proc. IEEE* **65**, 762 (1977).
9. E. Glaser and D. Ruchkin, "Principles of Neurobiological Signal Analysis." Academic Press, New York, 1976.
10. J. J. Eggermont, "The Correlative Brain." Springer-Verlag, Berlin, 1990.
11. D. H. Perkel, G. L. Gerstein, and G. P. Moore, *Biophys. J.* **7**, 419 (1967).
12. G. L. Gerstein, D. H. Perkel, and J. E. Dayhoff, *J. Neurosci.* **5**, 881 (1985).
13. G. L. Gerstein and A. M. H. J. Aertsen, *J. Neurophysiol.* **54**, 1513 (1985).
14. G. L. Gerstein and D. H. Perkel, *Biophys. J.* **12**, 453 (1972).
15. A. M. H. J. Aertsen, G. L. Gerstein, M. Habib, and G. Palm, with the collaboration of P. Gochin and J. Kruger, *J. Neurophysiol.* **61**, 900 (1989).
16. G. Palm, A. M. H. J. Aertsen, and G. L. Gerstein, *Biological Cybernet.* **59**, 1 (1988).
17. M. Abeles and G. L. Gerstein, *J. Neurophysiol.* **60**, 909 (1988).
18. C. Koch and I. Segev (eds.), "Methods in Neuronal Modeling: from Synapses to Networks." MIT Press, Cambridge, Massachusetts, 1989.

19. F. H. Eeckman, "Analysis and Modeling of Neural Systems." Kluwer Academic, Boston, 1992.
20. N. H. Goddard, K. J. Lynne, and T. Mintz, *Technical Report 233* (*revised*), Computer Science Department, University of Rochester, New York, 1988.
21. R. J. MacGregor, "Neural and Brain Modeling." Academic Press, New York, 1987.
22. G. L. Gerstein, P. Bedenbaugh, and A. M. H. J. Aertsen, *IEEE Trans. Biomed. Eng.* **36,** 4 (1989).
23. A. M. H. J. Aertsen and G. L. Gerstein, *in* " Neuronal Cooperativity" (J. Kruger, ed.), p. 54. Springer-Verlag, Berlin, 1991.

[5] Computer Methods in Nuclei Cartography

E. C. Hirsch, O. Lejeune, G. Colliot, G. Corkidi,
and M. Tajani

Introduction

Histological techniques, such as immunohistochemistry, enzymohisto-chemistry, and *in situ* hybridization, have been shown to be reliable analytical methods for studies in neuroscience (1, 2). For example, *in situ* hybridization allows modifications of gene expression to be quantified at both a regional and cellular level (3–5). The advantage of these techniques consists of a higher spatial resolution than "test tube" molecular biology or biochemical techniques (such as Northern blot or dot-blot procedures) commonly used for the study of genetic expression in the central nervous system (CNS). On the other hand, the sensitivity of histology may be lower than biochemical techniques using amplification procedures like the polymerase chain reaction.

Considerable interest has been focused on quantification procedures in histology because of the fundamental importance of numerical parameters in brain function. In line with this concept, the symptomatology of numerous neurological disorders has been related to the under- or overrepresentation of well-characterized neurotransmitters (6). Among neurological disorders, Parkinson's disease provides a good paradigm because, in the mesencephalon, only a selective population of neurons degenerates (7). Two important questions remain unanswered, however: (1) which specific lesion in the mesencephalon induces the symptomatology? (2) Why do the neurons degenerate? One possible means of solving this problem consists of analyzing the neurons which are preserved in the mesencephalon of patients that died with diagnosed Parkinson's disease, in order to identify the characteristics of the neurons which degenerate in this disease. This task has been greatly facilitated by the use of computer-based image analyzers, which allow the rapid quantification of thousands of neurons. The analysis of tissue from the CNS with a highly arranged cytoarchitectony causes some problems, however. During macroscopic analysis, at low magnification, different cell groups or regions of interest, such as the brain stem nuclei, can easily be analyzed. It is difficult, however, to perform a precise quantitative analysis of the cellular elements contained in the region of interest owing to the low magnification. By contrast, at higher magnification, the neurons can easily be counted or analyzed, but their precise location within the region of interest is difficult to

Methods in Neurosciences, Volume 10

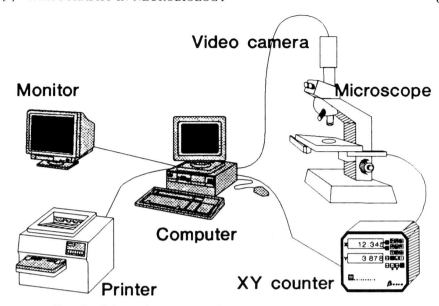

FIG. 1   Schematic representation of the image analysis system.

ascertain, since only a small portion of the region of interest can be visualized at a time. The problem is greater at the electron microscopy level, where analysis magnification is even higher.

A computer-based method is presented that allows microscopic and macroscopic quantitative examination of CNS tissue sections to be combined in a single image analysis system, which is now marketed by Biocom (Les Ulis, France). The method is suited to both light microscopy and electron microscopy using a transmission or scanning electron microscope. After a brief description of the equipment used and of the capabilities of the software, we shall focus on the application of the method to the analysis of the brain stem nuclei implicated in the pathophysiology of Parkinson's disease.

Equipment

Software application developed for nuclei cartography is implemented on an image analysis system, consisting chiefly of the acquisition and processing units (Fig. 1). The acquisition system is critically important. The success of neuroimaging depends principally on the quality of the primary images, but also on the acquisition device.

Acquisition Device

Because the size of the biological samples analyzed can vary from a few centimeters to a few millimeters, the observation of biological samples is performed with different types of acquisition devices. However, all systems require good spatial resolution, an intensity range suited to the initial image, and an absence of optical image distortion.

Macroscopic images are generated on a highly stable light box. For photometric measurements, stability of the light source is essential for good quality results. High-frequency fluorescent illuminators have been shown to meet this requirement (G. Corkidi, 1987, unpublished data). With a controlled power supply, at a constant 18°C, the variation in field illumination using an Osram (Paris, France) dulux L36W/21 lamp was shown to be less than 3% over 24 hr of continuous measurement.

Macroscopic images are generated on a light microscope equipped with a polarized incident light source for analysis. In microautoradiography, special attention needs to be given to the choice of microscope and lenses. The best results were obtained with either a Nikon Labophot (Charenton-le-Pont, France) or a Leica Laborlux (Rueil-Malmaison, France) microscope. More sophisticated microscopes often result in light interference owing to additional mirrors or lenses in the optical train.

Ultrastructural images are generated with a Philips 525M scanning electron microscope (SEM) (Bobigny, France) or a Jeol 1200SX (Rueil-Malmaison, France) transmitted electron microscope (TEM), although other microscopes yet to be tested may also be compatible with the system. Image processing in electron microscopy has been greatly simplified, as most electron microscopes are now controlled by microprocessors able to correct geometrical distortion caused by changes in magnification. Both microscopes used in the study are equipped with motorized stages, allowing a sample to be replaced in the same position to an accuracy of 50 nm.

The different image generators used are connected to image sensors which consist of a video camera, except for the SEM because the analogical signal directly generates an image in the microscope itself. Many types of video cameras can be used for image analysis but all must meet the following criteria: (1) high spatial resolution, with more than one pixel of the image covering the smallest elements to be analyzed; (2) a good signal-to-noise ratio; (3) a lack of geometrical distortion generated by the system; and (4) a sensitivity that is adapted to the image. Two types of cameras are generally used for image analysis: the CCD (charge-coupled device) and the CTV (cathode-tube video). For electron microscopy, the low intensity of the signal necessitates an SIT (silicon intensified tube) camera sensitive to 10^{-4} lux. CCD cameras have numerous advantages over CTV cameras: long life, high reliability, high sensitivity, low sensitivity to magnetic fields, and high resistance to vibration and shock. Moreover, the spatial distortions induced by CTV cameras are totally eliminated by the CCD chip, which represents a plane detection surface. In addition, they are popular because of the miniaturization of the photosensor network. For the present study, the following cameras were selected: a Cohu (San Diego, California) 4710 CCD camera (699 × 580 photosensors; sensitivity 0.32 lux) for light microscopy and a LH4036 Lhesa (Cergy-Pontoise, France) camera (Newvicon 25-mm tube; sensitivity 10^{-6} lux) for transmission electron microscopy.

Object Position Sensor

The main feature of the system is that the position of each field of view is continuously measured with reference to a virtual space of 6.4 × 6.4 cm. The measurements are performed using a Heidenhain (Traunreut, Germany) LS 107 X, Y optoelectronic sensor system mounted on the stage of the microscope. This allows the linear measurement of X, Y movements of the stage by reference to an absolute or relative landmark, to an accuracy of 1 μm. The X, Y sensors are hooked up to a Heidenhain VRZ 735 or VRZ 720 electronic reversible counter, which is connected to the personal computer (PC) by an asynchronous RS232 linkup (9600 bands). The same measurements are performed on the electron microscope equipped with a motorized stage, which uses step by step motors and a counter accurate to within 0.25 μm.

Image Digitization and Visualization

To be analyzed by the computer, the image produced by the camera is first digitized by an electronic board hooked up to the PC through the AT bus. The optimal configuration for light and electron microscopy differs, however, because in electron microscopy high background noise first needs to be filtered. With an SEM, the slower the scan (slow-scan mode), the more intense the signal and the better the quality of the image. In contrast, in video scanning, the signal/noise ratio is poor and has to be enhanced by integrating several digitized images (8). In this system, two different digitizing boards are used for light and electron microscopy. The Matrox (Dorval, Canada) PIP 1024 × 1024 electronic board used for light microscopy offers four main functions: digitization of standard video signals; storage of four 512 × 512 on 256 gray levels (8 bits) digital images; display of one of the four images in gray or pseudo color; and image processing (filters, histograms, etc.). The Matrox MVP-AT 1024 × 1024 digitizer provides some additional functions: electronic processing, color visualization, and integration of several digitized images (indispensable for the SEM and TEM).

Characteristics of Software for Nuclei Cartography

The goal in developing software for mapping and counting objects within tissue sections is to relieve the histologist of the tedious work of manually analyzing the objects and constantly switching between magnifications to locate them. Thus, the main characteristics of software application are the capacity to combine different microscope magnifications and provide a quantitative analysis of objects located within regions of interest.

Definition of Regions of Interest and Objects for Analysis

Two different types of elements constitute the "building blocks" of nuclei cartography: the nuclei, defined by their outer boundaries, and the objects located within the nuclei, which, in the CNS, may represent neurons or glial cells. Regions of interest and objects are each analyzed manually or semiautomatically. Manually, the objects or the regions are outlined with a mouse-controlled cursor on the video screen. This results in a graphic overlay specific to each object or region of interest. Semiautomatic detection of the objects is based on an edge-detection procedure, namely, an operation to identify regions of an image having similar properties. In existing software applications, edge detection is performed on an isodensity basis. The complete analysis of one section results in a map which is made up of the outlines of the section, the regions of interest, and the objects which may or may not be within the regions of interest.

The main characteristic of this software application is that analysis is not limited to the field of view visualized on the video screen, but can be extended to several fields of view up to a maximum of 6.4 × 6.4 cm. Thus, outlines of the regions of interest can be drawn on multiple fields of view. This is possible since the graphic overlay moves simultaneously with the video image on the screen. In fact, this software application can continuously read the position of the field within the section, owing to the displacement of the sensors mounted on the stage. Thus, the position of the image visualized on the video screen is always known, and the position of the graphic overlay recalculated accordingly. The realignment of the graphic overlay over the video image, following displacement of the microscope stage, is illustrated in Fig. 2a–c. One critical characteristic of this system is the refreshment time of the graphic overlay, since a microscope stage displacement should be immediately followed by an equivalent displacement of the graphic overlay. Using a 80386 Compaq (Les Ulis, France) microcomputer running at 25 MHz, the time necessary to display 2000 neurons is less than 1.5 sec. Moreover, this software application allows objects or regions of interest to be drawn at any

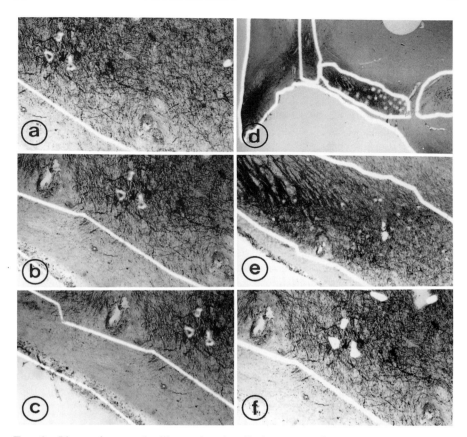

FIG. 2   Photomicrographs illustrating the displacement of the graphic overlay with the microscope stage movement and the management of the magnifications. All pictures were taken using a Leitz Orthoplan microscope equipped with a Microvid system which allowed the graphic overlay to be visualized in the microscope. (a)–(c) illustrate the way the graphic overlay follows the movements of the stage of the microscope. (d)–(f) illustrate the way the size of the overlay adapts to the new scale when the microscope magnification is changed. Note that the group of neurons shown in (f) is visible in (d) and (e).

microscope magnification. This software application manages the scale of graphic overlay in such a way that its size corresponds to the video image (Fig. 2d–f). To handle the different magnifications, the software application saves in a special file the different characteristics of the microscope lenses (magnification, size of field of view, optical axis adjustment). The procedure allows the region of interest to be drawn at low magnification and the outlines of the neurons at a higher magnification (Fig. 2).

In addition, the system enables the complete tissue section to be scanned, systematically or otherwise, without the same object being counted twice, since the graphic overlay moves with the object; without losing one's place in the section, since the system allows the field of view to be located within the map of the whole section; and without involuntarily exploring fields of view located outside the regions of interest, since the boundaries of the regions remain clearly visible at high magnification.

Realignment of Tissue Section with Cartography

Because there is often too much information to analyze in a section during a single analysis session or because successive double staining is performed, repeated analysis of the same section has to be performed. Thus, it is crucial to solve the problem of realignment of the map and the associated tissue section. The translation realignment is performed by freezing the graphic overlay of the section, moving the video image to the position of the graphic overlay, then realigning it with the video image. The standard refreshment of the graphic overlay with video image displacement is then reactivated.

Similarly the map of a given section can be realigned with the adjacent section, using rotation or even deformation of the graphic overlay. The adjustment algorithm is based on pairs of points (X, Y and X', Y') which enable the specific transformation between the graphic and the position of the section to be described:

$$(x', y', 1)^t = T(i, j)(x, y, 1)^t$$

where $(x, y, 1)^t$ is the initial position; $(x', y', 1)^t$ is the final position; and $T(i, j)$ is the 3×3 matrix of transformation.

Quantitative Results of Analysis

The number of objects of each type within each region of interest is determined by calculating the number of intersections between the vector (G_jG_i) and the boundaries of the region i (G_i and G_j represent the centers of gravity of the region i and the object j, respectively). An odd number indicates that the object belongs to the region, an even number that it is located outside the region of interest.

For each region of interest and each object, classified by region, numerous morphometric criteria are measured: perimeter, area, diameter (including

Feret's diameter in a given direction), orientation of the diameter, center of gravity, etc. Densitometric measurements can also be calculated for each region or object, from the Beer–Lambert relation, as follows:

$$OD = -\log(NG_c/NG_{max})$$

where OD is the optical density; NG_c is the gray level measured; and NG_{max} is the gray lever corresponding to maximal transmission. Quantitative results concerning the morphometric and densitometric parameters are displayed as tables or histograms. In addition, objects can be selected according to one or more of these criteria.

Silver Grain Counting for Histoautoradiographic Analysis

The use of radioactive probes in receptor binding studies, *in situ* hybridization, and immunoautoradiography has considerably increased the number of applications of histoautoradiography. The quantification procedure of histoautoradiography relates the amount of radioactive material present on the tissue section to the number of silver grains revealed in the photographic emulsion with which the section is coated. This procedure requires the use of standards which allow the amount of radiolabeled probe in the tissue to be related to the number of silver grains per cell. The use of the standard was extensively reviewed recently (4) and will not be discussed here. The grain-counting method is based on the principle that silver grains reflect polarized light whereas the tissue section does not (9) (Fig. 3a).

Using microphotometry, it has been demonstrated that the amount of light reflected by silver grains is directly proportional to the number of silver grains present in the emulsion (9). This characteristic was used to measure the optical density of cells using the image analysis system. Thus, a curve relating the optical density to the number of silver grains allows the number of silver grains per cell to be determined using a simple densitometric measurement. This technique has numerous advantages. First, it is very rapid since it is necessary only to outline the cell and perform a densitometric measurement to determine the number of silver grains per neuron. Compared to manual silver grain counting or edge-detection techniques to extract the outline of the silver grains, the analysis time is decreased 400-fold and 5-fold, respectively. Second, measurements are independent of the section counterstaining, since only silver grains reflect polarized light. Third, measurement is relatively independent of the focal plane, as illustrated in Fig. 3b–d. This is particularly important because the thickness of a cryostat section may vary from one part to another.

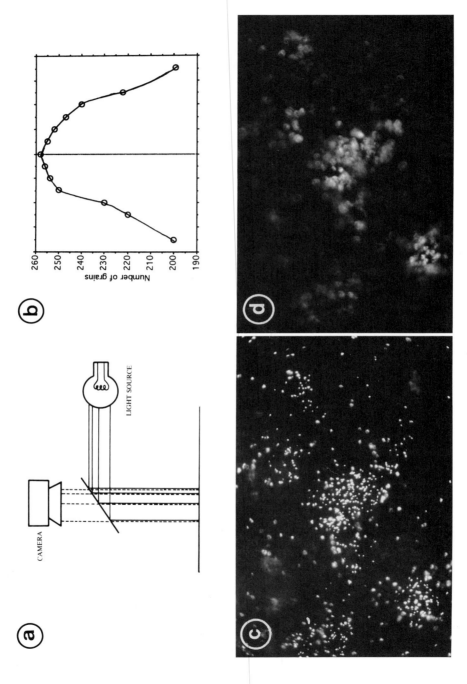

CAMERA

LIGHT SOURCE

Number of grains

260 250 240 230 220 210 200 190

(a)

(b)

(c)

(d)

Visualization of Data

Two-dimensional (2-D) maps locating the objects and the region of interest are generated on the screen of the video monitor (Fig. 5a). All or part of the maps can be edited with text and printed out using a Canon (Courbevoie, France) LBP8II laser printer (resolution 300 dpi). Synthetic images can also be constructed from the 2-D maps, relating the density of objects within a surface unit to a pseudo color scale. Thus the density of a given cell type within a region is associated with a color (blue, low cellular density; red, high cellular density). This method can also be used to compare the densities of two different cell populations A and B, since it enables a cell density image to be established for each population and images to be mixed. Thus, a "ratio image" representing the relative density of each population of cells can be calculated using the following formula:

$$F(i, j) = \{A(i, j)/[A(i, j) + B(i, j)]\}N$$

where $F(i, j)$ is the color of the pixel (i, j) in the resulting image; $A(i, j)$, the color of the pixel (i, j) in the population A image; $B(i, j)$, color of the pixel (i, j) in the population B image; N, normalization factor. $F(i, j) = 0$ if $A(i, j) = 0$ and $B(i, j) > 0$; $0 < F(i, j) \leq N$ if $A(i, j) \geq 0$; and $F(i, j) = N +$ background, if $A(i, j) + B(i, j) = 0$.

Three-Dimensional Modelization

The brain is a tridimensional structure and its representation by a series of 2-D maps rarely provides a realistic image of its anatomy. The present software application thus allows a 3-D structure to be created using maps of serial sections. The preliminary step in any 3-D reconstruction procedure is the realignment of the sections according to landmarks common to all sec-

---

Fig. 3   Schematic representation of the grain-counting method (a). The section, which is mounted on a glass slide and covered by a photosensitive emulsion, is analyzed under an intense polarized light source. Only the silver grains in the emulsion reflect the polarized light, which is measured by the image analysis system. (b)–(d) illustrate that the grain-counting procedure is independent of the focal plane. The curve relating to the number of grains detected by the system as a function of the focal plane is shown in (b). The zero value represents the "ideal focal plane" which is illustrated in photomicrograph (c) taken under polarized light with a Leitz Laborlux microscope equipped with a $50\times$ NPL Fluotar lens. The photomicrograph shown in (d) illustrates a 2-μm deviation from the ideal focal plane.

tions. Often, lasers are used to bore holes within the tissue. The realignment procedure used for the multiple analysis of sections is then applied, as described above, resulting in the collection of sections being displayed in a 3-D perspective with management of hidden surfaces (Fig. 5b).

Because the display of the contours in perspective is frequently unsuited to the quantitation or visualization of complete objects, it is necessary to build a framework in order to describe the surface of the biological material. The method used to create the 3-D object is called "polygonal wireframe" and enables a mesh composed of simple polygons, such as triangles, to be obtained. The next step in reconstruction is the shading process, which consists of filling each triangle of the mesh with a color which depends on the position of a modeled light source (Fig. 5c). This 3-D reconstruction method is classic and has been used by many authors (10). In the present case, the advantage is the realization of 3-D maps from the high-resolution 2-D maps obtained with this software application, as described above.

Application of Computer Method to Brain Stem Nuclei Analysis in Parkinson's Disease at Light Microscopy Level and Bacteria Detection at Electron Microscopy Level

The image analysis system has been evaluated at the light microscopy level for the analysis of the brain stem nuclei involved in the pathophysiology of Parkinson's disease. The purpose of the study was to determine which neurons degenerate in the mesencephalon in Parkinson's disease and to analyze the characteristics of neurons spared by the pathological process. The cardinal findings in our analysis of the nigral complex in Parkinson's disease are that (1) cell loss is nonuniform across cell types within the complex, (2) cell loss is nonuniform across the different subdivisions of the complex, and (3) surviving neurons may already be undergoing a degenerative process. At the electron microscopy level, a software application has been tested with an SEM for the study of adhesion bacteria to human respiratory epithelium cells in primary culture.

Detection of Different Types of Neurons within Different Mesencephalic Nuclei in Patients with Parkinson's Disease and in Controls

In Parkinson's disease, massive cell death occurs in the dopamine-containing substantia nigra (11). A link between the vulnerability of nigral neurons and the prominent pigmentation of the substantia nigra, though long suspected, has not been proved (12). Two sets of neurons were studied in autopsy material from four patients who died with a diagnosis of Parkinson's disease and from three people without known neurological disease. These were catecholamine-containing neurons identified by tyrosine hydroxylase (tyrosine 3-monooxygenase) immunohistochemistry and neuromelanin-pigmented neurons detected in adjacent hematoxylin–eosin stained sections. The numbers and locations of neurons of each type were recorded with the computerized plotting system for sets of serial sections through the midbrain.

First, the outlines of the sections (3 × 2 cm) were drawn at a magnification of 1.6 times. In brief, the operator used a mouse-controlled cursor on the image analysis video monitor to draw the portion of the outline of the section visible on one field of view. Then the microscope stage was moved in order to visualize the next field, while retaining on the screen a small portion of the section limit already drawn. This procedure was simple since the graphic overlay already drawn was "attached" to the section and moved with the microscope stage displacement. The procedure was repeated until the entire section was outlined.

Delimination of the outlines of the nuclei was difficult to perform on tyrosine hydroxylase-immunostained sections; outlining was therefore performed on the adjacent section stained for acetylcholinesterase activity using the same method as for the outline of the section. The main problems with this procedure were that the section rarely had the same orientation on the glass slide and tissue shrinkage occurred during acetylcholinesterase staining. A two-step realignment was therefore performed to match the outline of the acetylcholinesterase-stained section to the graphic overlay drawn on the tyrosine hydroxylase-immunostained section. The first step was a succession of translations and rotations in order to overlap the gravity center of the two sections; the second was a shrinkage of the graphic overlay in order to compensate for the atrophy of the section due to acetylcholinesterase staining. In this way, five catecholamine-containing regions were delimited: the substantia nigra pars compacta (SNpc), the substantia nigra pars lateralis (SNpl), the central gray substance (CGS), the ventral tegmental area, divided into a medial (M) and a medioventral part (Mv), and the remainder of the tegmentum dorsal to the medial lemniscus, corresponding to catecholamine-containing cell group A8 (A8).

Neuromelanin-containing or tyrosine hydroxylase-positive neurons were then mapped at a magnification of 10 times. First, the tyrosine hydroxylase-immunostained section was realigned with the graphic overlay and the magnification changed through the software application in order to superimpose the graphic overlay on the video image at a magnification of 10 times. A characteristic symbol specific to each type of cell (neuromelanin or tyrosine

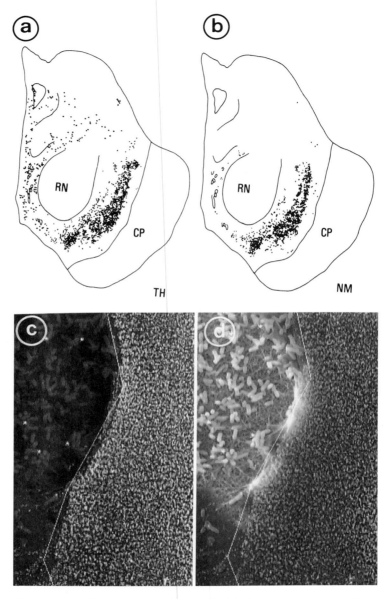

FIG. 4   Illustration of the 2-D maps obtained with the software application. (a) and (b) represent maps of adjacent sections from the human mesencephalon of a control patient stained to visualize tyrosine hydroxylase-positive (a) and neuromelanin-containing neurons (b). Note that the distribution of tyrosine hydroxylase-positive and

hydroxylase positive) was then placed, using the mouse-controlled cursor, over each neuronal perikaryon. The procedure was repeated for each field of view of the section. Double counting was avoided since the symbol marking a previously analyzed neuron was visualized on the graphic overlay when the neuron was present on the video screen.

Maps of neuromelanin-containing or tyrosine hydroxylase-positive neurons were thus obtained for the sets of serial sections (Fig. 4a,b). A perspective visualization of the different levels of analysis was obtained for each brain (Fig. 5). For each region delimited, the total number of neurons of each type was counted. These values were then plotted against the cumulative distance between the sections, and the surface area under this curve gave an estimate of the total number of neurons per region from its rostral to caudal extent (13).

Estimates of the total number of neurons within the substantia nigra pars compacta, ventral tegmental area (M + Mv), catecholaminergic region A8, and central gray substance of controls reveal that (1) most of the catecholaminergic neurons in the substantia nigra contain neuromelanin; (2) 98% of the catecholaminergic neurons in the central gray substance are devoid of neuromelanin; and (3) region A8 and the ventral tegmental area contain both types of neurons in the same proportion (13). In the parkinsonian midbrain, the loss of catecholaminergic neurons is severe in the substantia nigra (77%) but almost undetectable in the central gray substance (3%). The population of melanized catecholaminergic neurons is significantly decreased in catecholaminergic region A8, whereas the total population of tyrosine hydroxylase-containing neurons devoid of neuromelanin is not. This result can be generalized to all catecholaminergic areas analyzed in the parkinsonian mesencephalon and suggests that melanized catecholaminergic neurons are more sensitive to pathological processes than the nonmelanized catecholaminergic neurons (14).

---

neuromelanin-containing neurons differs in the dorsal mesencephalon. (c) and (d) illustrate use of the software at the electron microscopy level. The image obtained in backscattered electron mode (c) and in secondary electron mode (d) represents an epithelial cell labelled by a lectin/colloidal gold complex at a magnification of 4000 times. In backscattered electron mode, the labeled cell is more easily distinguishable from nonlabeled cells and from the background, but in secondary electron mode, the bacteria can be more easily counted. The contour of the cell and the symbol marking bacteria appear in white in the photograph.

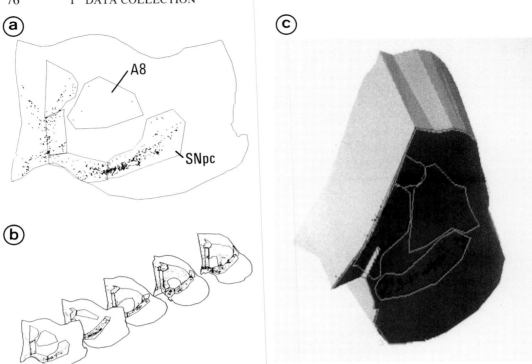

FIG. 5   Different modes of data visualization. (a) and (b) show maps of the catechol-aminergic neurons in the human brain stem and their location within the different mesencephalic regions. SNpc, Substantia nigra pars compacta; A8, catecholaminergic region A8. (a) shows single 2-D map; (b) shows a perspective view of a series of maps. The same series was used for the 3-D reconstruction shown in (c).

Densitometric Measurements within Neurons from Mesencephalic Nuclei

To analyze the functional state of the surviving catecholamine-containing neurons in the substantia nigra in Parkinson's disease, the level of tyrosine hydroxylase gene expression was studied by *in situ* hybridization histochemistry. The content of tyrosine hydroxylase messenger RNA in substantia nigra dopaminergic neurons was studied at the cellular level on emulsion-coated sections stained with hematoxylin. The number of silver grains per neuron was determined with the image analysis system at a magnification of 50 times on the section for which anatomical regions had already been drawn at a magnification of 1.6 times.

Grain counting was performed under polarized light, using the property of polarized light to be reflected only by silver grains (9). A standard curve of optical density as a function of an identified number of grains was established for each section. Through the standard curve, the computer converted the optical density of a neuron into a number of grains. Histological staining defined the boundaries of the cells. However, grain clusters usually overlapped the histological definition of the cell boundaries. Optical density per cell was measured over the area corresponding to the grain clusters overlying a neuron. Once the area corresponding to the grain cluster was delimited, the computer measured the surface of the area and the optical density over this area, which was then converted to a number of grains. Background grain density was estimated in the cerebral peduncle and in the substantia nigra at a distance from the dopaminergic neurons.

In the substantia nigra from parkinsonian patients, the number of neurons expressing tyrosine hydroxylase messenger RNA decreased compared to controls (5). In these patients, grain density per neuron was constantly subnormal. In no neuronal subpopulations did grain density exceed values in control subjects (mean decrease compared to controls was 63%). The decreased amount of nigral tyrosine hydroxylase messenger RNA, evidenced by the decrease in silver grains in *in situ* hybridization experiments, may reveal the suffering of dopamine-containing neurons still present at late stages of the disease.

Bacteria Detection at Electron Microscopy Level

The advantages of the computer method for light microscopy applications are even greater as the magnification of observation increases. Unfortunately, this software application has not yet been tested on specimens from the nervous system. However, one group of investigators (INSERM U314, Reims, France) has used it to study the adhesion of bacteria (*Pseudomonas aeruginosa*) to human epithelium cells in primary culture (15).

The adhesion of *Pseudomonas aeruginosa,* which can cause persistent respiratory infections, has been shown to be increased in the case of injury of the respiratory mucosa. The human nasal polyp primary culture constitutes a good model to study the respiratory epithelium since the cells migrate from the explant. The outgrowth so formed allows a good approach to be made to the study of epithelial cell adhesion during the process of migration, growth, and differentiation. Under these circumstances, two types of cells, ciliated or nonciliated, can be observed. The study revealed a high percentage of bacteria on ciliated cells, especially on those which migrated to the periphery of the culture.

The method of analysis consisted of defining, at a magnification of 3000 times, the outlines of the field of view and of the ciliated cells. The difference between the two surfaces corresponds to the surface of the nonciliated cells. Aggregates of bacteria in each region (nonciliated and ciliated) were counted and mapped at a magnification of 4000 times. Thus, the number of bacteria per square micrometer was obtained for each compartment. The resulting map allowed a quick and complete view of the complete distribution of the different cells in the outgrowth and of the bacteria among the different kind of cells.

The method was also used to analyze the same specimen under backscattered electrons. Indeed, the presence of specific sugar residues at the surface of epithelial cells is correlated to the maturation level of the cell. A cytochemical study was carried out using lectins conjugated with colloidal gold to identify the relationships between these sugar residues and bacteria adhesion. Under backscattered electrons, the labeled cells were more clearly distinguishable from the others (Fig. 4c,d).

In conclusion, this software application was developed to locate and count the bacteria within ciliated/nonciliated and labeled/nonlabeled cell populations using the map. In the future, two maps, one realized in SEM mode and the other in TEM mode, will probably be compared.

Conclusions

The computer method described extends the use of image analysis to more than one field of view and one working magnification. The procedure greatly simplifies the work of histologists since most of the routine work is now performed by computer. This software application has been used to count or map neurons or glial cells from the central nervous system but may help in the analysis of any histological tissue stained by various techniques. It also allows simple and complete storage of the characteristics of histological section. It differs from most other image analysis systems in that the objects or regions analyzed are not limited to the field of view. In contrast to sophisticated systems using motorized microscope stages, it enables a choice to be made of both the zones to be analyzed in the section, and the way the section is explored (random, interactive, or like most alternative systems, by the systematic selection of fields). Analysis is not restricted to only one magnification and several scans of the same section can be performed at different magnifications.

Our approach combines modern image analysis techniques with the advantages of X, Y plotters, which anatomists have been using for mapping since the 1960s. Given the high capacity and low cost of personal computers, imaging boards, and CCD cameras, the system may become more widely used than previously described systems based on specialized minicomputers (16, 17). Finally, although it was developed for tissue section analysis, this software application can be expected to find new applications in the study of more complex objects in three dimensions.

References

1. C. E. Gee, C.-L. C. Chen, J. L. Roberts, R. Thompson, and S. J. Watson, *Nature (London)* **306**, 374 (1982).
2. W. S. Young III, E. Mezey, and R. E. Siegel, *Neurosci. Lett.* **70**, 198 (1986).
3. M. Schalling, K. Seroogy, T. Hökfelt, S. Y. Chai, H. Hallman, H. Persson, D. Larhammar, A. Ericsson, L. Terenius, J. Graffi, J. Massoulie, and M. Goldstein, *Neuroscience* **24**, 337 (1988).
4. C. R. Gerfen, *Methods Neurosci.* **1**, 79 (1989).
5. F. Javoy-Agid, E. C. Hirsch, S. Dumas, C. Duyckaerts, J. Mallet, and Y. Agid, *Neuroscience* **38**, 245 (1990).
6. Y. Agid, *Lancet 1*, 1321 (1991).
7. C. Tretiakoff, *These Med.* (1919).
8. N. Bonnet, S. Lebonvallet, H. El Hila, G. Colliot, and A. Beorchia, *J. Phys. III* **1**, 1349 (1991).
9. J. C. Bisconte, J. Fulcrand, and R. Marty, *C. R. Seances Soc. Biol. Biol. Ses Fil.* **162**, 2178 (1968).
10. A. W. Toga, "3D Reconstruction." Raven, New York, 1990.
11. R. Escourole, J. De Recondo, and F. Gray, *in* "Monoamines ey Noyaux Gris Centraux et Syndrome de Parkinson" (J. de Ajuriaguerra and G. Gautier, eds.), p. 173. Fourth Symposium de Bel air, Geneva, 1970.
12. C. D. Marsden, *J. Neural Transm. Suppl.* **19**, 121 (1983).
13. E. C. Hirsch, A. M. Graybiel, C. Duyckaerts, and F. Javoy-Agid, *Proc. Natl. Acad. Sci. U.S.A.* **84**, 5976 (1987).
14. E. C. Hirsch, A. M. Graybiel, and Y. A. Agid, *Nature (London)* **334**, 345 (1988).
15. C. Plotkowski, N. Chevillard, D. Pierrot, D. Altemayer, J. M. Zahm, G. Colliot, and E. Puchelle, *J. Clin. Invest.* **87**, 2018 (1991).
16. X. Albe and J. C. Bisconte, *Innov. Tech. Biol. Med.* **3**, 691 (1982).
17. X. Albe, M. B. Cornu, S. Margules, and J. C. Bisconte, *Comput. Biol. Res.* **18**, 313 (1985).

[6]   Quantitative Computerized Immunocytochemistry: Tissue Preparation and Image Analysis Techniques

Sverker Eneström

Introduction

Quantitative immunocytochemistry (QIC) has become an important field in electron microscopy, made possible by the development of electron-dense markers (above all colloidal gold) and image analysis techniques. There are many advantages of QIC, such as unbiased, rapid, and comparable determination of immunoreactivity which can be described in numerical terms, that is, labeling density (gold particles/μm^2).

QIC is based on a rapidly progressing evolution of both hardware and software which will continuously bring about more sophisticated analytical programs. But the results of QIC are highly influenced by various tissue preparation steps as well as the immunocytochemical staining technique with impact on both specificity and labeling density. Therefore, this chapter will discuss different preparation protocols with their advantages and limitations in QIC. The suggestions given are based on our own experience, mostly using rat pituitary as a model. The basic principles are, however, generally applicable.

One important shortcoming of QIC today is the lack of simple reference systems to translate the labeling density to antigen concentration values. Some recent studies comparing antigen concentration and gold particle density in quantitative immunocytochemistry will, however, be mentioned. Studies of that kind will certainly be followed by others changing present semiquantitative assessments to real quantitative analyses.

Tissue Preparation Techniques and Influence on Specific and Nonspecific Labeling Density

Immunoelectron microscopy can be performed using pre- or postembedding staining. Only the latter technique is appropriate in QIC. It is thus essential to count the specific markers over the section surface only and not in deeper

Methods in Neurosciences, Volume 10

TABLE I   Epon Embeddings of Secretory Granules following Different Preparation Methods and Appearance in Ultrathin Sections

Preparation[a]	Epon infiltration of granules	Sensibility to H_2O_2 and water (in STEM)	Recommended incubation (min)	Epon surface appearance of granules (in SEM)	Immunogold-labeling density: specific (number/μm^2)	Nonspecific labeling (number/μm^2)
No fixation, CF-FD	Almost complete (83%)[b]	Very strong[c]	5 + 5 min (no H_2O_2)	Protruding, hollowed[c, d]	Very high but variable[b]	~1%
15 min, CF-FD	Almost complete[d]	Strong, general swelling[d]	60 + 60 min	Swollen, almost smooth[d]	Very high but variable[d]	<1%
No fixation, CF-FS	Almost complete	Low[e]	60 + 60 min	Short, low ridges[e]	High[e]	<1%
15 min, CF-FS	Almost complete	Low[e]	60 + 60 min	Short, low ridges[e]	High[e]	<1%
2 hr + OsO_4, chemical dehydration	Not complete (78%)[b]	Low[c]	60 + 60 min	Mountainlike (long, tall ridges)[c,f]	Moderate[d]	<1%

[a] Fixed in 2% glutaraldehyde. CF-FD, cryofixation and freeze-drying; CF-FS, cryofixation and freeze-substitution.
[b] S. Eneström and B. Kniola, *Biotechnic Histochem.* **67,** 100 (1992).
[c] Eneström and Kniola (6).
[d] Eneström and Kniola (4).
[e] Eneström and Kniola, *Biotechnic Histochem.* (1993) in press.
[f] Eneström and Kniola (9).

parts if the labeling is to be expressed as the number of gold particles per surface area of the object.

There are many conditions which influence immunostaining in electron microscopy. The most important are (1) the tissue preparation methods (type of fixation, dehydration, and embedding), (2) preincubation treatment (oxidizing agents), (3) immunolabeling, direct or indirect, and type of immunogold conjugate (protein A/G, immunoglobulins), (4) colloidal gold concentration and size as well as poststaining; and (5) the careful control of specific labeling.

1. Electron Microscopy Preparation Method

Significant differences in immunogold staining intensity may be due to the staining protocol and not solely to differences in antigen concentration. A survey of the effects of electron microscopy (EM) preparation methods on the appearance of ultrathin Epon sections and labeling density is given in Table I. These results are relevant for secretory granules in the adenohypophysis but can also be used as a general guide for other cell and tissue structures.

Apparently, cryofixation (CF) followed by freeze-drying (FD) can be highly recommended, giving both high specific and low nonspecific labeling. The avoidance of solvent extraction and denaturation of macromolecules, as seen after chemical fixation and dehydration, will optimally retain the antigens. These can be more or less lost and displaced during immunostaining. Therefore a short prefixation (for about 15 min) of glutaraldehyde (GA) might be necessary. Furthermore, a short fixation will protect the specimens from mechanical damage during dissection of the small tissue pieces and during freezing. Dehydration by freeze-substitution (FS), following cryofixation, is also superior to conventional fixation and chemical dehydration with regard to immunolabeling and can be performed using acetone or methanol. The fastest substituent is methanol, acetone being much slower (1). Aldehydes or osmium can be added to the substituent, improving morphological preservation but usually reducing labeling intensity (2). One disadvantage of freeze-substitution is the eventual probability of protein extraction and lipid loss which is claimed to preferentially occur with methanol (3).

Some comments must be added regarding cryofixation, which is most efficient using a damped slammer. It is hardly possible to avoid some deformation of cells which is detectable when the form factor of cell nuclei is measured by computerized morphological analysis (4). The optimally cryofixed tissue layer is only 10–15 μm thick at the side of contact with the copper mirror. The disadvantage of mechanical influence and a very narrow zone of well-frozen tissue seem to be much reduced using the high-pressure freezing machine, invented by Moore *et al.* (5). This technique permits much larger tissue pieces to be frozen homogeneously and without damage. Furthermore, the water will be frozen in a vitrified state or will appear as small cubic ice crystals. This drastically improves the subcellular morphology.

Embedding of CF–FD or CF–FS specimens can be performed in epoxy or acrylic resins depending on the type of investigation. Hydrophilic acrylics such as Lowicryl K4 M (Chemische Werke Lowi, Waldkraiborg, Germany) are often recommended for optimal immunocytochemistry, but good results can be achieved with hydrophobic acrylics such as Epon, Epon–Araldite, or Araldite, which are preferable when an optimal cell morphology is needed.

A promising technique for efficient immunolocalization of lipids has recently been described by van Genderen *et al.* (6a), combining aldehyde fixation and cryoprotection with freeze-substitution and low temperature embedding. The paper emphasizes the present author's opinion of freeze-substitution as the most appropriate dehydration technique in immunoelectron microscopy.

2. Effect of Preincubation Treatment on Labeling Density

Oxidizing agents, such as H_2O_2 or saturated aqueous solutions of sodium metaperiodate, are more or less standard in postembedding immunocytochemistry after glutaraldehyde–OsO_4 fixation. The immunostaining is enhanced about 3-fold after oxidizing the ultrathin Epon section with 10% (v/v) H_2O_2 for 10 min, but a good effect can also be achieved with water, which effectively removes osmium (6). Cell constituents such as secretory granules are sensitive to oxidizing agents as well as to water after CF–FD and are liable to disintegrate during an ordinary incubation time (60 + 60 min). It is possible, however, to obtain high immunolabeling after CF–FD even after a 5 + 5 min incubation time using a high concentration of colloidal gold conjugate (see below). It is necessary to know that CF–FD ultrathin sections are very sensitive to H_2O_2 and water even when embedded in epoxy resins. This effect could explain the easy access to antigens by antibodies in immunostaining, resulting in high labeling. A short prefixation in GA will still result in swelling of targets such as secretory granules facilitating the immunostaining.

How the different preparation methods influence the targets can most properly be studied in the scanning or scanning transmission electron microscope. The surface appearance of thin sections reveals the state of macromolecules after fixation, dehydration, and embedding. The surface appearance will also be influenced by further treatment steps such as oxidizing and immunostaining. Conventional preparation, namely, long GA fixation and chemical dehydration, aggregates and cross-links the proteins and gives a nonhomogeneous embedding. The surface is then deeply furrowed after postembedding treatments, and the gold particles will to a high degree be localized on the crests, thereby explaining the irregular distribution of particles as seen on electron micrographs (Fig. 1). A more even distribution of markers is seen after CF–FD, where the resin surface is protruding but less furrowed. CF–FS creates a fairly resistant surface with low ridges (Fig. 1). Thus, it is necessary to know the type of EM preparation in order to interpret the results of the immunolabeling.

3. Immunostaining

After oxidizing with 10% H_2O_2 for 10 min and washing in double-distilled water, the thin Epon sections (on bare nickel grids) are floated face down on drops of 1% (w/v) ovalbumin in phosphate-buffered saline (PBS) or a 10% (w/v) solution of heat-inactivated normal goat serum in 0.1% (w/v) bovine serum albumin/PBS. An incubation for about 5–10 min (at room temperature) will block nonspecific reactive sites, but this step should be omitted using protein A. The indirect technique of antigen labeling is recommended, being more sensitive and convenient than staining with gold-conjugated primary antibodies. The former method also has the advantage of much better penetration of the oxidized Epon resin by unconjugated antibodies to the target epitopes when compared to gold-conjugated antibodies. The secondary antibodies, adsorbed to the gold particles, are then able to attach to the primary antibodies more easily than in direct immunostaining. Alternatively, gold-

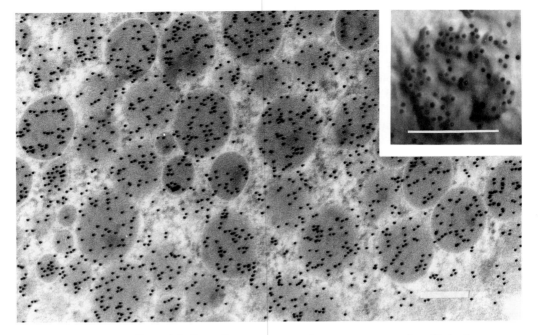

Fig. 1   Indirect labeling of rat somatotropic secretory granules. Note the uneven distribution of gold particles over the granules. This may be due to the topography of the Epon surface, with the gold particles being mostly confined to the ridges. *Inset:* TEM surface of a thin section which was sputter coated and tilted 35°. Fixation was for 15 min in GA, with CF–FS. Bars: 200 nm.

labeled protein A or G can be used, taking advantage of the affinity to Fc binding sites of the primary antibody.

The incubation time is dependent on the preparation protocol. Ultrathin, epoxy resin-embedded CF–FD tissue sections are very sensitive to water and are easily destroyed without a short prefixation for about 15 min, as mentioned above. Well-fixed tissue sections and those dehydrated by freeze-substitution are fairly resistant to 60 + 60 min incubation. Longer incubation times will alter targets, such as secretory granules, exaggerating their surface topographical changes already produced by the oxidizing treatment. Short washings between the incubation steps are mandatory, and this is most critical after the second step in order to remove unbound gold conjugate.

4. Colloidal Gold Concentration and Size; Poststaining

Colloidal gold conjugates with particle sizes of about 100 nm or less cannot sediment because the kinetic energy of the particles is greater than the force of gravity. The labeling intensity can then be assessed by the Einstein law of

Brownian motion according to the equation $q = n(kTt/12\pi yr)^{1/2}$ as given by Park *et al.* (7), where q is the theoretical value of the number of particles reaching a unit surface area, n is the concentration of gold marker, k is the Boltzmann constant (1.38×10^{-16}), T is the temperature in kelvins, t is time in seconds, y is the viscosity of the medium, and r is the radius of gold markers (cm). This means that we have to take into consideration parameters such as the concentration of gold particles, their sizes, as well as the temperature, which also will influence the viscosity. The most critical factor is the concentration of gold particles. It has to be high enough to permit optimal labeling within reasonable time. Using a gold conjugate which is 11–12 nm ($Au_{10} + Ig$) with incubation for 60 min at room temperature ($22°C$), a conjugate with about 8×10^{12} gold particles per milliliter is needed to obtain a theoretical q value of about 2000 particles/μm^2 (provided the antigens are evenly distributed and totally accessible). Comparing this value to the real count gives a probability factor (p value) that indicates the staining efficiency. This value must be compared to the nonspecific labeling ("background") in order to be sure about the specificity.

The second problem is to select the gold particle size which is optimal for specific staining and at the same time gives a low nonspecific labeling. It is important to include the preference maxima for gold probes in quantitative immunocytochemical analysis, together with the assessment of the labeling density. A high p value for a gold particle size means that it is favorable in that immunostaining situation. It is also necessary to use colloidal gold conjugates with correctly indicated gold particle size distributions as well as the percentage of gold particle aggregates, which is well specified by the manufacturer.

The ultrathin sections can be contrasted by poststaining with 2% (w/v) uranyl acetate in 50% ethanol for 5 min.

5. Controls

Standard negative controls include (1) omission of the primary antibody which is replaced with buffer, (2) use of an antibody against an antigen which does not exist in the tissue, (3) use of nonimmune serum from the same species or, preferably, the same animal which produced the specific antibodies, and (4) preabsorption of the primary antibodies with the target antigen. The "background" must be characterized and measured. It consists of (1) specific antibody staining outside targets (preferably assessed over cell nuclei), (2) nonspecific staining of the targets by a nonrelevant antibody, and (3) attachment of the gold conjugate to the embedding medium, which is measured over empty resin. The reason for the latter staining can be studied further by scanning–transmission electron microscopy (STEM), showing gold particles concentrated on protruding sites of the resin surface. This can be the main reason for background staining.

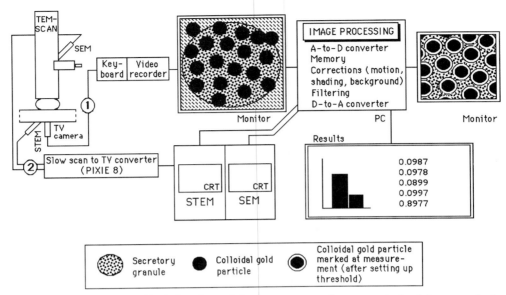

Fig. 2 Simplified drawing of the components for computerized immunoelectron microscopy. The image can be transferred to the computer and processed (1) through a TV camera connected to a video recorder with the image first appearing on the EM monitor (the images recorded on the video tape are then fed into the memory of the computer in a series of 90–100 images) or (2) through SEM or STEM detectors (the images, adjusted with the aid of the CRTs, are then transmitted to the computer). Note the rings around the gold particles after setting up the threshold and measuring them, as displayed on the second monitor. The results are presented on the third monitor as a histogram and tables.

Electron Microscopy

The immunolabeled section can be studied in transmission (TEM), scanning–transmission (STEM), or scanning (SEM) electron microscopical modes, depending on the purpose of the study (Fig. 2). Although TEM is intended for high-resolution work, SEM bridges the gap between light microscopy and TEM, offering opportunities to study supramolecular structures, such as membrane components, three dimensionally.

Good image contrast is essential in immunoelectron microscopy. In TEM this is determined by the absorption of scattered electrons, which increases by lowering the accelerating voltage; unfortunately, this also raises the risk of damage to the sections. Image contrast in SEM is mainly determined by the variation in the number of locally emitted secondary electrons, which in turn is dependent on the specimen composition and topography identified by differences in brightness. The amplitude of the video signal is controlled by the contrast circuit, which modifies the gain of the scintillator–photomultiplier. In that way, the image contrast can be increased. The secondary electron detector for SEM is placed above the objective pole piece to collect secondary and some backscattered electrons emitted from the irradiated specimen surface. A detector for transmitted electrons is situated at the lower end of the microscope in STEM. This detector will also produce signals which are sent to a video amplifier. They are amplified here and level-controlled to give an optimum contrast and brightness image on the monitor (cathode ray tube, CRT, Fig. 2).

Working with quantitative computerized immunoelectron microscopy (QCI) in the STEM mode is facilitated by the ability to enhance the contrast by changing the voltage to the photomultiplier. QCI also takes advantage of the video signal (slow scan) which is produced in the STEM mode. Imaging computer systems (Link Analytical Ltd., England; Cambridge Instruments Ltd., England; and others) provide programs (electron signal processing, ESP) for electron image acquisition from the STEM detector, including filtering to increase the signal-to-noise ratio. Without ESP the slow scan from the STEM detector can be converted to a TV scan by a digital image processor (PIXIE 8, Oxford Instruments Ltd., Oxford, England), which at the same time enhances the contrast by stored image manipulation and contrast inversion.

QCI in TEM mode is possible provided that the microscope is equipped with a TV camera. It is then easy to connect the camera to imaging hardware for signal digitization and processing. Such equipment will be described below.

Thin-Section Topography after Immunostaining

The result of the immunostaining is easy to check by examination of the section surface to look for the effect of preparation steps. For this purpose the grid-mounted sections are sputter coated with platinum using a magnatron sputter coater (Ion Tech Ltd., Teddington, England) installed in a freeze-drier and equipped with a quartz crystal thin film thickness monitor (Intellimetric, Glasgow, England). Sputtering of platinum is carried out in pure argon at 4×10^{-3} mbar, 1 kV, 20 mA; the metal is deposited at a thickness of 7 nm. Thin sections can also be coated with gold by thermal evaporation (JEE-4B/4C vacuum evaporator, Jeol Ltd., Tokyo, Japan). The metallized surfaces of the sections are examined in STEM mode operated at 100 kV. Photographs of sections coated with platinum are taken at a specimen tilt angle of 35°. For stereoscopy, stereopairs can be obtained by taking pictures of the same field at 0° and after tilting the specimen −6° from the original (0°) position. A three-dimensional image of the surface of structures is obtained by looking at properly adjusted pairs of such photographs with a stereoscopic binocular lens. The metallized resin surface can be more easily checked, and with acceptable results, in the TEM mode. The specimen is then tilted about 35° to have a good three-dimensional image (Fig. 1).

Computerized Analysis of Gold-Labeled Sections

Computer processing in image analysis has been discussed in a useful, practical way in Volume 3 of this series (8). The reader is referred to Chapter 11 in Volume 3 for detailed descriptions of basic principles of image processing and analysis. Briefly, a video signal from the TEM TV camera or from the slow scan to TV scan converter in the STEM mode is transmitted to the computer (Fig. 2). That means that the video image is captured and digitized by the analog-to-digital converter into an array of pixels, for instance 512 pixels in the x axis and 512 pixels in the y axis. Each point represents a gray level from black (0) to white (255) with varied shades of gray in between. The digitized image is stored in the frame memory of the computer for further processing. By way of a digital-to-analog converter the processed digital image can be displayed back to a color monitor.

Examples of Computerized Image Analysis in Scanning–Transmission Electron Microscopy Mode

AN 10000 Digital Imaging Processing (9)

The AN 1000 software can be used in the AN 10/50 computer (Link Analytical Ltd., Bucks, England) coupled to the video output of the video amplifier in the microscope. It comprises three main sections: (1) the electron signal processing (ESP) for acquisition, (2) the DIGI PAD for processing, and (3) the DIGISCAN/FDC for analysis. The ESP software takes signals from the scanning transmission electron microscope detector and from an area of a specimen by digitally controlling the position of the electron beam. The resulting map of intensity is displayed on the AN 10/50 screen as an image. The contrast of STEM images can be reversed electronically in the signal processor of the microscope (JEM 2000 EX and 1200 EX II electron microscope, Jeol Ltd., Tokyo, Japan) by flipping a toggle switch. Next, the contrast and dark levels are adjusted to generate an acceptable image in which the gold particles are visible as bright dots.

The ESP software is totally compatible with the DIGI PAD program so that STEM images can be stored on disk, converted to a binary version, and processed as required. Sometimes the gold particles are located close together. They can, however, be separated by a function from the process menu in the DIGI PAD program performing an erosion followed by expansion of the objects.

When STEM images are processed, the features can be detected and measured (using 20 projections) by the DIGISCAN/FDC program. Once all the data have been collected, a statistical summary of the results is printed and a histogram displayed. This summary gives the number, mean diameter, percent area of features, and size range specified by the user.

Quantitative Assessment

The labeling is assessed by computerized counting of the number of gold particles present over the surface of the targets. The area of the target profile (in μm^2) can be calculated according to the formula

$$A_{ta} = A_{sc} T_r \times 10^4/Mag^2$$

where A_{ta} is the true area of the target profile (in μm^2), A_{sc} is the area of the AN 10/50 screen (in mm^2), T_r is the percentage of the screen area occupied by the target, and Mag is the on-screen magnification.

Labeling density and background labeling (i.e., cytoplasm, mitochondria, and nucleus) are calculated by dividing the number of gold particles by the surface area. The results are usually expressed as the number of gold particles per 1 $\mu m^2 \pm$ SD. For each evaluation, a minimum of 20 images should be analyzed at magnifications of $\times 150,000$, $\times 200,00$, or $\times 250,000$.

Comments on System

We have used this image collecting and processing system for quantitative analysis of immunogold-labeled secretory granules in the adenohypophysis of rat and man (9). The magnification is set so that the image of the granule completely fills the observation CRT of the STEM observation unit. The granule area, measured as indicated above, will then always be the same, facilitating the measurements of labeling density. The system is convenient for studies of multiple parameters but is rather time consuming.

LabEye System

The LabEye system (Innovativ Vision, Linköping, Sweden) (Fig. 2) can be used in combination with the PIXIE 8 image processor which converts the slow scan video signal from the microscope to 625-line video standard fed into the computer. The processor gives recursive filtering for noise reduction and has a resolution of 512 (x) × 480 (y) pixels, whereas the LabEye system has the resolution of 512 × 512 pixels. The pixel array is loaded to the frame buffer, displayed, and processed by the imaging software. A series of about 50 images can be stored in the memory and then processed in succession. The image from the image processor is motion corrected by reducing the vertical resolution (to 256 pixels). A scale calibration is performed in advance using a standard to automatically obtain the measurements (in nm) which are easily transformed by the computer to areas in μm^2. Each stored image is identified by title and comments.

The analysis can be performed using the following steps: (1) Load an image from memory and adjust the contrast by performing a so-called histogram stretch, which ensures maximum contrast in the image. (2) The set up threshold function defines the gray level used to find the limit between the background and the object (gold particles) which will be measured. This can be automatically set or adjusted by the user. Measuring the object diameter will help find a suitable gray level, which is adjusted to fit the known diameter of, for example, 10-nm gold particles (Fig. 2). These adjustments are facilitated by pseudo coloring the particles in blue–red. (3) Set up object. *max* and *min* will define the object sizes between 0 and 32,000 of approximate image pixel perimeters which are measured; *min* is set to 10–25 pixels. Objects not fulfilling certain size criteria are automatically excluded when using the function *measure all*. This function locates and measures all the objects in the current image. Aggregated gold particles can be separated by manually drawing a line between them. Also the objects which hit the edges of the screen can be included in the measurements (by the function *edge hits*). (4) The show status/histogram function displays the results on the computer screen. Results can also be saved in result or text files and treated statistically.

Comments on System

Measurements of immunolabeled secretory granules in the adenohypophysis is done rather rapidly if the granule surface completely fills the screen area. The labeling density will then be easily assessed.

Computerized Image Analysis in Transmission Electron Microscopy Mode

Modern electron microscopes are equipped with TV cameras. The images can then be directly transferred to an image analysis system. We recommend, however, the use of a high-quality video recorder to store series of images. After adjusting focus and brightness, the images are displayed on the TV monitor screen via the recorder. The separate images are recorded and numbered (using the keyboard in the circuit) on a video tape. The video recorder and the computer are connected, and all the series of images are then fed into the computer (Fig. 2). The magnification is chosen at a level which will facilitate measurements of the target areas and the gold particles. A large number of images can be stored on the video tape and played into the computer independently of the microscopical work. The same image processing procedure can be used as mentioned above for the LabEye system.

Comments on System

We have found this system rapid and very convenient. At least 90 images of specific labeling and the same number of images of background staining are stored from the same specimen. The measurements are more rapidly performed when compared to the systems previously described. The present protocol also allows for the reexamination and processing of images many times without damage, which is inevitable during repeated direct communication between microscope and computer owing to exposure of the sections to the electron beam.

Computer-Assisted Indirect Quantitative Immunoelectron Microscopy

The predominant type of quantification after immunogold staining of EM sections, as mentioned in the literature, has been performed in three steps. First, electron micrographs are prepared which are then digitized to obtain the profile areas. Thereafter, the number of gold particles are counted, and the analyses are finally put into a computer with different kinds of software. One well-developed computer program for biological specimens was particularly designed to aid in the measurements of labeling density (10). It has been used in quantitative immunoelectron microscopy of neuroactive amino acids (11) (see below).

Comparison between Immunogold-Labeling Density and Concentration of Antigens

Only a few attempts have been made to establish the relationship between immunogold labeling and relative antigen concentration in the tissue. This has been tentatively performed for neuroactive amino acids (11, 12) because of the possibility of raising antibodies against a complex of glutaraldehyde–bovine serum albumin–amino acid (13). The antiserum will show a specific immunoreactivity to the comparative glutaraldehyde-fixed amino acid in the brain. By including reference sections with conjugates of known graded concentrations of the fixed amino acid bound to a brain homogenate, a rough comparison can be made between labeling density and concentration (in mmol/liter) of fixed amino acid. The question is, however, whether the antigen concentration at the exposed, oxidized, and incubated resin surface really corresponds to the biochemically assessed concentration in the conjugate. It is also obvious that the measurements pertain only to glutaraldehyde-fixed amino acids.

Another approach to the quantification of tissue antigen is the use of reference concentrations of target antigens included in a matrix such as gelatin or microcellulose membranes. By application of the same fixation and embedding the tissue together with the standard, the material will be cut at the same thickness and treated equally at the immunolabeling steps. This procedure has been developed for ultrathin cryosections of different tissues (14). The procedure can give a rough estimate of antigen concentration, but a complete comparison is not possible due to the difference in macromolecular structure between cell constituents and the standard.

Concluding Remarks and Recommendations

Even the most elaborated computerized image processing system is meaningless if the tissue preparation protocol is inadequate. It has to be very carefully tested for specificity and sensitivity. All steps must be analyzed to be aware of how they influence antigen preservation and recognition. This procedure should include the following:

1. A short prefixation for small tissue sections (2–3 mm) which are then dissected into smaller pieces while still in the fixative; 15 min suffices for this initial fixation step.
2. Cryofixation of tissue pieces not larger than 1×1 mm using a damped slammer, chilled with liquid nitrogen.
3. Dehydration by freeze-drying or freeze-substitution.
4. Embedding in acrylic resin after incubation in diluted and then 100% plastic overnight at room temperature.
5. Oxidation of ultrathin sections with 10% H_2O_2 for 10 min and blocking for nonspecific staining.
6. Indirect immunostaining where the colloidal gold conjugate has a known and optimal concentration and size of gold particles. Poststaining.
7. Adequate controls.
8. Checking the metal-coated resin surface (in SEM or TEM modes) of the ultrathin sections in order to distinguish topography and labeling pattern.

9. Measurement of the specific and nonspecific labeling density in TEM, preferably by the use of a video cassette recorder coupled to the TV camera of the electron microscope.
10. Processing the images by automatically measuring the number of gold particles over an object surface which fills the area of the monitor. If this is calibrated the results in labeling density per 1 μm^2 of object area will be obtained.
11. Rough determination of the antigen concentration relative to a standard curve using some simple reference system (e.g., antigen included in a matrix which is comparable with the object).

Acknowledgments

I would like to thank Barbara Kniola for excellent technical assistance. Research was supported by the Swedish Research Council, Project No. 6536.

References

1. R. A. Steinbrecht and M. Müller, *in* "Cryotechniques in Biological Electron Microscopy" (R. A. Steinbrecht and K. Zierold, eds.), pp. 149–172. Springer-Verlag, Berlin, 1987.
2. G. Nicolas, *J. Electron Microsc. Tech.* **18**, 395 (1991).
3. B. Ph. M. Menco, *Cell Tissue Res.* **256**, 275 (1989).
4. S. Eneström and B. Kniola, *Biotechnic Histochem.* in press (1992).
5. H. Moore, G. Bellin, C. Sandri, and K. Akert, *Cell Tissue Res.* **209**, 201 (1980).
6. S. Eneström and B. Kniola, *Biotechnic Histochem.* **66**, 246 (1991).
6a. I. L. van Genderen, G. van Meer, J. W. Slot, H. J. Geuze, and W. F. Voorhout, *J. Cell Biol.* **115**, 1009 (1991).
7. K. Park, S. R. Simmons, and R. M. Albrecht, *Scanning Microsc.* **1**, 339 (1987).
8. A. J. Smolen, *Methods Neurosci.* **3**, 208 (1990).
9. S. Eneström and B. Kniola, *Stain Technol.* **65**, 263 (1990).
10. T. W. Blackstad, T. Karagülle, and O. P. Ottersen, *Comput. Biol. Med.* **20**, 15 (1990).
11. O. P. Ottersen, *Anat. Embryol.* **180**, 1 (1989).
12. N. Zhang, F. Walberg, J. H. Laake, B. S. Meldrum, and O. P. Ottersen, *Neuroscience* **38**, 61 (1990).
13. J. Storm-Mathisen, A. K. Leknes, A. Bore, J. L. Vaaland, P. Edminson, F. M. S. Haug, and O. P. Ottersen, *Nature (London)* **301**, 517 (1983).
14. J. W. Slot, G. Posthuma, C. Lin-Yi, J. D. Crapo, and H. J. Geuze, *Am. J. Anat.* **185**, 271 (1989).

[7] Confocal Fluorescence Microscopy in Three-Dimensional Analysis of Axon Terminal Distribution, Neuronal Connectivity, and Colocalization of Messenger Molecules in Nervous Tissue: Computerized Analysis

B. Ulfhake, K. Mossberg, K. Carlsson, and U. Arvidsson

Introduction

With the introduction of powerful tools for tracing (for review, see Heimer and Záborsky, 1989) and immunohistochemistry (Coons, 1958; see also, e.g., Hökfelt *et al.*, 1983), studies on neuronal connectivity have greatly expanded our knowledge of the organization of the nervous system. In nervous tissue, structure and function are intimately linked (Cajal, 1909), and research has by necessity involved a great deal of quantification, as to, for example, number of axon terminals, spatial distribution, as well as content of neuroactive molecule(s). Quantitative studies have utilized either conventional light microscopy (LM) or electron microscopy (EM). The limited resolving capability of LM and its inability to provide 3-dimensional (3-D) data, however, impose severe constraints on quantitative analysis. In practice only fairly large elements can be analyzed, and then only after mechanical sectioning of the tissue. An alternative, giving superior resolution, is EM. The drawbacks with this technique are, however, obvious in that it demands even more laborious mechanical sectioning and for practical reasons only permits analysis of quite small tissue regions. With the introduction of confocal microscopy (CM) a new instrument for this type of analysis was provided, bridging to some extent the gap between LM and EM (for review, see Wilson and Sheppard, 1984; Wilson, 1990). Although the principles for CM date back to the late 1950s (Minsky, 1961), instruments have been commercially available only since the 1980s. The reasons for this are that not until recently have suitable lasers and sufficiently powerful computers been readily available.

The aim of this chapter is to provide an introduction to confocal fluorescence microscopy and the use of this technique in quantitative analysis of immunofluorescently labeled axons. One section is devoted to multicolor fluorescence recording and analysis. Although some general principles for

Methods in Neurosciences, Volume 10

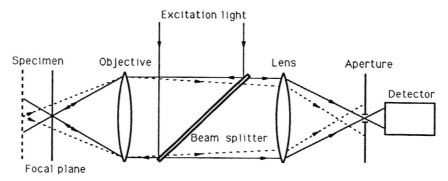

FIG. 1   The ray path in a confocal microscope. The specimen is illuminated by a focused laser beam, one point at a time (point illumination), and the fluorescent light emitted from the focal plane (solid lines) is focused on an aperture in front of the detector (point detection) while light from other specimen layers (dashed lines) is suppressed.

fluorescence microscopy are touched on, the main focus is on issues related to image analysis of nervous tissue subjected to immunofluorescence and/or fluorescent tracer studies.

Confocal Microscopy and Image Recording

Confocal microscopy is a point illumination/point detection technique, allowing collection of high-resolution 3-D data. The principle is depicted in Fig. 1. The diaphragm (aperture) inserted in front of the detector in the ray path suppresses light deriving from out-of-focus tissue regions. In most CM instruments one point at a time is illuminated, and, in order to get an image, scanning of the specimen or the illuminating beam must be performed. In the tandem scanning microscope, multiple specimen points are illuminated/ detected simultaneously by use of a Nipkow disk (Petran *et al.,* 1968). However, we will only deal with confocal scanning laser microscopy in fluorescence studies using single-spot illumination/detection.

The Confocal Microscope

Figure 2 shows the PHOIBOS confocal scanning laser microscope (CSLM), developed as an add-on device to a conventional epifluorescence microscope (Carlsson and Åslund, 1987; Carlsson and Liljeborg, 1989). In principle it is not different from commercially available CSLMs. Image recording is obtained by illuminating (exciting) the specimen with a focused laser spot [laser

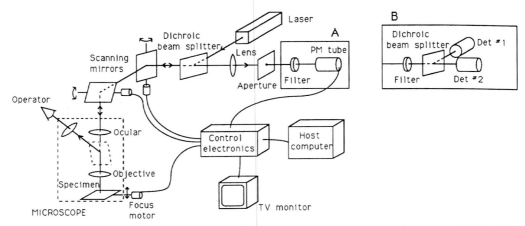

FIG. 2   (A) Schematic diagram of the confocal scanner laser microscope PHOIBOS. The laser beam is reflected into the microscope by a beam splitter and two mirrors that perform the two-dimensional scanning. The fluorescence light from the specimen goes the opposite way, passes the beam splitter, and is focused on a pinhole aperture in front of the detector. The scanner is controlled by a microprocessor which controls the movements of the mirrors and of the specimen stage, as well as the detection of the PMT signal. The imaged data are transferred to a computer where they are stored and analyzed. (B) Dual-detector arrangement.

beam diameter is $\sim$0.4 μm at wavelengths around 500 nm with a 1.4 numerical aperture (N.A.) objective]. The emitted (fluorescence) light is focused on a pinhole aperture in front of the detector, a photomultiplier tube (PMT). Two-dimensional scanning is managed by two moving mirrors that steer the laser beam in a raster pattern across a square-shaped specimen area. The scanned specimen area varies with objective magnification from 1 mm $\times$ 1 mm ($\times$10 objective) down to 100 μm $\times$ 100 μm ($\times$100 objective). Each part of the specimen within the scanned area is illuminated by the laser beam during a very short period of time (referred to as the integration time, during which photons emitted from the specimen are recorded by the detector). The recorded specimen field can be divided into a maximum of 1024 $\times$ 1024 pixels (picture elements). The emitted fluorescence light is detected by the PMT. The signal from the PMT is analog to digital (A/D) converted to 8 bits of data, corresponding to 256 gray levels for each pixel, and stored in a computer memory.

Owing to the point illumination/point detection technique, only a thin section of the specimen (down to <1 μm), is imaged by each scan. By repeatedly scanning the specimen and refocusing the microscope (shifts down to 0.1 μm in focus can be managed by a computer-controlled stepping motor; Fig. 3) a "stack" of optical sections are recorded from which 3-D specimen

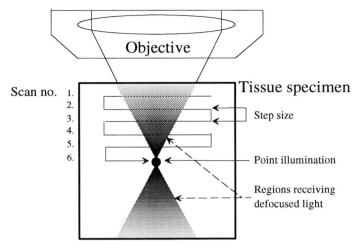

FIG. 3   Schematic illustration of the distribution of focused and defocused laser light during scanning. Note that out-of-focus tissue regions will also be illuminated during scanning. Thus, fading will occur in defocused depths of the specimen. The graded shading of the defocused light indicates that light intensity will fall off with distance from the focus plane. However, the defocused regions will be illuminated over longer times; thus, the integrated excitation light exposure will be equal for all tissue regions. Between scans the specimen stage is moved one "step" in order to record successive focal planes. The step size can be set down to 0.1 μm.

information can be extracted and calculated (Carlsson *et al.*, 1985). The image data can be processed by computer and then be inspected from different angles or in stereo viewing, revealing the detailed spatial arrangement of fluorescently labeled objects. In Fig. 4 (color plate), a fluorescein-labeled spinal cord specimen revealing serotonin-immunoreactive axon terminals is shown as it appears in a conventional epifluorescence microscope and in images recorded with a CSLM.

For recording and analysis of two-color fluorescence, a version of PHOIBOS equipped with dual detectors may be employed (see Fig. 2b; Carlsson, 1990a). In this instrumental setup the light passing through the detector aperture is split into two spectral bands by a beam splitter. If the two fluorophores to be recorded have sufficiently separated emission bands, this instrument allows simultaneous imaging of both fluorophores (see below; for general discussion on multicolor fluorescence, see, e.g., Wessendorf, 1990).

Before proceeding to applications we shall discuss some properties and parameter settings of the CSLM, as well as specimen considerations critical for high-quality recordings, which may be useful to the user of CSLMs in fluorescence studies.

Resolution

The lateral point-to-point resolution in fluorescence microscopy, based on the Rayleigh criterion, is about 30% higher in a confocal microscope than in a conventional microscope. In the special case of equal excitation and emission wavelengths (λ), the resolution is given by

$$R_f = (0.46\lambda)/N.A. \tag{1}$$

where $N.A.$ is the numerical aperture of the objective. However, more important in confocal microscopy is the depth discrimination, that is, the suppression of light intensity from out-of-focus tissue regions, a property that is not present in conventional microscopy. The depth discrimination in the optical section can be estimated by studying the intensity distribution along

Fig. 4   (A) Epifluorescence micrograph taken in a Nikon microscope with a ×20/0.75 dry objective showing serotonin-immunoreactive axon fibers in the cat spinal cord ventral horn. (B) Optical section through the center of the same specimen recorded with a CSLM using a ×63/1.4 oil immersion objective, optical section thickness approximately 0.8 μm. (C, D) Stereo pair ($-4°$ and $+4°$) of the entire stack of optical sections through the same specimen. Colors in (B)–(D) reflect intensity levels, where black $=$ 0 and white $=$ 255.

Fig. 13   (A) Look-through projection of a sample specimen showing FITC-fluorescent axonal fibers containing thyrotropin-releasing hormone-like immunoreactivity in the cat spinal cord ventral horn. The specimen was recorded using a ×63/1.4 N.A. oil immersion objective, with a pixel size of 0.16 μm and a vertical step size of 0.4 μm. The lower right region of (A) (framed) is also shown in (B) and is used to illustrate the different steps of the computer program. Colors in (A) and (B) indicate intensity levels (black $=$ 0 and white $=$ 255). The panel side equals 80 and 40 μm in (A) and (B), respectively. (C) Identified 3-D objects in the sample specimen. (D) After splitting into subobjects. (E) Following merging of subobjects. (F) Following removal of objects that are too small or nonellipsoidal. The remaining objects are regarded to be axon terminals and are subjected to quantification. Colors in (C)–(F) indicate objects/subobjects, except white which is used to indicate overlap of objects in this viewing direction.

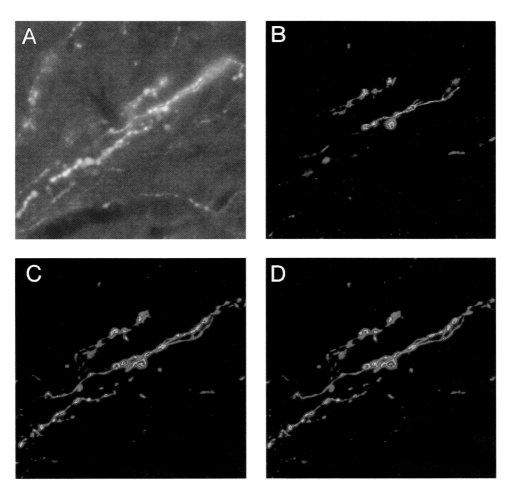

Fig. 4

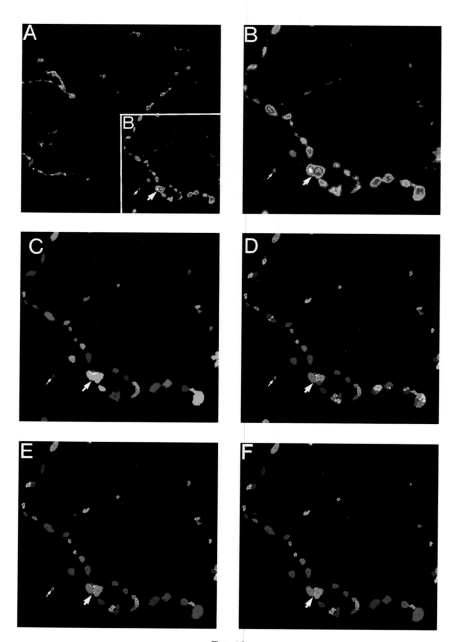

FIG. 13

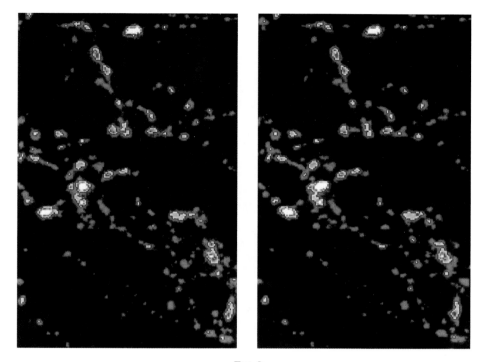

Fig. 9

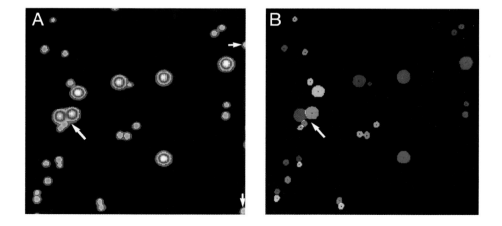

Fig. 16

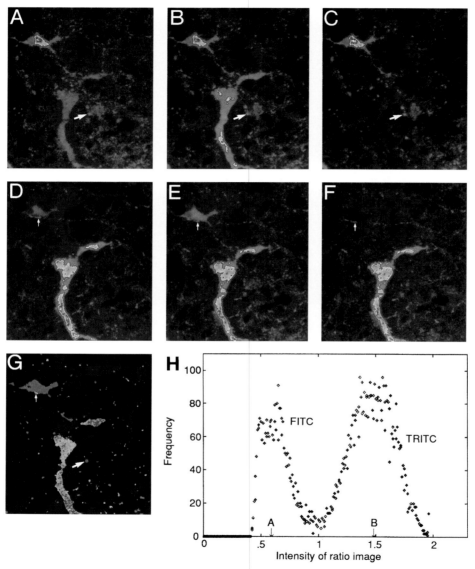

FIG. 17

the optical axis from a very thin layer of fluorescent dye, as shown in Fig. 5. The full-width half-maximum (FWHM) value of that curve can be used as a measure of the optical section thickness. Theoretically, the FWHM is given by

$$R_d = (8.5\lambda)/[8\pi n \ \sin^2(\alpha/2)] \tag{2}$$

where n is the refractive index of the immersion medium and α is obtained from $N.A. = n \sin \alpha$. n is equal to 1.0 for air and 1.5 for oil. Thus, for a $\times 25/0.65$ dry objective and a $\times 63/1.4$ oil immersion objective, R_f is 0.32 and 0.15 μm, respectively, and R_d is 1.3 and 0.32 μm, at a wavelength of 450 nm.

FIG. 9   Stereo pair ($0°$ and $+4°$) of a rat ventral horn spinal cord specimen excited in the UV (351 nm) using a $\times 63/1.25$ (Zeiss) oil immersion objective, showing AMCA fluorescence of substance P-immunoreactive axonal fibers.

FIG. 16   (A) Look-through projection of a sample specimen containing fluorescent latex beads of two sizes (1.1 and 2.4 μm). Colors indicate intensity levels as in Fig. 13. (B) The objects found by the computer program. A perfect match with the image in (A) is evident. Colors in (B) indicate objects. The large arrow indicates a cluster of large and small objects resolved by the program. Small arrows indicate some of the objects touching the boundaries of the recorded image stack, and these are excluded by the program.

FIG. 17   Look-through projection of two neurons and axon terminals distributed in the surrounding neuropil immunoreactive to substance P (FITC; A, B, C, G) and serotonin (TRITC; D, E, F, G), respectively, in the lateral brain stem of the cat. In (A, FITC) and (D; TRITC) the recordings have been done separately using one detector, and between recordings both the excitation wavelength and filters were changed in accordance with the fluorophore to be detected. (B) and (E) show the corresponding images from simultaneous recording using a dual-detector system (see text). Note the more extensive breakthrough between labels with this method. (C) and (F) show the corresponding images after linear subtraction. Note the pronounced improvement with respect to breakthrough without any noteworthy reduction of the specific fluorescence intensity to be detected. Large arrows in A, B, C, and G indicate a FITC-positive axon terminal; small arrows in D, E, F, and G indicate a TRITC-positive axon terminal apposing the FITC-positive neuron. Colors refer to intensity levels as in Fig. 13. In (G) a ratio image obtained by dividing the FITC image by the TRITC image recorded with dual detectors is shown (in this case only a single optical section is displayed). (H) Histogram of the ratio image shown in (G), from which the subtraction coefficients can be determined from the two visible peaks (A and B on the abscissa).

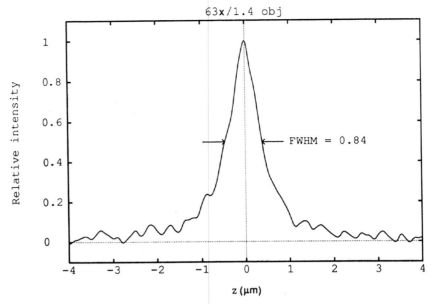

FIG. 5   Depth discrimination curve for a ×63/1.4 oil immersion objective, obtained by the step response recorded when entering a layer of fluorescein solution (FITC dissolved in H_2O/NaOH, pH 13; excitation light 497 nm). To make the response similar to that obtained from an infinitely thin dye layer, the step response recorded was differentiated. The full-width half-maximum (FWHM) is in this case 0.84 μm. In a conventional microscope there is no intensity tail-off, instead the intensity will remain at a high level.

At a wavelength of 550 nm, the corresponding values are 0.39 and 0.18 μm for R_f and 1.6 and 0.40 μm for R_d.

Equations (1) and (2), however, apply only in the special case of an infinitely small detector aperture and for optics that have no aberrations (Wilson and Carlini, 1987, 1989). In practice none of these criteria are fulfilled, and the resolution will therefore be somewhat lower. Two important conclusions that can be drawn from Eqs. (1) and (2) are, however, that R_f is inversely proportional to $N.A.$, whereas R_d is approximately inversely proportional to the second power of $N.A.$ For high numerical apertures ($N.A. = 1.3$–1.4), R_f will be a factor of 2–3 better than R_d; for smaller numerical apertures, this factor will be even larger.

The finite resolution of the microscope is determined by the point-spread function (PSF), which will cause objects imaged to be extended beyond their real size. The PSF will distort smaller objects more than large objects, and

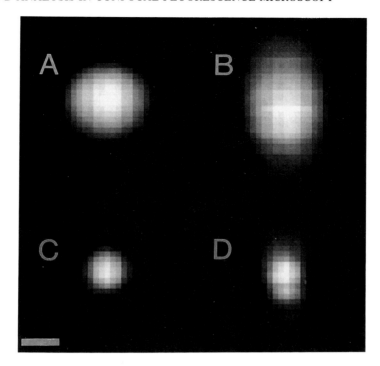

FIG. 6   Look-through projections of fluorescent latex beads of two different sizes (A, B, 2.4 μm; C, D, 1.1 μm) imaged with a ×63/1.4 oil immersion objective, as they appear viewed in the lateral plane (A, C) and along the optical axis (B, D). Note that the objects have been substantially extended in depth (B, D) owing to microscope PSF; this property will distort the smaller object (C, D) the most. For a sample of such beads the average optical-to-lateral axis ratio was 1.12 (2.4-μm beads) and 1.43 (1.1-μm beads), instead of the ideal 1.0. By use of the PSF correction algorithms (see Appendix), the bead axes could be recalculated, and the thus obtained ratios were 1.03 (2.4-μm beads) and 1.01 (1.1-μm beads). Bar, 1 μm.

it will be most evident along the optical axis (Fig. 6). Thus small tissue objects, with sizes near or below the FWHM, will appear unproportionally elongated in depth. Theoretically, the image objects can be recalculated to their "true" size by deconvolution techniques (e.g., Bertero *et al.*, 1987a,b; Castleman, 1979; Shaw and Rawlins, 1991). However, such methods demand rather noise-free image data, which according to our experience often is not possible to produce, and involve massive calculations which in many cases is impractical. The object distortion is, however, troublesome for small tissue objects such as axon terminals. Axon terminals are spherical-to-ellipsoidal

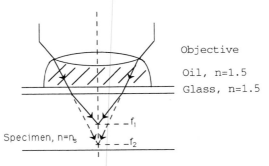

FIG. 7   Influence of specimen refractive index. If the refractive index of the specimen/embedding medium is equal to oil (n_s 1.5; dashed line), then the light will focus in plane f_2; however, if the refractive index is closer to water, for example, (n_s 1.33; solid line), the light will focus on plane f_1. Thus, if the specimen stage is moved Δz then f_2 will be moved Δz but f_1 will be moved $(n_{water}/n_{oil})\Delta z$.

objects in the size range 0.5 to 4 μm (Peters *et al.*, 1976), but when imaged they will appear asymmetrically enlarged with an average optical-to-lateral axis ratio of about 1.4 instead of 1.0 (using a $\times 63/1.4$ oil immersion objective). This error may be acceptable in many applications. However, to permit high-resolution quantitative analysis as in our applications, the effects of the PSF need to be compensated for. For this purpose an alternative method was devised for the special case of objects that have an approximately ellipsoidal shape (e.g., axon terminals). The method is described in the Appendix.

Effects of Specimen Retractive Index

If the refractive index of the specimen embedding medium/tissue differs from that of the immersion medium of the objective, this may cause focal displacement resulting in a compressed or elongated depth scale (Fig. 7). If the refractive index is the same throughout the specimen, the depth scale in the image can be compensated by a scale factor equal to the ratio of the refractive indices of the specimen and the immersion medium (Carlsson, 1991). The scale factor can be obtained, for example, by vertical scanning of a reflecting specimen, as a piece of aluminum, in which a step of known depth has been filled with embedding medium (Carlsson, 1991). The compressed or elongated depth scale due to the change in refractive index can be measured directly from the recorded depth response. We commonly use phosphate-buffered saline in glycerin (1 : 3) containing 1% p-phenylenediamine as embedding medium, and this solution has a refractive index (n) of 1.44 (n_{oil} = 1.52). Also, the depth resolution (FWHM) will be lower when refractive indices mismatch (Carlsson, 1991; Sheppard and Cogswell, 1991; see also Fig. 11), and, especially with thick specimens (e.g., water 300 μm), the FWHM may be several times larger at the bottom than at the top of the specimen. With thin tissue specimens (<20 μm), however, this error is small.

Optics

To best utilize the confocal instrument in fluorescence microscopy, color-corrected objectives of high quality and with the largest possible numerical aperture should be used, i.e., as a rule, oil immersion objectives. For use within visible wavelengths objectives from a number of manufacturers are available. The objectives should be tested with respect to resolution along the optical axis, as many objectives are not aberration free and cannot produce the resolution expected on basis of their N.A. Thus, objectives of the same brand with identical N.A. may differ significantly with respect to R_d (Carlsson *et al.*, unpublished observations; see also Brakenhoff *et al.*, 1990). As indicated above, R_d may be estimated by measuring the FWHM of the recorded light intensity along the optical axis from a thin fluorescent dye layer (see, e.g., Ulfhake *et al.*, 1991).

Longitudinal chromatic aberration will cause a focus shift for light of different wavelengths (Fig. 8A). In fluorescence imaging, where the excitation and emission wavelengths are different, this aberration will result in a lower depth discrimination and also a decrease in intensity (e.g., Carlsson *et al.*, 1992). It is therefore important to choose color-corrected objectives of high quality for confocal fluorescence microscopy. Even for such objectives a slight shift in focal plane can often be measured between wavelengths. Although this shift may be small and without significance for the operation of the instrument in many applications, it is of importance when two-color fluorescence from small tissue objects such as axon terminals are imaged (see below). Therefore, objectives to be used for quantitative analysis of two-color fluorescence should be tested for the wavelengths to be employed. A way to measure this directly is to select a tissue object labeled with both fluorophores and measure its depth position with microscopical setting for detection of respective fluorophore.

For wavelengths where color-corrected objectives are not available, as in

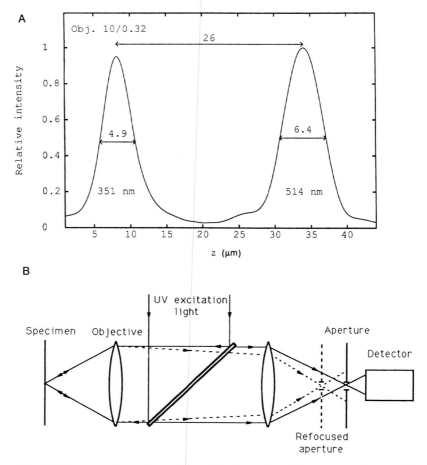

FIG. 8   (A) One way to analyze chromatic aberration is to scan a mirror with two wavelengths simultaneously and record the focal shift between wavelengths. This example shows depth discrimination curves for a × 10/0.32 dry objective when scanning a reflecting specimen vertically with both UV (351 nm) and visible light (514 nm) simultaneously. The distance between the peaks (26 μm) is a measure of the focus shift between wavelengths. The FWHM of the UV curve is narrower by about 25% owing to the shorter wavelength. (B) The influence of chromatic aberration when using UV illumination. The solid lines indicate the ray path for reflected UV light that will be focused on the pinhole aperture. The visible fluorescent light will, owing to the chromatic aberration, be focused at some distance in front of the aperture (dashed line). To detect the visible light, the aperture can be refocused. An alternative is to use convergent UV light as the illuminating light.

the UV, chromatic aberration may cause a focal shift between excitation and emission light of such a magnitude that it forces a change of the position of the illumination or detector ray paths in order to enable confocal imaging at all (Fig. 8B). An alternative in such cases is to use reflective (i.e., mirror objectives) rather than refractive objectives (Ulfhake *et al.*, 1991), avoiding all problems with chromatic aberration. However, this type of objective is very expensive and does not have as high of a numerical aperture as a refractive immersion objective with corresponding magnification.

Lasers

Lasers are powerful light sources, which can be used for either single- or multiple-wavelength excitation. Single-wavelength excitation is advantageous in cases where large focal shifts between different wavelengths are present owing to chromatic aberration as, for example, in near-UV excitation (Ulfhake *et al.*, 1991). Multiple-line excitation may be selected when the different lines can be used to excite several fluorophores with different absorption spectra simultaneously. Separation of the emission light is then dependent on differences in emission spectra of the fluorophores.

As light source in confocal microscopes, small argon ion lasers providing wavelengths between 458 and 514 nm have commonly been used. Such lasers are restricted to blue/green excitation. More recently a krypton–argon laser has been developed which enables the use of fluorophores excited in yellow (568 nm) and red (636 nm) light as well.

At Physics IV a confocal microscope equipped with a laser providing lines in the UV has been developed (Ulfhake *et al.*, 1991; Carlsson *et al.*, 1992), enabling the use of fluorophores such as aminocoumarin derivatives (AMCA; Fig. 9), paraformaldehyde vapor-induced conversion of monoamines to fluorescent compounds (Falk–Hillarp technique; Falk *et al.*, 1962), as well as calcium indicators such as Indo-1 and Fura-2 (Tsien, 1989). An example recorded with UV laser excitation is shown in Fig. 9 (color plate). As expected from theory, moving from visible light to the UV improves the resolution by about 20–30% (Ulfhake *et al.*, 1991), provided that the chromatic aberrations can be compensated for (see section on optics). This may be critical for imaging of very small objects. One drawback is that the autofluorescence contribution becomes more pronounced. Perhaps the largest obstacle in UV confocal microscopy is the current lack of color-corrected high-resolving immersion objectives needed to overcome the problems caused by chromatic aberration (Ulfhake *et al.*, 1991). Nevertheless, the recent introduction of new types of lasers for CSLMs has broadened the range of applications in neurobiology.

Signal Detection

The quality of the recorded images is determined to a large extent by the signal-to-noise ratio (S/N). The signal strength is proportional to the number of photons detected by the PMT. The noise in signals detected from fluorescent specimens is largely photon noise, that is, statistical fluctuations due to the quantum nature of light (Carlsson, 1988; Carlsson *et al.*, 1989). If the average number of photons detected per pixel is N, the signal-to-noise ratio is equal to $N^{1/2}$. Thus, S/N can be improved by increasing the number of photons reaching the detector from the fluorescent profiles. This may be achieved by increasing the photon emission from the tissue by increasing the excitation light, or by increasing the integration time during which photons are collected. Both methods may cause disturbing increases in fluorophore fading. Increased numbers of photons may also be achieved by increasing the volume from which the photons are collected for each pixel. This can be obtained by increasing the pixel size, an option that is possible in most instrumental designs. Another way to proceed is to increase the size of the aperture in front of the detector (Fig. 1). The size of the aperture determines the degree of suppression of light from out-of-focus regions of the tissue specimen and thus the amount of light reaching the detector. Apertures of very small diameter will provide excellent confocality but demand a very strong specimen signal. Thus, a weak signal may be imaged rather noise-free with a large aperture, but the imaging of small objects will then suffer from the large-aperture decrease in resolution.

By using a higher PMT voltage, the detector sensitivity can be increased and weak signals more easily detected. However, there will be an equally large increase in image noise; thus S/N in the image will not improve. In the case of immunofluorescence, or fine tracer-labeled tissue profiles, the situation is quite unfavorable, since the signal is often rather weak and emitted from quite small objects. In our application a S/N of about 10 still allows the image data to be useful in quantitative analysis (see also Carlsson *et al.*, 1989). However, it is hardly meaningful to give any general recommendation for an acceptable S/N, rather trial and error will determine this level for the particular type of specimen to be used in the image analysis.

The last step in image recording involves A/D conversion of the PMT signal to 8 or more bits of data. Instruments are commonly equipped with an 8-bit A/D conversion corresponding to 256 gray levels, and this is sufficient for most of our applications. However, in specimens containing a wide range of signal intensities, the dynamic range may not allow for detection of weak signals while still avoiding saturation of strong signals.

It is also important to be aware that the detection sensitivity of PMTs varies with photon wavelength. Thus, for the Hamamatsu R1463 PMT, used in PHOIBOS, the efficiency is 12% at 500 nm, 7% at 600 nm, and 2% at 700 nm. Finally, light is also lost in the ray path, owing to absorption/reflection by optical components. Needless to say, if one works with material containing weak fluorescence, an instrumental design with as large as possible transmission of the emission light should be favored.

Fluorophores

Fluorophores are a heterogeneous family of compounds in which light absorbed by the chemical may, through a chain of events, cause the molecule to fluoresce, that is, to emit photons. The emitted photons will vary in energy and thus wavelength, giving rise to an emission spectrum. The emitted photons will have less energy than the exciting photons and thereby longer wavelengths. Also, the absorption of light for a fluorophore is dependent on the wavelength of the excitation light; hence the photon yield, that is, the intensity of the fluorescence, will vary with excitation wavelength. The peak yield occurs when excitation is at absorption maximum. Available data seem to indicate that shifts in the excitation wavelength do not cause any large displacement of the emission spectrum (Wessendorf, 1990). The excitation/emission spectra for some commonly used fluorophores in immunolabeling have been compiled in Fig. 10. We will not present the physics of fluorophores in depth, but some practical points will be considered below.

Fluorescence properties may differ for the free fluorophore dye and the protein-conjugated counterpart (Hansen, 1967; McKinney and Spillane, 1975); therefore, when selecting fluorophores for immunofluorescence it is preferable to have absorption/emission spectra at hand for the antibody-conjugated and not the free dye. Furthermore, fluorophores may change their properties in response to environmental factors such as pH and solution composition (e.g., Nairn, 1969; Hansen, 1967). Also, the concentration of the fluorophore may cause slight displacement of the emission spectrum, as measured for the free dye (Carlsson and Mossberg, 1992; Prenna *et al.*, 1974).

A particular problem with fluorophores is photobleaching, or fading, owing to the finite capacity of the fluorophores to give off photons on excitation. This will show as a fluorophore signal attenuation which largely appears to be a function of integrated excitation energy, excitation wavelength, and heat exposure (e.g., Kaufman *et al.*, 1971; Mossberg *et al.*, 1990; Nairn, 1969; Schauenstein *et al.*, 1978; Wells *et al.*, 1990). The rate of fading varies among fluorophores, which may make one compound more desirable over other alternatives. In the early 1980s several compounds were found to retard fading if they are included in the tissue embedding medium (e.g., Johnson

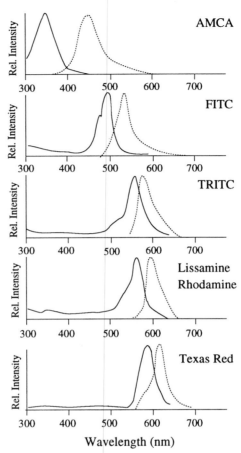

FIG. 10 Compilation of approximate absorption (solid curve) and emission (dotted curve) spectra for AMCA (7-amino-4-methylcoumarin-3-acetic acid; Kahlfan *et al.*, 1986; Wessendorf *et al.*, 1990a), fluorescein (FITC; Hansen, 1967), tetramethyl rhodamine isothiocyanate (TRITC isomer *R*; Hansen, 1967), lissamine rhodamine B (Hansen, 1967), and Texas Red (Titus *et al.*, 1982), conjugated with immunoglobulins.

et al., 1982; Platt and Michael, 1983; Valnes and Brandtzaeg, 1985), and a compound that has been widely used for this purpose is *p*-phenylenediamine (Johnson and De C Nogueira Araujo, 1981). Although useful, the action of these compounds remains obscure.

A further problem which may be encountered is fluorescence absorption by other chemicals in the specimen. This is often referred to as quenching.

If the compound that absorbs the fluorescence is itself a fluorophore, scintillation may occur; that is, the fluorescence of the compound we aim to detect may be absorbed by a second fluorophore which then emits undesirable fluorescence.

In summary, we may conclude that the selection of fluorophores must be based on their absorption/emission spectra. In the context where more than one fluorophore is used it is particularly important that the emission spectra should be separate enough to allow for safe individual detection by use of pass filters in front of the detector(s). Finally, compounds prone to fading should be avoided in order to decrease the amount of compensation needed in quantitative analysis. Some useful combinations based on the fluorophores in Fig. 10, are AMCA (Khalfan *et al.*, 1986) and FITC (Nairn, 1969), when UV/blue excitation is employed. With a krypton–argon laser, FITC and Texas Red (Titus *et al.*, 1982) is a useful pair of fluorophores. Also, cyanin derivatives (Ernst *et al.*, 1989; Mujumdar *et al.*, 1989; Southwick et al., 1990), such as allophycocyanin (peak absorbance ~650 nm, peak emission ~660 nm; not shown), could be combined with the former two, allowing rather safe triple-color detection (cf Staines *et al.*, 1988; Wessendorf *et al.*, 1990a; Schubert, 1991; see also Wessendorf, 1990). Fluorophores excited at longer wavelengths may be preferable if the contribution from autofluorescence to the recorded image is troublesome. The drawback with using longer wavelengths is that the resolution will be lower [see Eqs. (1) and (2)]; also, the quantum efficiency of the PMT is less for red wavelengths (see above). For the argon ion laser, the widely used combination of FITC and TRITC (e.g., Hansen, 1967) will cause problems with emission signal separation. It would perhaps be better to use lissamine rhodamine B (LRB; Nairn, 1969; Hansen, 1967) instead of TRITC together with FITC, since the FITC/LRB emission spectra do not overlap to the same extent as FITC/TRITC (Fig. 10).

Tissue Specimen and Embedding Medium

In immunofluorescence studies, tissue samples typically consist of a single or a series of rather thin (6–20 μm) sections cut on a cryostat and thaw-mounted on pretreated (with gelatin or equivalents) glass slides. For immunofluorescence labeling, there exists a large number of protocols for single- as well as double- and triple-labeling procedures (for references, see above); however, we will not deal with the basic issues of immunolabeling techniques here. Following labeling, the sections are embedded in a fluid medium such as a mixture of buffer and glycerin. This type of embedding medium allows for a photobleaching retarding agent, such as *p*-phenylenediamine, to be added and often (e.g., with fluorescein) such an agent cannot be omitted since fading otherwise will be severe during analysis. Thin glass slips (a thickness of 0.17 mm is often assumed in lens design) are used to cover the specimens.

The tissue preparation described is widely used and is among the best available for immunofluorescence labeling. However, if the specimens are to be used in quantitative confocal analysis there are some points that need to be considered. First, it is critical that the tissue be fully adherent to the slide and flat-embedded, since any movement of the tissue severely impedes analysis performed on the "submicron" scale. Second, the refractive properties of the embedding medium should be determined to allow for numerical compensation as needed (see section on optics). Third, the tissue should be preferably mounted in the center of the glass slide in order to enable rigid clamping of the slide onto the specimen stage of the microscope. Fourth, coverslips should be compatible with the type of objectives used (see also Sheppard and Cogswell, 1991). Also, all four edges of the coverslip need to be glued to prevent specimen movement or leakage of embedding medium during analysis. It would be advantageous if a solid embedding medium such as Entellan (Merck, D-61 Darmstadt, Germany) could be employed, since such media often have well-defined refractive properties and will prevent movement of the specimen. A drawback, however, is that solid embedding procedures include dehydration of the tissue and will thus invariably cause tissue shrinkage. We have tried to use Entellan, but this type of embedding appears to suppress the fluorescence; also, antifading measures do not appear to work well.

In tracing studies, when axon pathways or whole cells are intracellularly labeled with a fluorescent dye such as Lucifer Yellow (Stewart, 1978), thick tissue sections are often used for practical purposes. Because fading seems to be less of a problem with these types of markers, a solid embedding medium has commonly been employed, which is an advantage. However, with thick specimens three other tissue properties need to be carefully considered, namely; tissue refractive aberrations (see section on optics), light absorption, and light scattering. Tissue will always cause a certain amount of light scattering, which will vary with tissue type and will be more disturbing with increasing depth. Light absorption and bleaching may also cause considerable problems, as illustrated schematically in Fig. 11.

Light Absorption

If absorption can be assumed uniform, the intensity of the excitation light, I_{exc}, as a function of the penetrated depth (z) is equal to

$$I_{exc} = I_0 \, e^{-\alpha z} \tag{3}$$

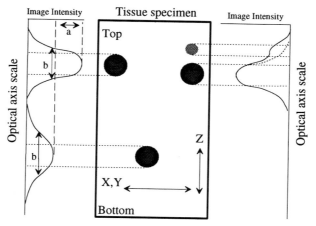

FIG. 11   Schematic illustration of the influence of light absorption, fading, tissue refractive index, and nearby fluorescent objects on the imaged size and intensity of the object. The tissue specimen contains three equally large and intense objects and one smaller and less intense object. As can be read from the image optical axis scale, objects located deep in the specimen will have a larger FWHM (*b*), due to tissue specimen refractive properties, and a lower intensity (*a*), due to both the specimen retractive index and to light absorption. Absorption is increased because the traveling distance to and from this object will be longer. If the tissue specimen is scanned from top to bottom, the number of scans will be much larger before the focus plane will be at the deeply located object compared to that located at the top, and fading may have become more severe. To the right, the problem with imaging closely located fluorescent objects is shown. The light intensity distributions of the objects will influence each other, and if one of the objects is much smaller and/or weaker in signal it may be quite difficult to resolve it from its larger/stronger emitting neighbor.

where I_0 is the incident light intensity and α the extinction coefficient for the excitation light. Likewise, the fluorescent light is attenuated on its way to the surface of the specimen. α depends on the specimen tissue. The light absorption by the tissue can be estimated by comparing recorded light intensities of a fluorescence standard from the top and the bottom of the specimen. If the background label is sufficiently even, it will suffice for this purpose. An alternative is to implant fluorescent beads (commercially available in a variety of sizes) into the top and bottom surfaces of the specimen. The obtained measurement will be a combined value of incident and emission light absorptions, and can be used to compensate the recorded image intensity values. In more complicated situations where there are good reasons to assume an uneven absorption, more elaborate methods to achieve the same objective have been suggested (Åslund *et al.*, 1991; Visser *et al.*, 1991).

Fading

With thick specimens, fading becomes a considerable problem, since each scan of the laser light will cause fading at all depths of the specimen and not only the focal plane imaged by the scan (Fig. 3). The degree of fading is sensitive to laser light intensity, exposure time, and excitation wavelength. We have tested FITC, a fluorophore prone to fade, and AMCA, which is rather resistant to fading, in situations identical to those of our applications (Mossberg *et al.*, 1990; Ulfhake *et al.*, 1991). The results indicated that with FITC the fading decreased the fluorescence intensity by about 5% when the number of scans was 10–20, and about 30% after 100 scans (see also Mossberg *et al.*, 1990); the same exposure for AMCA caused only marginal signal attenuation (Ulfhake *et al.*, 1991). By analogy with light absorption, fading can be compensated for, within reasonable limits, by recording the fluorescence attenuation factor for the number of scans and the instrumental settings to be used in the specific analysis. This can be done on the same specimen used for image analysis. However, since the buildup of heat in the specimen during scanning may also cause undesirable effects on fluorophores in the nearby tissue regions, we recommend that an adjacent tissue section be used for fading estimates. In many cases with low laser excitation energies in the specimen, the relationship between the degree of fading and the exposure time appears to be close to linear (Mossberg *et al.*, 1990), making compensation rather easy.

Photometry

Quantitative fluorescence analysis often involves recording of photometric data. The recorded signal for each pixel is proportional to the number of photons detected. However, the interpretation of the signal is far from straightforward. If the signal strengths from two equivalent objects in an imaged tissue volume are to be compared, consideration must be taken of variations in light absorption if the objects are located at different depths as well as of amount of fading due to variations in excitation–exposure time (Fig. 11). Suggestions for how these flaws may be compensated for have been given above. Also, the presence of fluorescent objects in adjacent optical layers may contribute to the fluorescence signal of the object under study (Fig. 11). This may be difficult to resolve, and the only simple and complete remedy for this is to select nonclustered objects for the analysis. Even when these constraints have been overcome, the photometric data should only be used for comparative analysis of fluorescent objects or object populations prepared and imaged under similar circumstances. In many applications, however, the main aim is to compare populations of objects, and in such situations minor errors will tend to be of little importance.

With the immunofluorescence technique, a tissue antigen, for example a transmitter substance, is labeled by antibodies conjugated to a fluorophore (direct method) or by a sequence of a primary antibody followed by a secondary antibody (indirect method). If information on fluorophore concentration or a measure of tissue content of an antigen is desired, the situation is indeed complicated. First, we need to know the relationship between the fluorophore concentration and the recorded signal intensity. In measurements of free FITC dye layers, a close to linear relationship (Carlsson and Mossberg, 1992) was obtained between dye concentration and light intensity for concentrations ranging from 0.001 to 0.1 mg dye/ml solution. Still, the relevance of this estimate for the protein-conjugated dye is not clear. Thus, prior to applying this procedure such measurements should be performed since with high fluorophore concentrations a pronounced nonlinearity has been observed (Carlsson and Mossberg, 1992). Second, the fluorophore-to-protein binding ratio in the antiserum needs to be estimated. Wessendorf and collaborators (1990b) have shown that this ratio can be estimated fairly accurately for an antiserum. Thus, for fluorescent signals that are not too strong it may be possible to estimate the amount of antiserum protein in a specimen object, and this measure may suffice. However, to deduce the approximate tissue antigen concentration we must also have a reliable estimate of the binding ratio of the tissue antigen to the antiserum protein, which is a formidable task since the binding coefficient is influenced by a large number of factors.

Quantitative Analysis of Immunofluorescence-Labeled Axon Terminals in Nervous Tissue

Studies on neuronal connectivity and colocalization of messenger molecules in axonal pathways would be greatly enhanced if quantitative analysis of transmitter-identified axon varicosities could be automated. So far, such data have only been obtained by very laborious counting, often performed directly in the microscope (e.g., Arvidsson et al., 1990; deLima et al., 1988; Tuchscherer et al., 1987, 1989) or by deploying two-dimensional image analysis systems. For the purpose of enabling reliable 3-D quantitative analysis of axon terminals labeled as to their content of messenger molecule(s), a computer-based system was developed which provides data on (a) number of axon terminal-like objects, (b) individual size and shape of axon terminal-like objects, (c) their individual positions, and (d) the integrated fluorescent signal strength of individual axon terminal-like objects in tissue volumes recorded with a CSLM (Mossberg *et al.,* 1990; Mossberg *et al.,* in preparation; Mossberg and Ulfhake, 1990). The computer-based method is described below.

To be able to identify an image object as an axon terminal, the procedure must accomplish a separation of axon terminals from axon fibers and noise. Axons have a spheroid-shaped terminal enlargement (swelling) and also often a varicose preterminal region, evident in the light microscope. These swellings correspond to the (synaptic) boutons of the axon as revealed by electron microscopy (e.g., Lagerbäck *et al.,* 1979; Peters *et al.,* 1976; Ulfhake *et al.,* 1987a). Also, the fluorescence signal is stronger in the swellings than in the parent axon in specimens labeled with antiserum raised against the transmitter substance, since the vesicles containing the signal substance(s) are concentrated in these swellings (Peters *et al.,* 1976; Ulfhake *et al.,* 1987a). Taken together, axon boutons may be discriminated from the parent axons, as well as from noise, by size, shape, and also, when occurring *en passant* (or in clusters) by the presence of local fluorescence maxima. These parameters, namely, a minimal size, an ellipsoidal shape, and local photometric intensity maxima, were deployed as discriminating criteria.

The program is based on a split–merge analysis of image objects in the process of testing identified objects against the constraining criteria. To enable this type of small-object image analysis, the recordings are made with high-resolution oil immersion objectives ($\times 63/1.4$ N.A.) and 0.1, 0.2, or 0.4 μm refocusing steps between scans, so that the smallest image object to be analyzed consists of over 10 pixels. Excitation power should be adjusted so that the recorded images have a signal-to-noise ratio that is as high as possible, while at the same time fading of fluorophores is acceptable. This can be established rather fast empirically by trial and error (for discussion on this topic, see above and Mossberg *et al.,* 1990). Rather thin sections (8–16 μm) are commonly employed; thus, fading and light absorption are, as a rule, not serious problems. When using fading-prone fluorophores such as FITC, or if the number of scans needed is large, or if high laser energies are used to improve S/N in weakly labeled specimens, compensation for fading is often needed (see section on tissue specimen and embedding medium). We have previously devised a method to estimate background contribution by light intensity histogram analysis directly on data recorded on the tissue specimen used for analysis (for details, see Mossberg *et al.,* 1990). This method was originally designed for thresholding but can, when applied to sections recorded at different depths, be used to extract data on light absorption and fading (see also above). When the refractive index of the specimen differs from that of the immersion medium, the scale error must be measured and the depth coordinates in the volume compensated. Finally, image distortions owing to microscope PSF must be compensated for; the algorithms used to recalculate the size of the axon terminal-like objects in this particular application are described in the Appendix.

Program Processing Steps

A flowchart of the program is shown in Fig. 12, and the stepwise effects on a recorded image volume of immunofluorescent labeled axons are shown in Fig. 13 (color plate). First, the recorded stack of images is filtered with a 3×3 median-value filter in order to reduce noise. This step can be omitted if the noise level in the recordings is low, and unnecessary loss in resolution is thereby avoided. If there is a background level in the images, for example, due to a photometric offset in the instrument, a constant value can be subtracted from the entire image volume. This value can be determined by calculating the mean intensity of a part of the image that contains only background. From here onward, the image data are considered as a 3-D volume and not as a stack of 2-D images. All local intensity maxima in the volume are identified in order to find the fluorescence maxima of the axon terminals. To avoid sampling the numerous minor maxima caused by variations in the nonspecific background label, maxima to be sampled must have a minimum intensity, indicated by the operator. For each identified intensity maximum, all surrounding pixels with an intensity value above the half-maximum intensity are tagged and assembled to an object (Fig. 13B,C). In an earlier version of the program we chose to tag all pixels with intensity values above a given fixed threshold value (Mossberg *et al.*, 1990; Mossberg and Ulfhake, 1990). That method has the disadvantage of being sensitive to the applied threshold and will also give high-intensity objects an unproportionally large size (see Fig. 14). The method described here, where the threshold applied is proportional to the maximum intensity of the object, has the weakness of being sensitive to noise that may influence the maximum intensity value. The filtering step in the beginning of the processing is therefore often necessary in noisy images.

At this stage of the program processing, an image object consists of a group of adjacent pixels (Fig. 13C) that may include several intensity maxima. In some cases this is due to the fact that the object represents a cluster of several axon terminals, whereas in others it is due to uneven distribution of the fluorescence in the terminal. The objects are therefore split into as many subobjects as there are maxima. The splitting is made along the "intensity valleys" between the maxima (Fig. 13D; large arrow Fig. 13B–D). Owing to remaining noise and the uneven distribution of fluorescence in the axon terminals, this step will produce a rather large number of subobjects, some

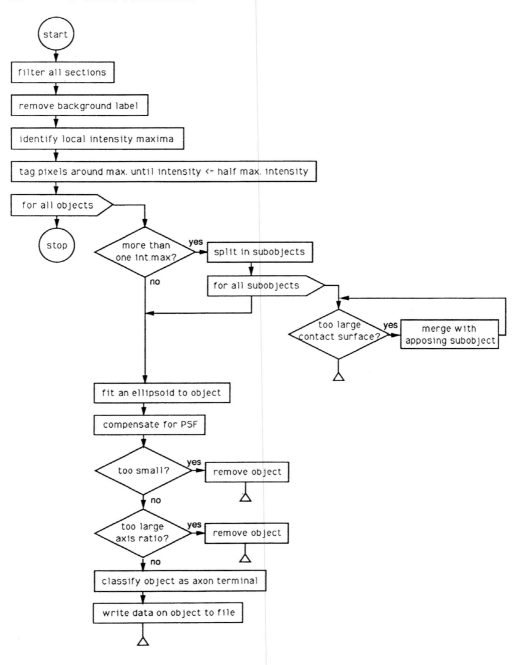

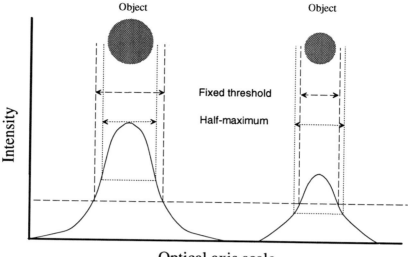

FIG. 14   Schematic illustration showing the effects of applying an intensity threshold (fixed threshold) versus an intensity half-maximum on the size of the objects. A fixed threshold will cause differences in size between strongly and weakly fluorescent objects to be exaggerated. Although this diagram only ilustrates the effect along the optical axis, the same applies to the lateral plane as well.

of which clearly do not represent axon terminals on their own. In the next processing step, therefore, such sobobjects are merged. The criterion used for merging of subobjects is to examine the largest contact surface that a subobject has to other subobjects. If this contact surface area exceeds a user-defined fraction of the total surface area (we have used values of 20–33%), it is merged with the object to which it has the largest contact area (Fig. 13E; large arrow in Fig. 13B,D,E). The remaining objects are regarded as axon terminals or axon fibers, and, in addition, there is often some noise remaining.

As mentioned earlier, the microscope PSF will make small objects elongated beyond their true size, in particular, along the optical axis (Fig. 6), and this effect is clearly noticeable in our application. To compensate for this we first fit an ellipsoid to each object. When calculating the size and direction

FIG. 12   Flowchart of the program. The software is written in C and has been implemented on an ordinary SUN workstation and a Silicon Graphics computer.

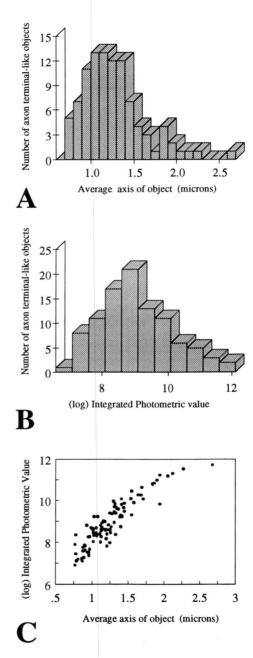

of the axes of the ellipsoid, the axis size is recalculated in order to compensate for the PSF of the microscope (see Appendix). The size of this compensation will vary with the size of the specimen objects and the PSF of the objective used. This compensation should not be confused with depth scale compensation performed for the refractive index of the embedding medium (see section on optics). This program step will improve the estimation of object size (see Fig. 6).

To remove remaining axon fibers and noise, all objects are tested with respect to their size and shape. Noise is eliminated by removing all objects that have at least one axis whose length is less than a predefined lower limit for objects to be regarded as axon terminals. We have used 0.5 μm (Fig. 13E,F; small arrow) which is reasonable considering the resolution of the microscope. Axon fibers are, as a rule, very much more elongated than the terminals. Thus, if the ratio between the largest and the smallest axis lengths exceeds a preset value, the object will be regarded as a fiber and therefore removed. Different ratio values may be tried; we have used 3 : 1 based on empirical tests and quantitative electron microscopic data on axon terminal diameters in the literature (e.g, Conradi, 1969; Lagerbäck et al., 1981; Ulf-hake et al., 1987a).

The remaining objects are considered to represent axon terminals and are subjected to quantification (Fig. 13F). The output file consists of the number of objects, their individual size and integrated intensity value, as well as their 3-D coordinates. In Fig. 15, the output from the sample volume used to illustrate the program steps is shown. One important property of this method, illustrated in Fig. 15C, is that axon terminals of equal size but with different fluorescence intensities are recognized by the program. Thus, not only can variances in densities, sizes, and spatial distributions of axon terminals be studied, but differences in fluorescent signal strength can also be detected. We have previously shown that the quantitative output of this program yields consistent results with estimates done directly in the microscope (Mossberg et al., 1990). Tests with known standards such as latex beads provide a useful means to validate the performance of this type of software image processing. We have also applied the program to image recordings of fluorescent latex

---

FIG. 15   Output data from the computer program. (A) Histogram of average axis size of the objects found in the sample specimen shown in Fig. 13A. (B) Histogram of the integrated intensity values of the objects. (C) Plot of object average axis versus object integrated signal strength. It is clear that these two variables covary in this sample but also that the program recognizes objects of similar size but with different fluorescent signal strengths.

beads with known diameters and obtained consistent results with respect to both accuracy and reproducibility. An example of such a test is shown in Fig. 16 (color plate).

Imaging and Analysis of Multicolor Fluorescence in Analysis of Neural Connectivity and Colocalization of Messenger Molecules

Studies on the distributions of classic transmitters, and neuropeptides, have revealed that several messenger molecules frequently coexist in the same axon terminal (e.g., Hökfelt, 1987). However, quantitative studies of colocalization are scarce. Immunofluorescence techniques used to visualize transmitter-identified axon terminals can be combined with intracellular labeling of, for example, target neurons (see, e.g., Ulfhake *et al.*, 1987b; Wallén *et al.*, 1988). Furthermore, by combining the above-mentioned labels with markers for synaptic contacts (e.g., Siekevitz, 1987), such as postsynaptic receptors (Triller *et al.*, 1990), a detailed analysis of neuronal connectivity can be made directly. This type of information is much wanted, and with a high-resolving microscope as the CSLM and the multicolor immunofluorescence technique this seems to be possible to obtain. However, a key issue in this context is the capacity to distinguish the different fluorophores with a high degree of accuracy.

For the purpose of multicolor fluorescence imaging, several approaches are available. We will focus on techniques we have used and also refer to other options.

Using a CSLM with a single detector (Fig. 2A), two-color fluorescence can be recorded by scanning the specimen twice, once for each fluorophore (Mossberg *et al.*, 1990). Between the scans, the laser light wavelength and filter setting are changed according to the fluorophore to be detected. With the instrument we have used, it has been most practical to scan the entire stack of images separately for each fluorophore. With access to multiline laser excitation, however, the instrument can be modified to allow fast switching of excitation and pass filters between image scans, or even between scan lines. The advantage of the latter procedure is that the risk of displacement of the specimen between fluorophore recordings is reduced. With the former approach alignment of the image stack must be checked using reliable landmarks in the specimen.

Another potential source of image error which needs to be assessed when detecting two-color fluorescence is that the focal length of the objective may differ for different wavelengths, causing a displacement along the optical axis between the pair of recorded image (see section on optics). This is due to the chromatic aberration of the objective and stresses the importance of selecting objectives that are color-corrected for the wavelength bands to be used.

To achieve an acceptable separation of the two fluorophores, the exciting wavelength and barrier filters must be chosen carefully. We will exemplify this with the widely used combination of TRITC/FITC, where both the excitation and emission spectra show substantial overlap, resulting in difficulties in separation of the recorded fluorophores. We have found that FITC is best detected by using a shorter (458 nm) than optimal (i.e., 495 nm; see Fig. 10) excitation wavelength, since the comcomitant excitation of TRITC is minimized (cf. Fig. 10). The weaker FITC signal thus obtained can be compensated, as needed, by accumulating the signals from several scans. In addition to a long-pass filter (500 nm; Corion), a short-pass filter (550 nm; Nikon) is also inserted in the emission ray path to suppress TRITC fluorescence. TRITC is detected by exciting at the longest available wavelength, that is, at 514 nm, and a long-pass filter (LP 610 nm; Schott) is employed to suppress undesired FITC fluorescence. With these settings it was found that in most cases breakthrough between fluorescence labels could be kept to 5% or less of the total image intensity (Mossberg *et al.,* 1990; Mossberg *et al.,* unpublished observations). To estimate the coexistence, the (FITC and TRITC) volume images are compared pixel by pixel. A two-color fluorescence specimen (FITC blue excitation/green emission; TRITC green excitation/red emission) where the signal separation is rather poor (Fig. 17A,D) (color plate) will be used here as an example when comparing methods. In this case the two cells contain only one of the labels. There is a low-intensity breakthrough evident in both images, amounting to 15% for FITC and 33% for TRITC, which can be reduced further by inserting additional filters. This will, however, also reduce the signal strength of the fluorophore to be detected, which may be critical if the specimen fades quickly.

With a CSLM equipped with dual detectors (Fig. 2B) it is possible to detect two emission wavelength bands simultaneously. This enables detection of two fluorophores on the basis of differences in emission spectra between the compounds. This is a faster procedure than the method described above and avoids the problems associated with repeated scanning of a tissue specimen as well as fading owing to repeated exposures. With a single-line laser, both fluorophores must be excited with the same wavelength. This may call for a compromise with respect to excitation efficiency of the respective fluorophore. In the example used above, namely, a FITC/TRITC-labeled tissue specimen, the wavelengths 448, 497, and 514 should be tested. The line that gives comparable intensity strength of the two fluorphores, depending on the relative strength of the two labels, should be selected. However, the breakthrough between labels (see Fig. 17B,E) will be a greater problem relative to the situation where the excitation wavelength was changed between fluorophore. In our example, the amount of breakthrough was 46% for FITC and 64% for TRITC. Thus, simultaneous two-color fluorescence recording demands good separation of the emission spectra which is rarely the case.

To reduce the problem with breakthrough between labels, a method based on linear subtraction of the images has been used (Carlsson and Mossberg, 1992). In brief, a fraction of the images recorded in one of the spectral bands ("FITC" detector) is subtracted from the image recorded in the other spectral band ("TRITC" detector), and vice versa. The fraction of the FITC image to be subtracted from the TRITC image is determined by dividing the image recorded by the two detectors. Such a "ratio image" is shown in Fig. 17G, where the blue color (low ratios) corresponds to regions stained with FITC and red color (high ratios) to regions stained with TRITC. A histogram of the ratio image (Fig. 17H) shows two peaks corresponding to the different ratio values (arrows on the abscissa in Fig. 17H). The ratio value corresponding to the FITC peak gives directly the fraction of the FITC image to be subtracted from the TRITC image, while the FITC image is subtracted by a fraction of the TRITC image corresponding to the inverted value of the TRITC peak (B on the abscissa in Fig. 17H) in the ratio image. The result of these subtractions can be seen in Fig. 17C,F. The "image subtraction" method can remove breakthrough if (a) the detector response is linear and (b) the spectral properties of the fluorophores are the same in the entire specimen. The first requirement is met with PMTs as detectors, while the spectral properties of the fluorophore often do not strictly adhere to the requirements. Empirical tests (Carlsson and Mossberg, 1992) such as the one shown in Fig. 17, however, show that although the fluorophore requirements are not met, a considerable decrease in cross-talk can be obtained. It is critical, however, that a single-labeled object for each fluorophore be present to allow calculations of the appropriate ratio values. This method can also be applied with the single-detector imaging method for two-color fluorescence described above.

The problem with breakthrough in double-labeled specimens can only be circumvented completely by choosing a combination of fluorophores with sufficiently separated excitation/emission spectra. If multiline excitation can be employed, the situation becomes more favorable since one can choose from a broader range of fluorophores. When emission spectra become sufficiently separate, the problem with breakthrough diminishes. Suitable combinations of fluorophores would be FITC/Texas Red (cf. Fig. 10) or FITC/cyanin derivatives to be used with a krypton–argon ion laser. AMCA/FITC should be useful with a laser providing wavelengths between 350 and 450 nm. When combining immunofluorescence with fluorescent traces such as fluorogold (Schmued and Fallon, 1986) or Lucifer Yellow (Stewart, 1978), the best choice is probably a fluorophore emitting in blue, for example, AMCA (Ulfhake *et al.*, 1991), or far-red, for example, cyanine derivatives (for references, see section on fluorophores). An alternative is to use a blue tracer, for example, True-blue (Bentivoglio *et al.*, 1980), in combination with fluorophores emitting in red, such as lissamine rhodamine or Texas Red (Fig. 10). In some cases two fluorophores are readily excitable at the same wavelength but differ considerably in emission spectra (e.g., fluorogold and AMCA), allowing safe separation only by varying the pass filters (Ulfhake *et al.*, 1991).

Concluding Remarks

With the introduction of the confocal microscope we have a tool enabling light microscopic analysis of neuronal connectivity on the axon terminal level in three dimensions. Although CSLM instruments have come in general use only recently, this technique has been deployed in a number of neurobiological applications, revealing information previously not possible to obtain directly. Thus, CSLM has been used for 3-D reconstruction of neurons intracellularly labeled with Lucifer Yellow (Wallén *et al.*, 1988), for recording of two-color fluorescence to reveal the connectivity between identified neurons (Brodin *et al.*, 1988; Ulfhake *et al.*, 1991) as well as colocalization of signal molecules in axon terminals (Mossberg *et al.*, 1990; Mossberg and Ulfhake, 1990), monitoring changes in the content of signal substances caused by imposed lesions (Lindå *et al.*, 1990; Örnung *et al.*, 1989), and for measuring the size and form of glycine receptors in neuronal membranes (Triller *et al.*, 1990).

The main aim of this chapter has been to describe the potential of CSLMs in fluorescence microscopy, exemplified by the procedure developed for quantitative axon terminal analysis, but also to stress that there are clear limitations. Thus, noncritical use of CSLMs in these types of applications may lead to misinterpretations of the data and possibly false conclusions. We can envisage that various types of software packages will become available for the CSLM user in the near future, not only providing routines for quantitative 3-D analysis but also helping with a number of the problems encountered in CSLM imaging, such as light signal attenuation, PSF, and refractive index errors.

Among the most exciting recent developments in CM is the introduction of new lasers, enabling the instrument to be applied in UV fluorescence microscopy (Arndt-Jovin *et al.*, 1990; Puppels *et al.*, 1990; Ulfhake *et al.*, 1991). Provided that high-resolution objectives color-corrected for the UV become available, CSLMs should be ideal for the rapidly expanding technique of measuring cytosolic free calcium by UV-excited fluorescent indicators such as Fura-2 and Indo-1 (Tsien, 1989). Another important future option is the deployment of "two-photon" excitation by the use of pulsed lasers (Webb, 1990). This technique will strongly reduce fading of tissue regions out of focus.

Appendix*

When an object (f) is imaged by a confocal microscope, the PSF (h) will result in a distorted image (g). The image that represents the object convolved with the PSF of the microscope can be written as

$$g = h \circ f \tag{A1}$$

If the object is one dimensional the following relation is valid:

$$\sigma_g^2 = \sigma_h^2 + \sigma_f^2 \tag{A2}$$

where σ is the standard deviation of the intensity distribution (Cramer, 1966). In three dimensions this relation can be expanded:

$$\sigma_{(g,xy)}^2 = \sigma_{(h,xy)}^2 + \sigma_{(f,xy)}^2 \tag{A3}$$

and analogously for other combinations of x, y, and z. σ_{xy} can be calculated as

$$\sigma_{xy}^2 = [1/\sum_{x,y,z} I(x, y, z)] \sum_{x,y,z} I(x, y, z)(x - \bar{x})(y - \bar{y}) \tag{A4}$$

$$\bar{x} = [1/\sum_{x,y,z} I(x, y, z)] \sum_{x,y,z} I(x, y, z)x \tag{A5}$$

with the similar expression for $\bar{y}$. $I(x, y, z)$ is the intensity of the pixel with coordinates (x, y, z).

If an object has an ellipsoidal shape, the size and direction of its axes can be calculated by determining the eigenvalues and eigenvectors of the covariance matrix $\mathbf{C}$ (Cramer, 1966):

* Prepared by K. Mossberg, B. Ulfhake, and Johan Philip, Royal Institute of Technology, Stockholm, Sweden.

$$\mathbf{C} = \begin{bmatrix} \sigma_{f,xx} & \sigma_{f,xy} & \sigma_{f,xz} \\ \sigma_{f,yx} & \sigma_{f,yy} & \sigma_{f,yz} \\ \sigma_{f,zx} & \sigma_{f,zy} & \sigma_{f,zz} \end{bmatrix} \tag{A6}$$

Equation (A3) gives $\sigma^2_{(f,xy)} = \sigma^2_{(g,xy)} - \sigma^2_{(h,xy)}$, and the matrix elements in Eq. (A6) can therefore be replaced by $[\sigma^2_{(g,xy)} - \sigma^2_{(h,xy)}]^{1/2}$.

If we assume that the PSF (h) is symmetric around the x, y, and z axes, then $\sigma_{h,xy} = \sigma_{h,yz} = \sigma_{h,zx} = 0$ and Eq. (A6) can then be written as

$$\mathbf{C} = \begin{bmatrix} [\sigma^2_{(g,xx)} - \sigma^2_{(h,xx)}]^{1/2} & \sigma_{g,xy} & \sigma_{g,xz} \\ \sigma_{g,yx} & [\sigma^2_{(g,yy)} - \sigma^2_{(h,yy)}]^{1/2} & \sigma_{g,yz} \\ \sigma_{g,zx} & \sigma_{g,zy} & [\sigma^2_{(g,zz)} - \sigma^2_{(h,zz)}]^{1/2} \end{bmatrix} \tag{A7}$$

The eigenvalues and eigenvectors calculated from Eq. (A7) give information about the size and direction of the axes in the original object f. Thus, the volume and orientation of the original object can be calculated. A prerequisite for the compensation is that the standard deviations, laterally and along the optical axis of the PSF, $\sigma_{(h,xx)}$, $\sigma_{(h,yy)}$, and $\sigma_{(h,zz)}$, have been measured, and that the different combinations of σ_g can be calculated from the digital volume for each object. This method was tested on a specimen containing fluorescent latex beads of known sizes (1.1 and 2.4 μm). For a sample of such beads the optical-to-lateral axis ratios were on average 1.12 (2.4-μm beads) and 1.43 (1.1-μm beads), respectively, instead of the ideal 1.0. By recalculating the axes using Eq. (A7), the optical-to-lateral axis ratios were 1.03 (2.4-μm beads) and 1.01 (1.1-μm beads), respectively.

Acknowledgments

This study was supported by grants from the Swedish Medical Research Council (12X-6815, O4X-2887), Goljes Minnesfond, ELFAs Forskningsford, Anders-Otto Swärds Stiftelse, Marcus and Amalia Wallenbergs Minnesfond, Riksföreningen för Åldringsforskning, the Swedish Research Council for Engineering Sciences, and the Karolinska Institutet. We are grateful to Professor R. Elde (Dept. of Histology and Neurobiology, Karolinska Institutet, Stockholm, Sweden) for constructive criticism on the manuscript.

References

Arndt-Jovin, D. J., Robert-Nicoud, M., and Jovin, T. M. *J. Microsc.* **157**, 61 (1990).
Arvidsson, U., Cullheim, S., Ulfhake, B., Bennett, G. W., Fone, K. C., Cuello, A. C., Verhofstad, A. A. J., Visser, T. J., and Hökfelt, T., *Synapse* **6**, 237 (1990).

Åslund, N., Liljeborg, A., Oldmixon, E. H., and Ulfsparre, M. *Proc. SPIE Int. Soc. Opt. Eng.* **1450,** 329 (1991).

Bentivoglio, M., Kuypers, H. G. J. M., and Catsman-Berrevoets, C., *Neurosci. Lett.* **18,** 19 (1980).

Bertero, M., Brianzi, P., and Pike, E. R. *Inverse Prob.* (*Proc. SPIE-Int. Soc. Opt. Eng.*) **3,** 195 (1987a).

Bertero, M., De Mol, C., and Pike, E. R. *J. Opt. Soc. Am.* **4,** 1748 (1987b).

Brakenhoff, G. J., Visscher, K., and van der Voort, H. T. M. *in* "Handbook of Biological Confocal Microscopy" (J. B. Pawley, ed.), p. 87. Plenum, New York, 1990.

Brodin, L., Ericsson, M., Mossberg, K., Hökfelt, T., Ohta, Y., and Grillner, S., *Exp. Brain Res.* **73,** 441 (1988).

Cajal, S., and Ramon, Y. "Histologie du Systeme Nervaux de l'Homme et des Vertebres." Maloine, Paris, 1909.

Carlsson, K., Thesis, Physics IV, The Royal Institute of Technology, TRITA-FYS4010 (1988).

Carlsson, K., *J. Microsc.* **157,** 21 (1990a).

Carlsson, K., *J. Microsc.* **163,** 167 (1991).

Carlsson, K., and Åslund, N., *Appl. Opt.* **26,** 3232 (1987).

Carlsson, K., and Liljeborg, A., *J. Microsc.* **153,** 171 (1989).

Carlsson, K., and Mossberg, K., *J. Microsc.* in press (1992).

Carlsson, K., Danielsson, P. E., Lenz, R., Liljeborg, A., Majlöf, L., Åslund, N., *Opt. Lett.* **10,** 53 (1985).

Carlsson, K., Wallén, P., and Brodin, L., *J. Microsc.* **155,** 15 (1989).

Carlsson, K., Mossberg, K., Helm, J. P., and Philip, J., *Proc. SPIE* 1660 (1992).

Castleman, K. R. "Digital Image Processing," p. 351. Prentice-Hall, New Jersey, 1979.

Conradi, S., *Acta Physiol. Scand.* **332**(Suppl.), **5** (1969).

Coons, A. H., *in* "General Cytochemical Methods" (J. F. Danielli, ed.), p. 399. Academic Press, New York, 1958.

Cramer, H. "Mathematical Methods of Statistics." Princeton Univ. Press, Princeton, New Jersey, 1966.

Ernst, L. A., Gupta, R. K., Mujumdar, R. B., and Waggoner, A. S., *Cytometrics* **1,** 3 (1989).

Falck, B., Hillarp, N. A., Thieme, G., and Torp, A., *J. Histochem. Cytochem.* **10,** 348 (1962).

deLima, A. D., Bloom, F. E., and Morrison, J. H., *J. Comp. Neurol.* **274,** 280 (1988).

Hökfelt, T. *in* "Synaptic Function" (G. M. Edelman, W. E. Gall, and W. M. Cowan, eds.), A Neurosciences Institute Publication, p. 179. Wiley, New York, 1987.

Hökfelt, T., Skagerberg, G., Skirboll, L., and Björklund, A., "Handbook of Chemical Neuroanatomy, Methods in Chemical Neuroanatomy" (A. Björklund and T. Hökfelt, eds.), Vol. 1. (1983). Elsevier, Amsterdam.

Hansen, P. A., *Acta Histochem.* Suppl. **7,** 167 (1967).

Heimer, L., and Záborszky, L., "Neuroanatomical Tract Tracing Methods 2, Recent Progress." Plenum, New York and London, 1989.

Johnson, G. D., and De C Nogueira Araujo, G. M., *J. Immunol. Methods* **43,** 349 (1981).

Johnson, G. D., Davidson, R. S., McNamee, K. C., Russell, G., Goodwin, D., and Holborow, E. J., *J. Immunol. Methods* **55,** 231 (1982).

Khalfan, H., Abuknesha, R., Rand-Weaver, M., Price, R. G., and Robinson, D., *J. Histochem.* **18,** 497 (1986).

Kaufman, G. I., Nester, J. F., and Wasserman, D. E., *J. Histochem. Cytochem.* **19,** 469 (1971).

Lagerbäck, P.-Å., Ronnevi, L.-O., Cullheim, S., and Kellerth, J.-O., *Brain Res.* **207,** 247 (1981).

Lindå, H., Cullheim, S., Risling, M., Arvidsson, U., Mossberg, K., Ulfhake, B., Terenius, L., and Hökfelt, T., *Brain Res.* **534,** 352 (1990).

McKinney, R. M., and Spillane, J. T., *Ann. N.Y. Acad. Sci.* **254,** 55 (1975).

Minsky, M. Microscopy Apparatus. U.S. Patent 3,013,467 (1961).

Mossberg, K., and Ulfhake, B., *Trans. R. Microscop. Soc.* **1,** 413 (1990).

Mossberg, K., Arvidsson, U., and Ulfhake, B., *J. Histochem. Cytochem.* **38,** 179 (1990).

Mossberg, K., Philip, J., and Ulfhake, B., in preparation (1992).

Mujumdar, R. B., Ernst, L. A., Mujumdar, S. R., and Waggoner, A. S., *Cytometry* **10,** 9 (1989).

Nairn, R. C. *in* "Fluorescent Protein Tracing" p. 95. E&S Livingstone, Edinburgh and London, 1969.

Örnung, G., Ulfhake, B., Arvidsson, U., Cullheim, S., Hökfelt, T., Mossberg, K., and Visser, T. J., *Eur. J. Neurosci., Suppl.* **2,** 206 (1989).

Peters, A., Palay, S. L., and Websters, E. F., "The Fine Structure of the Nervous System: The Neurons and Supporting Cells." Saunders, Philadelphia, Pennsylvania, 1976.

Petran, M., Hadravsky, M., Egger, M. D., and Galambos, R., *J. Opt. Soc. Am.* **58,** 661 (1968).

Platt, J. L., and Michael, A. F., *J. Histochem. Cytochem.* **31,** 840 (1983).

Prenna, G., Leiva, S., and Mazzini, G., *J. Histochem.* **6,** 467 (1974).

Puppels, G. J., deMul, F. F., Otto, C., Greve, J., Robert-Nicoud, M., Arndt-Jovin, D. J., and Jovin, T. M., *Nature (London)* **347,** 301 (1990).

Schauenstein, K., Böck, G., and Wick, G. *in* "Immunofluorescence and Related Staining Techniques" (W. Knapp, K. Holubar, and G. Wick, eds.), p. 81. Elsevier/North Holland, Amsterdam, 1978.

Schmued, L. C., and Fallon, J. H., *Brain Res.* **377,** 147 (1986).

Shaw, P. J., and Rawlins, D. J., *J. Microsc.* **163,** 151 (1991).

Sheppard, C. J. R., and Cogswell, C. J., *Optic* **87,** 34 (1991).

Shubert, W., *Europ. J. Cell Biol.* **55,** 272 (1991).

Siekevitz, P. *in* "Encyclopedia of Neurosciences" (G. Adelman, ed.), Vol. 2, Birkhäuser, Boston, Basel, and Stuttgart, 1987.

Southwick, P. L., Ernst, L. A., Tauriello, E. W., Parker, S. R., Mujumdar, R. B., Mujumdar, S. R., Clever, H. A. Waggoner, A. S., *Cytometry* **11,** 418 (1990).

Staines, W. A., Meister, B., Melander, T., Nagy, J. I., and Hökfelt, T., *J. Histochem. Cytochem.* **36,** 145 (1988).

Stewart, W., *Cell (Cambridge, Mass.)* **14,** 741 (1978).

Titus, J. A., Haugland, R., Sharrow, S. O., and Segal, D. M., *J. Immunol. Methods* **50,** 193 (1982).

Triller, A., Seitanidou, T., Franksson, O., and Korn, H., *New Biol.* **12,** 637 (1990).

Tsien, R. Y., *Annu. Rev. Neurosci.* **12,** 227 (1989).

Tuchscherer, M. M., Knox, C., and Seybold, V. S., *J. Neurosci.* **7,** 3984 (1987).

Tuchscherer, M. M., and Seybold, V. S., *J. Neurosci.* **9,** 195 (1989).

Ulfhake, B., Arvidsson, U., Cullheim, S., Hökfelt, T., Brodin, E., Verhofstad, A., and Visser, T., *Neuroscience* **23,** 917 (1987a).

Ulfhake, B., Cullheim, S., Hökfelt, T., and Visser, T. J., *Brain Res.* **419,** 387 (1987b).

Ulfhake, B., Carlsson, K., Mossberg, K., Arvidsson, U., and Helm, P. J., *J. Neurosci. Methods* **40,** 39 (1991).

Valnes, K., and Brandtzaeg, P. (1985).

Visser, T. D., Green, F. C. A., and Brakenhoff, G. J., *J. Microscop.* **163,** 189 (1991).

Wallén, P., Carlsson, K., Liljeborg, A., and Grillner, S., *Neurosci. Methods* **24,** 91 (1988).

Webb, W., *Trans. R. Microscop. Soc.* **1,** 445 (1990).

Wells, S. K., Sandison, D. R., Strickler, J., and Webb, W. W., *in* "Handbook of Biological Confocal Microscopy" (J. B. Pawley, ed., p. 27. Plenum, New York, 1990.

Wessendorf, M. W., *in* "Handbook of Chemical Neuroanatomy" (A. Björklund, T. Hökfelt, F. G. Wouterloud, and A. N. van den Pool, eds.), Vol. 8, Analysis of Neuronal Microcircuits and Synaptic Interaction, p. 1. Elsevier, Amsterdam, 1990.

Wessendorf, M. W., Appel, N. M., Molitor, T. W., and Elde, R. P., *J. Histochem. Cytochem.* **38,** 1859 (1990a).

Wessendorf, M. W., Tallaksen-Greene, S. J., and Wohlhueter, R. M., *J. Histochem. Cytochem.* **38,** 87 (1990b).

Wilson, T., *J. Microsc.* **154,** 143 (1989).

Wilson, T. (ed.), "Confocal Microscopy." Academic Press, London and New York, 1990.

Wilson, T., and Carlini, A. R., *Opt. Lett.* **12,** 227 (1987).

Wilson, T., and Carlini, A. R., *J. Microsc.* **154,** 243 (1989).

Wilson, T., and Sheppard, C., "Theory and Practice of Scanning Optical Microscopy." Academic Press, London, 1984.

[8]   Computer-Assisted Quantitative Receptor Autoradiography

Louis Segu, Pierre Rage, Robert Pinard, and Pascale Boulenguez

Introduction

Autoradiography

Anatomical methods generally use a visual reaction to identify chemicals in tissue sections. To achieve a quantitative analysis it is necessary to have a reproducible method, a standard, and a relation between the standard and the physical measurement of the product of the reaction. Radioactivity is detectable at very low concentrations, is easy to quantify, and allows the determination of the absolute concentration of a radiolabeled molecule of known specific activity.

The radioactive emission can promote a latent image on a photographic emulsion. After development, a silver image of the radiation with darkened surfaces over the radioactive source is obtained, that is, an autoradiogram. Thus, if a tissue section contains a radiolabeled molecule, the molecule may be localized by apposing an emulsion to the section. Autoradiography is used frequently in biology to detect molecules linked to biochemical reactions and cell functions (1).

Autoradiography is a nonspecific method; only the labeled molecule and the way in which it is used are specific for the function to be studied, the molecule being a marker of one of the steps of the mechanisms of interest. For example, several neurons of the central nervous system have the capacity to take up specifically and accumulate biogenic amine neurotransmitters (2). By using ^{3}H-labeled monoamines it is possible to determine by autoradiography a regional distribution of the uptake function, specific for the amine used.

Quantitative Autoradiography

Determination of silver grain density (3) allows quantification at the subcellular level of axon terminals able to take up the amines (4). This is an example of the use of autoradiography to determine relative densities of the molecules and hence of their anatomical site. A similar method was used to detect the binding sites of neurotransmitter receptors (5). Again, the specificity is based on the labeled molecule or tracer used, such as [^{3}H]serotonin (5-HT), and the conditions of incubation employed, specific for the 5-HT$_1$ receptors (6). Autoradiograms show a regional distribution of the intensity of labeling of certain specific anatomical structures. This distribution is different from that of serotonin uptake cited above. The optical densities of autoradiograms can be determined, but they are not directly related to absolute concentrations of the source, since the response of the emulsion to the intensity of the radiation is not linear. A relationship between the physical measurement of the reaction (optical density of emulsion) and radioactivity must be established. This can be achieved by using calibrated radioactive standards as reference for every type of radioactive emission commonly used, namely, tritium (7) and iodine-125 (8). A relationship between the known radioactive atom concentration of the standards and the optical density obtained with these same standards allows one to calculate absolute concentrations of the molecule used if its specific activity is known. The quantitative autoradiographic method has been employed to determine the anatomical distribution of binding sites of neurotransmitter receptors (9) and to measure cerebral metabolic rates of glucose with 2-deoxyglucose (10).

Computer-Assisted Quantitative Autoradiography

Quantitation of autoradiograms requires the measurement of optical densities or of the level of gray in small areas of the emulsion. This can be done with manually driven photocells (11) or by densitometry with mechanical scanners (12). The optical density value and the coordinates of the defined point where measurements are done can be stored in the memory of a computer. This is an improvement considering that the number of measurements that have to be performed for statistical significance in each photograph has to be high and could not be done manually. Autoradiograms can also be digitized with a tube camera or a solid-state camera (13). This is probably the most convenient procedure (14). Charge-coupled device (CCD) elements are used in current cameras; they permit good resolution and high-speed image acquisition.

Once digitized, an image is analyzed by using a special program which allows the selection of regional values with appropriate tools, automatically transforms gray levels into drug concentrations, and performs statistical treatment of the data.

We describe here the components of a computer device for quantitative autoradiography and the characteristics and properties of each component. A program for image analysis as well as a rationale for its utilization is described. Finally, we show a typical application of this method for analysis of the distribution of specific binding sites of neurotransmitters. The device presented here is currently functioning in our laboratory (15), but there are other setups and commercially available components with similar characteristics. What is important is to understand which technical characteristics are essential for a convenient device and how to use the system.

System for Quantitative Analysis

Hardware

The camera has a photosensor (CCD) defined by the dimension of each element (11 × 11 μm), the number of elements (756 × 604), and the sensing area (8.8 × 6.6 mm). The resolution depends on these characteristics and on the enlargement of the picture by the objective; in the case of rat cerebral sections, for example, we use a 12.5–75 mm zoom, with an aperture ranging from f1.6 to f16 (Lens, Japan). The camera gives a video signal (CCIR standard) with a determined voltage (1 V) and impedance (75 Ω).

Digitization of the images is done with a frame grabber (Photon, Orkis, Marseille, France) which is designed for either black and white or color images. Picture definition of the frame is defined by the number of horizontal and vertical quantal picture elements (pixels) which should, of course, correspond to the number of elements of the photosensor so as to preserve the initial resolution. Optical transmission values are digitized at least to 8 bits, representing 256 possible levels of gray for each pixel. Although higher definition is obtainable for color grabbers, it is not essential in view of the sensitivity of the photographic emulsions used in our experiments. By convention, gray level 0 corresponds to maximum brightness and 255 to maximum darkness. The time of acquisition of an image (40 msec) and of the transfer to the computer (133 msec) is not an essential point for the analysis of the static images which we use, but it must not be too slow as the number of images analyzed is important. Regarding analysis, the most time-consuming step is the pointing of the structures and collecting the data. This depends, of course, not only on the speed of the operator, but also on computer capacity (68020 processor, 16.5 MHz) and on the software.

The digitization card is inserted into the extension of a computer (Macintosh II, Apple, U.S.A.). The computer should have a quick processor ($\geq$68020), more than 5 megaoctets of random access memory, a 40 megaoctets of internal hard-drive, and, if possible, external hard-drives (Megatek or Syquest, Fremont, U.S.A.), to stock simultaneously all pictures corresponding to one experiment, or at least those exposed at the same time as the same standard scale.

The other components of the setup are a light box (Orkis, Marseille, France) to illuminate the autoradiograms, a standard video monitor (Visionor, France) to control the image before digitization, a high-resolution color monitor (Apple, U.S.A.) for image display (640 × 430), and a printer (Image Writer II, Apple, U.S.A.) (Fig. 1).

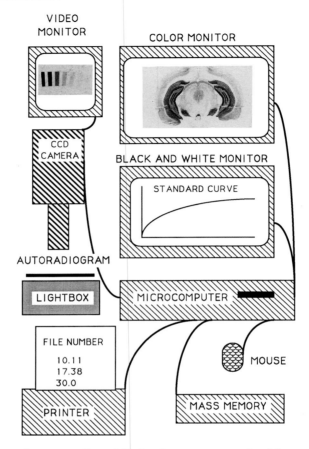

FIG. 1   Schematic presentation of the hardware components of the computer device for quantitative receptor autoradiography.

Accuracy of Measurements

The functional characteristics of the setup have to be determined in a real situation. Each element integrated in the device contributes to each of the parameters determining the accuracy of final gray level measurements. These characteristics may be furnished by the supplier of the elements, but it is important for the user to test the characteristics.

Dynamics of Camera

It is possible to vary the illumination with neutral density filters of known optical density (OD). The illumination obtained is measured with a lux-meter

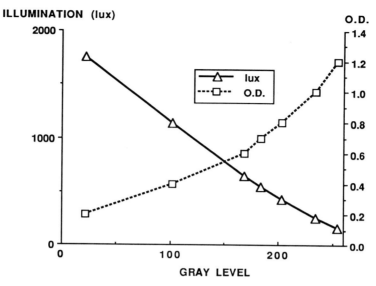

FIG. 2   Dynamic characteristics of a camera.

module (Chauvin-Arnoux, France). The gray level obtained with each filter in the digitized image is noted. The plot of illumination versus gray level values gives the dynamic response of the camera. With a good camera this plot is rectilinear, as shown by solid line of Fig. 2 (13). On the contrary, the plot of OD versus gray level is nonrectilinear; in most cases, it can be fitted with a fourth degree polynomial function (16). This function should be used to convert gray levels to OD allowing comparison with data obtained using true microdensitometers (12, 17).

Temporal Characteristics

Temporal characteristics depend on the camera and the light box. As CCD cameras have high sensitivity, it is possible to test the long-term stability with constant illumination. Once the temporal stability of the camera has been determined, generally over a very short transient period, the light box is tested over several hours. If the light source is a filament lamp, heating produces a time-dependent variation. With fluorescent tubes, the stability of illumination is attained in 1 hr. Long-term drifts are important because the elements are generally used throughout a working day. We prefer to wait for stability and digitize all the autoradiograms referred to the same standard over a short time so as to diminish temporal variability.

   Short-term variability, say, over a 20-min period, is tested with two neutral density filters. Gray levels are measured every 30 sec. The variability is expressed by the relative standard error of the gray levels [standard error of the mean divided by the mean gray level (18)]; values of 0.4 to 0.6% are obtained.

Photometric Uniformity

Photometric uniformity depends both on the light source and on the camera (14). To test photometric uniformity of both components it is necessary to digitize the image of the light source (i.e., in absence of autoradiogram). Mean gray levels are measured in different regions of the surface. The nonuniformity is expressed by the relative standard deviation (standard deviation of the mean divided by the mean gray level); under our conditions we found it to be 6.36%.

To test the photometic uniformity of the camera it is necessary to digitize the images obtained with several neutral density filters and to subtract from every image the image of the source previously obtained. Random measurements over the surface show a relative standard deviation ranging from 2.44 to 0.35%, depending on the filter used. The mean standard deviation is equal to one gray level. In conclusion, the shading of the sensor is very low, and that of the light box is diminished by subtracting the image of the light source from the images studied.

When a mask is placed around the field of view, light scattered by the edges of the mask reduce the measured density of a given region. This effect is called flare (14). The flare effect should be tested by noting the difference between the gray level values obtained in a 1×1 mm surface alone and the same surface surrounded by a mask set 2×2 mm apart. The flare effect is expressed by the relative error (mean flare errors divided by 256 gray levels); under our conditions, a value of 0.012% was obtained.

Camera Geometry

Geometrical uniformity is determined by measuring, in pixels, the digitized image of an objective micrometer (1 cm) in the horizontal and vertical positions. Geometric distortion is determined by the ratio between the difference in length obtained in the two positions and the total length of the micrometer. This determination must be carried out properly when the surface of the sensor is strictly parallel to the light source.

Spatial resolution depends on the surface of the elements of the CCD and on the overall magnification. For macroautoradiography we used 10-fold magnification to digitize rat brain sections. A 3 cm length object is digitized into 768 pixels on the grabber, giving a theoretical resolution of 40 μm.

Software

The software is composed of three types of programs: one to digitize autoradiograms, one to perform image analysis, and others for statistical analysis and presentation of the results.

Image Digitization

In a program specifically designed for image digitization (Flash 1.0, Orkys, France), there is a choice of functions to digitize the autoradiograms: (1) the number of gray levels (2, 4, 16, 256; 256 levels are needed for proper autoradiogram analysis); (2) the encoding format (SATIE, PICT, TIFF); (3) the compression of the image, permitting reduction of the space memory occupied by the image; (4) the precise selection of the surface of the image under study, since this is not done by the camera; and (5) the registration number (reference of the image to be stored).

Autoradiogram Quantitation

A program is specifically designed for image analysis, namely, selection of regions of interest, mean gray level value determinations, and conversion of molecular density by a transfer function. We developed a modular program, BIOLAB, written in PASCAL with the help of the Macintosh Programming Workshop (MPW).

The region of interest (ROI) is selected on the digitized image in the color monitor with the mouse either by manual delineation, or by superimposing on the ROI one or more rectangular windows of varying surfaces. The choice of the pointing tool depends on the structure under analysis. It is difficult to delineate small anatomical structures; it is easier to point the structure with multiple square measurements. Moreover, the mean gray values obtained by delineation are modified by the opacity of the neighboring structures. The reproducibility is identical for the two methods, and by using small squares there is no variation in the delineation by the user of the boundaries of brain structures.

Standard Curve

To transform the gray level values into radioligand concentrations, it is necessary to use a reference. This is done with microscales of known content in radioactivity. The microscales are made with a polymer containing a radioactive source; each strip corresponds to a defined concentration of radioactivity expressed in nanocuries per milligram of tissue equivalent, as determined by comparison with a mashed tissue preparation. The calibrated standards are different for tritium and for iodine-125; since the half-lives of the radioelements are different, even if under strictly defined conditions, it is possible to use tritium standards for iodinated ligands (16). Of course, as the half-life of ^{125}I is 60 days, the sections of polymer cannot be used for very long. Moreover, the penetration of the iodine emission in tissues is high, and the thickness of the polymer standard sections has to be identical to that of the brain tissue sections. This is not the case for tritium, because radiation from only the first 2–3 μm of the section is able to reach the apposed photographic emulsion; thus the thickness of tritium standards is not important in this case.

The standards are exposed simultaneously with the tissue sections, and after development one obtains the image of the strips. The radioactivity of each strip (1.3–33 nCi/mg tissue equivalent for Amersham (Les Ullis, France) tritium microscales, for example) is transformed into radioligand concentration by dividing the values by the specific activity of the radioligand under study (Ci/mmol). It is essential to take into account the decay of the radioactivity, for the standards as well as for the radioligand, especially when iodine is used as it has short-lived radioactivity. The equation for this correction is

$$N = N_0\, e^{-(t-t_0)\,\ln 2/T}$$

where N is the specific activity at date t, N_0, the specific activity at date t_0 (manufacture), and T the half-life of the radioelement. Generally the program should execute this correction; the worker need enter only the data for the standards, specific activity, and respective dates of use.

The stored image of the microscales is recalled; the image of the light box is subtracted, as explained in the section on hardware. Then the background measured around the image is also subtracted. A rectangular window is apposed to each of the strips in order to measure the gray level values; these values are compared to the radioligand concentration using the standard curve. The plot of this curve could be done with different coordinates, namely, linear/linear, log/log, log/linear. Several modes should be used to fit this curve, such as linear interpolation, Bezier's function, and linear regression. A polynomial fit is often used, but it presents sharp changes in slope difficult to accommodate to low gray level values. The biexponential fit seems to be more regular in the delineation and closer to the real values (Fig. 3). This fit is used as a transfer function and allows the transformation of gray level values into femtomoles of radioligand per milligram of tissue equivalent. The values and the fit are stored in a special "standard file," which may be used for measuring the tissue autoradiograms.

The sensitivity of the measurements made by this method is defined by the first value of the standard giving a difference of one gray level unit from the preceding value (3–5 fmol/mg tissue equivalent). Relative sensitivity is

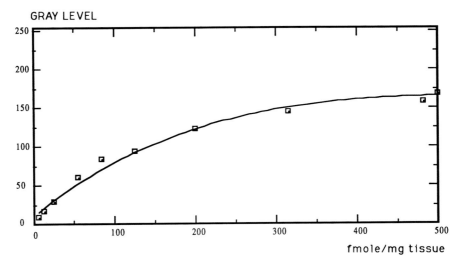

FIG. 3   Gray level values of the strips of standard scale plotted as a function of serotonin concentration. A biexponential function was used to fit the curve.

defined by the sensitivity divided by the maximum value (0.8%). This parameter depends on the exposure time of the emulsion and can be improved by increasing the exposure time. This corresponds to the use of the entire range of the 256 gray levels for digitization and the emulsion density.

As described in the preceding section, the physical accuracy of the hardware was within one gray level unit. As the darkening of the emulsion is a saturable phenomenon, the radioligand concentration, equivalent to a difference of one gray level unit, varied with the radioligand concentration in the tissue, as shown by the standard curve. The relative accuracy of the measurements is higher near the saturation part of the curve (10%) than in the most linear part (2–5%). This means that the asymptotic part of the curve should be avoided for measurements. The best way to obtain accurate quantitation is to adapt the exposure time to the level of radioactivity expected. For that, it is necessary to expose the same sections twice: once for a short time for anatomical structures with high radioactive concentrations and then for a long time for low radioactive structures.

The radioligand concentration is related to the mass of tissue equivalent (19), theoretically allowing the comparison with membrane binding data that are expressed in milligrams of protein. This comparison is difficult since the protein content is different from one anatomical structure to another, and the true values are not known. The unit area would be a good reference for radioligand concentration (16) when comparisons are to be made with data from histological experiments on the densities of cellular or subcellular elements carrying the receptors or the neurotransmitters (4).

Rationale of Autoradiogram Measurement

The digitized autoradiographic image of a rat brain section is recalled in the analysis program. As for the image standards, the image of the light box and the value of the background surrounding the image under study have to be subtracted. Each anatomical structure is defined by a reference number. Mean gray levels are measured within a square tool placed on the structure. The mean gray levels, the corresponding radioligand concentration, the standard error of the mean of this last value, and the surface covered by the square are calculated, stored in a file, and presented on the screen of the monitor. We can measure several squares of the same section and the same anatomical structure of the same autoradiogram; all of the values obtained for each square are stored with the same reference number in consecutive lines of the same file. We can then change the reference number for another region and/or change the file for measurements in another section. The mean value for a defined region on one or several sections should be calculated automatically by the program. These values are then transferred to other programs for statistical analysis or other data processing.

Reproducibility of Experimental Measurements

The variability of the measurements in one anatomical structure from a given autoradiogram is just above the range of accuracy compatible with the physical characteristics of the device, the difference being mainly due to the transfer function of the autoradiographic process. The main variability in the results should be due to the processing of sections and to the biological intraspecific variability (15). This aspect has to be tested by the investigator by comparing the results obtained in successive sections of the same animal; this tests the quality of section and the reproducibility of the incubation conditions for binding site detection. The same studies have to be done for sections coming from different animals of the same species.

Receptor Autoradiography

Determination of the regional density of neurotransmitter receptors in the central nervous system is essential for understanding the function and plasticitiy of the neurotransmitters and receptors (17, 20). The method of obtaining an autoradiogram of neurotransmitter receptors has been extensively described (9). We want to emphasize some points essential for a good quantitative analysis.

Choice of Radioligand

The most important point is the specificity of the labeled molecule for the type of receptor binding site to be analyzed. For example, [^{3}H]5-HT labels all serotonin 1 receptor subtypes (5-HT$_1$) (21), whereas 8-[^{3}H]hydroxydi-*n*-propylaminotetraline labels only the 5-HT$_{1A}$ subtype (22). Once the molecules have been picked, we can choose the radioactive element, namely, tritium or iodine-125. Tritium has the advantage of not modifying the chemical properties of the molecule in which it has been inserted. Iodine is a large atom which may modify the affinity of the ligand for the receptor, but it has the advantage of not being absorbed by white matter, unlike tritium. This phenomenon, called quenching, results in a modification of the darkening of emulsion for a given activity of the radioactive element. The optical density observed with tritium has to be modified by a coefficient which takes into account the proportion of white versus gray matter of each anatomical structure studied (19).

There is also a difference in the spatial resolution of the two radioelements. The radioactive emission of iodine is intense (specific activity ~2000 Ci/mmol) as compared to tritium (10–50 Ci/mmol). For very thin emulsions, iodine shows better resolution than tritium (3). This is similar for thicker emulsions (hyperfilm ^{3}H, Amersham), although for long exposure times iodine gives a "foggy" image. Iodine has the advantage of needing shorter exposure times than tritium.

Quality of Emulsion

The tritium-sensitive film is suitable for anatomical work with autoradiography. The response of the emulsion to radiation has a small range of density (1.5 OD values), but it is possible to re-expose the same sections to obtain significant labeling for weakly emitting structures. The response of the film is not good during the first hours, and thus it is unsuitable for short exposures for quantitative studies; this is important for iodine and could invalidate saturation studies designed to determine the affinity constant of a new ligand.

The emulsion developer is important. D19 (Kodak, Rochester, NY) is a good developer, since it shows a large range of gray levels and results in good proportionality between the observed optical density and the radiation emission. As we have standards and a transfer function, the correspondence

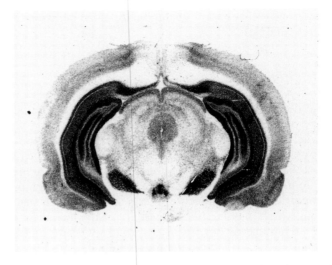

FIG. 4   Autoradiogram of rat brain section showing distribution of serotonin 1 binding sites (Magnification, × 4.5).

between the two parameters can be established. It is possible to develop with Microdol (Ilford, France), which gives a fine grain and thus a more precise localization of the emitting source; however, Microdol has a short range of optical densities, so it is not very suitable for quantitative macroautoradiographic studies.

Example of Receptor Autoradiography

Figure 4 shows the autoradiography of 5-HT$_1$ receptor binding sites of the rat brain (23). Sections were prepared as described by Segu *et al.* (17). One section was incubated with the radioligand [^{3}H]5-HT alone (TB, total binding), and an homothetic section was incubated in the same medium supplemented with 10 μM 5-HT in order to determine the nonspecific binding (NSB) (24). Sections were apposed to hyperfilm ^{3}H together with tritium microscales. After development, the autoradiograms were analyzed with the image analysis device as described above.

Complementary Tools

Some tools or special programs can help in the analysis of the autoradiograms and in their presentation.

Cellular Analysis

The camera may be coupled to a microscope to give greater enlargement of the autoradiogram. The limit of resolution is the granularity of the emulsion. The user has to choose between measuring the gray level (related to the optical density) and counting the number of individual grains on the emulsion; the latter situation is most often found when sections are dipped in liquid emulsions, resulting in a thin layer closely apposed to the section and presenting very fine grains which can only be viewed at magnifications ranging from ×250 to ×400 with either dark- or bright-field illumination. In this case, the digitizing system is functioning in the same way, but the analysis is different; it measures the density of silver grains showing a defined gray level, above the background (counting module of Biolab, LNB, Marseille, France).

Geometric Analysis

It is possible to obtain a histogram of the relative proportion of the gray values in the regions of interest either delineated or pointed with a rectangular frame. The region of interest can be defined by a threshold value of gray level. A zone of the screen should be predefined as the region of interest. Measurements will be made in this zone for all successive autoradiograms. This tool allows the rapid accumulation of a set of measurements.

Image Modification

It is possible to change the histogram of the gray levels by shifting of the scale, equilization, or exponential filtering. Contrast and luminosity of the image may also be changed. Of course, after these modifications, the data should not be used to measure gray level values. Several images may be presented simultaneously on the screen and labeled for presentation. It is also possible to select some parts of different images and present them together in the same final image. Gray levels of the image can be transformed into color-coded images with a possible three-dimensional representation. This translation affects the standards, so that the colors correspond to a given concentration of radioligand, determined by the transfer function.

Conclusion

The detection of neurotransmitter receptor binding sites by autoradiography is an important method for the study of the function of receptors in the physiology and pathology of the brain. Such work can be done only with the help of quantitative analysis in order to determine the biochemical, pharmacological, and anatomical properties of the receptors.

The computer is essential to assist the manual work in studying receptor density distribution. The accuracy of a computer device may be determined readily. There are now several commercially available setups for such analyses, and software programs are available individually, allowing one to construct a system with a personal computer and camera. Use of the software is an important point for the efficacy of the work. The system may also be used for other work, particularly in immunohistochemistry (18) or *in situ* hybridization of mRNA (25) either at the macroscopic or microscopic level.

References

1. C. Pilgrim and W. E. Stumpf, *J. Histochem. Cytochem.* **35,** 917 (1987).
2. A. Calas and L. Segu, *J. Microsc. Biol. Cell.* **27,** 249 (1976).
3. M. M. Salpeter, L. Bachmann, and E. E. Salpeter, *J. Cell Biol.* **41,** 1 (1969).
4. L. Segu and A. Calas, *Brain Res.* **153,** 449 (1978).
5. M. J. Kuhar, *Trends Neurosci.* **4,** 60 (1981).
6. A. Biegon, T. C. Rainbow, and B. S. McEwen, *Brain Res.* **242,** 197 (1982).
7. J. R. Unnerstall, D. L. Niehoff, M. J. Kuhar, and J. M. Palacios, *J. Neurosci. Methods* **6,** 59 (1982).
8. W. Rostène and C. Mourre, *C.R. Acad. Sci. Ser. III* **301,** 245 (1985).
9. M. J. Kuhar and J. R. Unnerstall, *in* "Methods in Neurotransmitter Receptor Analysis" (H. I. Yamamura, ed.), p. 177. Raven, New York, 1990.
10. G. M. Alexander, R. J. Schwartzman, R. D. Bell, and A. Renthal, *Brain Res.* **223,** 59 (1981).
11. J. I. Choca, F. M. Clark, R. D. Green, and H. K. Proudfit, *J. Neurosci. Methods* **24,** 145 (1988).
12. C. Gooche, W. Raiband, and L. Sokoloff, *Ann. Neurol.* **7,** 359 (1980).
13. G. M. Alexander and R. J. Schwartzman, *J. Neurosci. Methods* **12,** 29 (1984).
14. L. McEachron, C. R. Gallistel, J. L. Eibert, and O. J. Tretiak, *J. Neurosci. Methods* **25,** 63 (1988).
15. L. Segu, P. Rage, and P. Boulenguez, *J. Neurosci. Methods* **31,** 197 (1991).
16. A. P. Davenport and M. D. Hall, *J. Neurosci. Methods* **25,** 75 (1988).
17. L. Segu, J. Abdelkefi, G. Dusticier, and J. Lanoir, *Brain Res.* **384,** 205 (1986).
18. R. R. Mize, R. N. Holdefer, and L. B. Nabors, *J. Neurosci. Methods* **26,** 1 (1988).
19. W. A. Geary and G. F. Wooten, *Brain Res.* **336,** 334 (1985).
20. J. M. Palacios and A. Pazos, *in* "Quantitative Receptor Autoradiography" (C. A. Boast, E. W. Snowhill, and C. A. Altar, eds.), Vol. 19, p. 173. Alan R. Liss, New York, 1986.
21. M. Hamon, J. M. Cossery, U. Spampinato, and H. Gozlan, *Trends Pharmacol. Sci.* **7,** 336 (1986).

22. M. Marcinkiewicz, D. Vergé, H. Gozlan, L. Pichat, and M. Hamon, *Brain Res.* **291,** 159 (1984).
23. A. Pazos and J. M. Palacios, *Brain Res.* **364,** 205 (1985).
24. G. A. Weiland and P. B. Molinoff, *Life Sci.* **29,** 313 (1981).
25. L. Maroteaux, F. Saudou, N. Amlaiky, U. Boschert, J. L. Plassat, and R. Hen, *Proc. Natl. Acad. Sci. U.S.A.* **89,** 3020 (1991).

[9] Measuring Spatial Water Maze Learning

Victor H. Denenberg and Gerry H. Kenner

Introduction

Several years ago Richard Morris (1) developed a simple yet elegant method for measuring spatial learning in rodents. The apparatus consisted of a large circular tub (1.3 m diameter) partially filled with water heated to $26 \pm 1°C$ and a circular platform (0.11 m diameter) just below the surface of the water. Milk was added to the water to make it opaque, thus reducing the likelihood that the rat would be able to see the submerged platform. In such a situation the standard procedure is to place the platform in one of the four quadrants of the tub (arbitrarily called NW, NE, SW, or SE), some distance away from the edge (so the animal cannot find the platform by swimming around the outside). The animal is released, on successive trials, from the four arbitrary compass points (N, E, S, and W), and the task is to find the platform.

Morris found that the rats quickly learned the location of the platform. In a series of experiments he showed that the animals were not using information from within the tub to find the platform, and concluded, "... rats can rapidly learn to approach a spatially fixed object that they can never see, hear, or smell" (p. 257). The rats located the platform by means of extramaze information. The tub was in a room with asymmetrically placed distal cues, including a window, a door, shelves on a wall, and a cupboard. These were used as relative references by the rat to find the platform. An analogous situation would be the car driver who uses a distinctive landmark (such as a very tall building) as a reference to locate his objective.

Performance on the spatial learning task has been shown to be degraded in rats with hippocampal lesions (2), medial frontal or orbital frontal lesions (3), or a severed corpus callosum (4), as well as in autoimmune mice with cortical ectopias (5). Brandeis *et al.* (6) have recently reviewed apparatus, procedures, and findings.

Several procedures have been used to measure behavior in this apparatus. The simplest is to record the time it takes the animal to reach the platform after being released. Morris did this in the original study. He also videotaped the animals and later transcribed the paths they followed to determine path length. It is apparent that these are labor-intensive procedures which do not gather all the information present in the test environment. Since then various procedures have been developed to make data collection and analysis more

Methods in Neurosciences, Volume 10

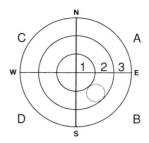

FIG. 1   Screen display of the Morris maze, consisting of the bull's-eye, the cross hairs, and the location of the platform within the second annulus. The numbers 1, 2, and 3 specify the annuli; the letters A, B, C, and D specify the quadrants. The smaller letters N, W, S, and E indicate where the animal is released into the maze. These numbers and letters are not part of the screen display.

efficient. Denenberg *et al.* (7) have recently developed a relatively simple computer-assisted procedure for recording the behavior of the animal that yields a number of important parameters. We describe this procedure below.

Apparatus

We use a circular black tub, 123 cm in diameter, partially filled with water kept at an ambient temperature (20–22°C). We have found that rats and mice swim well at this temperature with no ill effects afterward and this temperature is most convenient for the researcher.

Our platform is an inverted clear Pyrex jar, 23 cm in diameter, covered with black wire mesh. The water is approximately 1.3 cm above the top of the escape platform. We have found that neither rats nor mice are able to see the platform under our test conditions; therefore we do not use milk or any other substance to make the water opaque. The apparatus is in a room that contains several distinguishable extramaze cues.

Description of Computer Program

Our program, called Spatial Maze, has been developed for the Macintosh computer. The procedure involves tracing the path of the animals on a template of the maze (Fig. 1) which is attached to a digitizing tablet. We use a paper template, and the digitizing stylus contains a pen so that the researcher obtains a hard copy of the path of the animal.

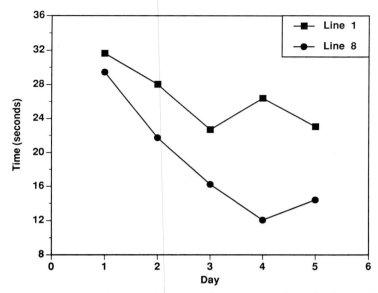

FIG. 2   Mean daily time in seconds to find the escape platform in the spatial maze for NXRF Lines 1 and 8.

The program first asks for the radii of the tub and the platform. Next, a template of the maze is put on the digitizing tablet, and the stylus is placed at the upper left-hand corner of the template and then in the center of the platform. This locates the template on the tablet and the platform in its proper location in the tub. The computer then uses the two radii to generate a screen image of the template which is proportional to the actual template.

The computer then asks for the following information: subject ID, test day, trial number, trial length (seconds), and amount of time (seconds) the subject is to remain on the platform after reaching it. This last bit of information is needed because the researcher records the number of rears by the animal while on the platform (discussed below). The current date is entered automatically.

When the animal is put into the water, the experimenter starts the program and begins tracing the path of the animal. When the subject reaches the platform, the experimenter turns off the stylus, thereby stopping this portion of the program. The experimenter then strikes the X key, and this starts a "platform timer." We have set this for 10 sec, but that can be changed by the researcher. We record the number of rears by the animal during the 10 sec by striking the R key each time a rear occurs. The researcher can, of course, use the R key to record some other behavior of interest.

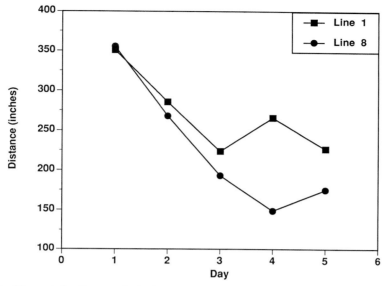

FIG. 3   Mean daily distance in inches to find the escape platform in the spatial maze for NXRF Lines 1 and 8.

When the platform time is done, the trial is over. The screen shows the template pattern seen in Fig. 1 and the path of the animal. This image can be saved as a MacPaint document.

Data Files

The computer samples the path information approximately 60 times/sec and translates the information into XY coordinates. Two sets of data files can be obtained. The first set includes the following measures: the distance traveled in inches; the time from entry into the pool until the platform is found; swimming speed, obtained by dividing distance by time; the absolute amount and percentage of time spent in annuli 1, 2, and 3; the absolute amount and percentage of time spent in quadrants A, B, C, and D; and the number of times the subject reared up while on the platform.

The second data set is optional and is based on the XY data points. The researcher can (1) retain each XY coordinate set so that the path taken can be reproduced later, if desired; (2) ask the computer to determine swimming angles (this is discussed below); or (3) not retain any of the XY data. The first

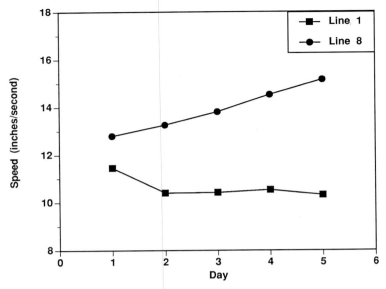

FIG. 4   Mean daily speed to find the escape platform in the spatial maze for NXRF Lines 1 and 8.

option uses more memory than the other two and takes the most time to process, whereas the third option uses least memory and is the fastest. The computer is set to store all the *XY* data, but this can be reversed by the researcher. If one wishes to store the *XY* data, it is possible to choose a sampling rate less than 60/sec.

The measure of swimming angles has been used as an index of spatial orientation and learning by several research groups. The default mode for this program is as follows. When the animal has traveled 4 inches from the starting point, an angle is determined. The angle is defined by the line from the position of the subject (after traveling 4 inches) to the starting point and the line from the starting point to the center of the escape platform. When the animal has traveled an additional 2 inches from its prior position, a second angle is calculated. The location of the subject at the time the first angle was determined is used as the reference, and the angle is constructed by projecting a line from that point to the center of the escape platform and a second line to the location of the subject which is now 2 inches away. A total of 5 angles are obtained in this manner. The data are stored both as true angle (range 0–360°) and as absolute deviation from the zero line (range 0–180°). The experimenter can change any of the three default parameters: the initial

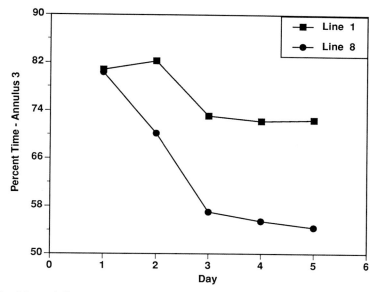

FIG. 5   Mean daily percentage of time spent in annulus 3 in the spatial maze for NXRF Lines 1 and 8.

distance traveled (default = 4 inches), the distance to be traveled thereafter (default = 2 inches), and the number of angles to be recorded (default = 5). At the end of each trial, the data are sent to Excel for statistical processing.

Example

Wimer *et al.* (8) derived four recombinant inbred lines by crossing the NZB and the RF mouse strains. We have investigated the behavior and biology of these animals. The animals were given the Morris maze test when 70 days old. They were given four trials a day (from the N, E, S, and W positions, randomly determined) for 5 days (a different random sequence was used each day) with the platform located in quadrant B. The data were averaged over the four daily trials to yield a single daily score. For illustrative purposes the maze data on Lines 1 ($N = 20$) and 8 ($N = 20$) are shown in Figs. 2–7. The two lines differ markedly, with Line 8 showing much better learning than Line 1. Both lines have essentially the same time and distance scores (Figs. 2 and 3) on the first day, and both lines show decreasing scores over time, which is indicative of learning, but the rate of change is much greater in Line

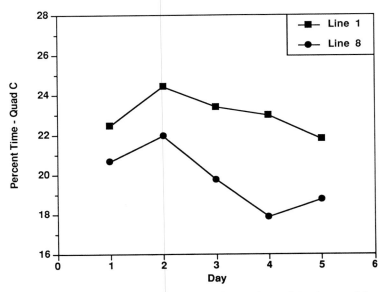

FIG. 6   Mean daily percentage of time spent in quadrant C in the spatial maze for NXRF Lines 1 and 8.

8. Note that the time and distance curves for each line are similar in form, which is often the case.

Figure 4 shows speed scores and reveals that the mice from Line 8 increased their speed of swimming each day, whereas Line 1 animals swam at the same speed, or even slowed down slightly, over the 5 days of testing.

Figures 5 and 6 help illuminate the strategies used by the mice. To find the platform, it is necessary to spend minimum time in annulus 3 (the outermost ring). As the animals got to know the location of the platform, time in this ring decreased markedly for the Line 8 mice but only minimally for the Line 1 animals. Another index of learning efficiency is the amount of time spent in the quadrant opposite the one containing the escape platform. An examination of Fig. 1 reveals that an animal who has learned that the platform is located in quadrant B need only spend a slight amount of time in the diagonally opposite quadrant (quadrant C). Therefore, as learning proceeds, time in quadrant C is expected to decrease. Figure 6 shows this to be the case for Line 8 but not Line 1. Finally, Fig. 7 shows that neither group spent much time rearing up on the escape platform.

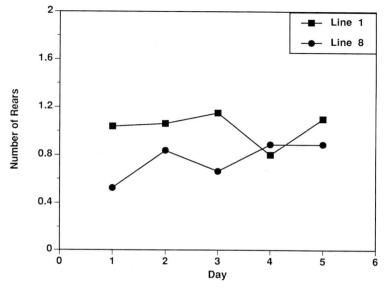

Fɪɢ. 7   Mean number of daily rears while on the escape platform for NXRF Lines 1 and 8.

Conclusions

For an animal to be able to adapt in any ecology, it is necessary to learn the spatial features of the environment in which it lives. The Morris maze appears to measure some critical features of this process. In addition, behavior on this test has been shown to be correlated with brain anomalies and with immune deficiencies. Thus, this test is a valuable tool to help the neuroscientist in understanding the complex relationships between brain and behavior.

Acknowledgments

The research reported in this chapter was supported, in part, by Grant HD 20806 from the National Institute of Child Health and Human Development, NIH.

References

1. R. Morris, *Learning Motivation* **12,** 239 (1981).
2. R. Morris, P. Garrud, and J. Rawlins, *Soc. Neurosci. Abstr.* **11,** 237 (1981).

3. B. Kolb, R. Sutherland, and I. Whishaw, *Behav. Neuroscience* **37,** 97 (1983).

4. A. Adelstein and D. Crowne, *Behav. Neurosci.* **105,** 459 (1991).

5. L. Schrott, V. Denenberg, G. Sherman, N. Waters, G. Rosen, and A. Galaburda, *Dev. Brain Res.* in press (1992).

6. R. Brandeis, Y. Brandys, and S. Yehuda, Int. J. Neurosci. **48,** 29 (1989).

7. V. Denenberg, N. Talgo, N. Waters, and G. Kenner, *Physiol. Behav.* **47,** 1027 (1991).

8. R. Wimer, C. Wimer, C. Winn, N. Ravel, L. Alameddine, and A. Cohen, *Brain Res.* **534,** 94 (1990).

[10] Measuring Discrimination Learning

Victor H. Denenberg and Gerry H. Kenner

Introduction

A simple, yet powerful, way to measure one aspect of learning in animals is via the discrimination method. The method is almost as old as Pavlov's conditioned reflex. Munn (1) reports that Yerkes used a discrimination procedure in 1907 to study vision in the dancing mouse.

Discrimination learning is a form of associative learning, independent of spatial learning (see [9] in this volume for a description of a way to measure spatial learning). The animal is presented with two distinctly different stimuli which are equally attractive. Approaching one stimulus is followed by reinforcement (reward) while approaching the other stimulus has no consequences, or may be followed by punishment. The animal is given a series of trials per day over a number of days, and the most common score is the number of correct choices made daily. Initially, the animal performs at a chance level of 50%. As practice continues, the stimulus which is reinforced comes to be preferred, and the percentage of correct choices increases in a gradual fashion, often reaching an asymptote of 85–95% correct. This process is depicted via a learning curve which plots trial blocks on the X axis and percent correct responses on the Y axis.

The most common form of discrimination learning involves the simultaneous presentation of two visual stimuli. This will be the paradigm discussed in this chapter. However, discrimination methods can use successively presented stimuli and are not restricted to the visual modality. For example, two different auditory tones can be presented successively in a semirandom order, with the animal required to turn to the right for one tone and to the left for the other.

Description of Visual Discrimination Learning Experiment

We will use the example of mice learning a simple black/white visual discrimination to illustrate the principles involved in running a discrimination experiment. A T-maze is commonly used in visual discrimination studies, especially with rodents. The animal is placed at the base of the stem and runs (or swims, depending on the apparatus) toward the T-junction. As the animal nears the junction, the two stimuli come into view, and the task of the animal is to turn into the alley containing the correct stimuli.

We have mice swim through a T-maze with the water temperature at 20–22°C. Mice and rats are very effective swimmers, and water is a simple way to motivate them. The stem of the T-maze is painted a neutral gray; one arm is painted white and the other black. Rather than being a straight T, the arms actually curve back toward the stem (2). This prevents the mouse from seeing the end of the alley. This is critical because there is a wire mesh ladder at the end of the positive alley which the mouse uses to climb out of the maze. It is important that the mouse make the choice of which alley to enter based on the black/white distinction, not because it sees wire mesh at the end of one alley and not at the end of the other. After climbing out, the animal is placed into a nest box with dry shavings and an overhead heat light. It remains there until the other members of the squad have been tested, and is then given its second trial. This continues until each animal has been given 10 trials. The procedure is repeated for 5 days.

We arbitrarily assign black as the positive stimulus for half the animals and white for the other half. In this way any stimulus bias is counterbalanced (e.g., rodents typically prefer a dark chamber to a lighted one and thus may favor black over white). In addition, one can run a statistical test to determine whether the learning curve of one stimulus group differed from that of the other.

In a discrimination learning experiment the positive stimulus occurs equally often on the right and the left. This prevents the animal from using spatial information to solve the problem. It is necessary that the left–right order from trial to trial be done in a semirandom manner within a day; and that different semirandom orders be used from day to day. Gellerman (3) has generated 44 semirandom orders based on a 10-trial sequence. Half the trials are left (L), the other half right (R). In no instance does a sequence contain more than three successive choices in the same direction (i.e., LLL or RRR) because a sequence of four or five is liable to cause the animal to adopt a side strategy.

We have a symmetrical and bottomless maze, and to reverse the left–right position of the two stimuli, we simply turn the maze over. If more complex visual stimuli are used (e.g., geometrical forms), these can be put on metal or wooden plates that can be hung at the T-junction at the entrance to the two alleys. These can easily be reversed from trial to trial.

Computer-Aided Recording and Scoring of Discrimination Learning

Denenberg *et al.* (4) have developed a computer program which (1) aids the researcher in setting up a discrimination experiment, (2) directs the re-

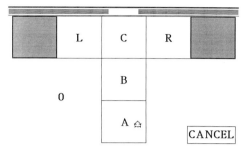

FIG. 1   Picture of computer screen showing T-maze for discrimination learning. [Reprinted with permission from V. H. Denenberg, N. W. Talgo, L. Schrott, and G. H. Kenner, *Physiol. Behav.* **47,** 1032 (1991), Pergamon, Elmsford, NY.]

searcher during the course of the study, (3) collects and processes the data, and (4) sends the processed information to a spreadsheet at the completion of a session. The following description is adapted from the original paper.

The program is called DLT (Discrimination Learning T-Maze) and runs on a Macintosh computer. The 44 semirandom Gellerman series are stored in the computer. Since these are 10-unit series, each animal will be given 10 trials per day. A different Gellerman series is selected for each test day, and the complete set is entered into the computer file prior to the start of the experiment. The trial duration has a default time set for 60 sec, but this can be changed by the experimenter. The computer assumes that black is the positive stimulus; this can be easily changed to white by the experimenter. The date, trial number, and positive location are automatically entered and displayed on the screen. The positive location is an R or an L and represents the Gellerman position for that particular series and trial number, and it specifies the location of the positive stimulus for that trial.

An important feature of the program is that the researcher, before the start of the study, can prepare files for each animal for each day of testing. A file will contain the ID of the animal, the positive stimulus (black or white), and its location (left or right) for every trial.

The researcher is now ready to start the experiment. An icon of a T-maze appears on the screen (see Fig. 1). The timer for each trial is in the lower left quadrant, and the zero there means the trial has not started. The cursor is placed into box A. When the animal is released at the stem of the T, the computer mouse is clicked to start the trial. The researcher then tracks the progress of the animal on the screen. As it swims forward, box B and then box C are clicked. (It is necessary to move the cursor to the new location and also click the mouse to enter information into the computer.) The L or

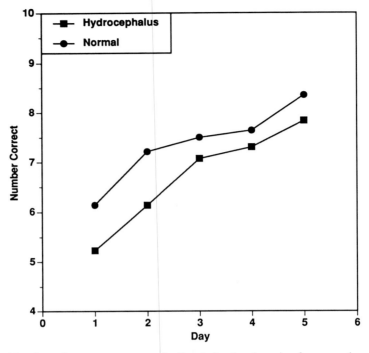

FIG. 2   Number of correct responses in discrimination learning for normal and hydro-cephalic MRL/MP *lpr/lpr* mice.

R box is clicked only after the animal has definitely entered one of the two arms. The shaded box is clicked only after the animal has made contact with the escape ladder. If the animal goes partway down an alley and then reverses, its path is continuously tracked by the researcher. A trial is terminated when the animal reaches the ladder (i.e., the correct shaded box is clicked), or when time is up. The researcher can then continue with the next trial for that animal, or shift to a new subject. This procedure is followed until all subjects are given their 10 trials for the day. The following information is obtained after each trial.

Left Choice
If the initial choice of the subject at the T-junction was to the left, a score of 1 is entered; if to the right, a score of 0. This enables one to determine whether an animal has a left–right position bias since an unbiased animal will have an average score of 5.

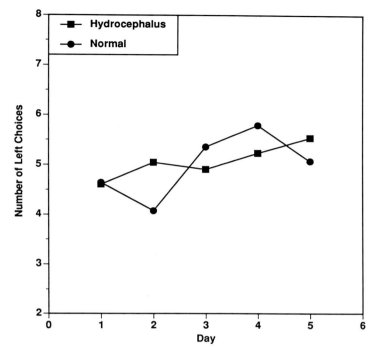

FIG. 3   Number of initial left choices in a discrimination learning task for normal and hydrocephalic MRL/MP *lpr/lpr* mice.

No Choice

On occasion an animal will not leave the stem of the T. After the trial is over the animal is forced to make a left or right choice.

Correct Choice

If the subject gets to the escape ladder without entering the incorrect alley, it is credited with a correct choice. It is permissible to swim back and forth in the correct alley, and/or to re-enter the stem of the T. However, if any portion of the incorrect alley is entered, the subject gets no credit.

Learning Score

The correct choice score is all-or-none. This is generally a sufficient measure for most experiments. However, there are occasions when animals behave in odd ways, and a measure of partial learning is needed. This is the purpose

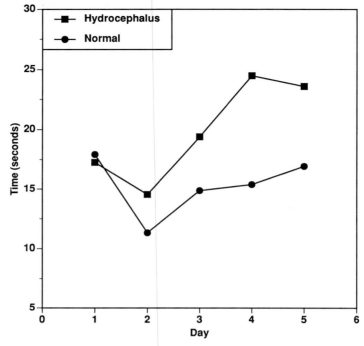

FIG. 4   Time to reach the escape ladder in a discrimination learning task for normal and hydrocephalic MRL/MP *lpr/lpr* mice.

of the learning score. Because this is not a common statistic, it will not be further discussed. See Denenberg *et al.* (4) for details.

Time

The actual time is recorded for each trial, with the default value being the maximum possible.

Example

Recently, Denenberg *et al.* (5) reported that the autoimmune MRL/Mp *lpr/lpr* mouse has a high incidence of hydrocephaly. Three measures from our discrimination learning program are plotted in Figs. 2–4 for 14 mice with normal brains and 43 with moderate or severe hydrocephaly. Those with hydrocephaly are significantly impaired ($p < 0.02$) in their ability to make the black/white discrimination (Fig. 2). Neither group had a left or right bias on their initial turns (Fig. 3). The time curves are rather odd in that they drop from Day 1 to Day 2, as expected, but then start increasing again (Fig. 4). Both groups showed the same pattern and did not differ significantly from each other.

Comment

A black–white discrimination is one of the simplest problems to present. The task can be made more difficult by changing the nature of the stimuli. For example, black and white horizontal stripes versus black and white vertical stripes is a moderately difficult discrimination. For a discussion and review of the classical literature on discrimination learning, see Munn (1).

Another way to extend this task is to reverse the positive and negative stimuli after the animals have reached a preset learning criterion. Thus, animals that had been run with black positive now must shift to white positive. The time it takes to make the shift has been used as an index of the strength of memory of the original learning.

The discrimination method is a very effective way to measure simple associative learning processes unrelated to spatial learning. As such this procedure complements the techniques used to measure spatial learning, discussed in Chapter [9] in this volume.

Acknowledgments

The research described in this chapter was supported, in part, by Grant HD 20806 from the National Institute of Child Health and Human Development, NIH.

References

1. N. Munn, ''Handbook of Psychological Research on the Rat.'' Houghton Mifflin, New York, 1950.
2. R. Wimer and S. Weller, *Perceptual Motor Skills* **20,** 203 (1965).
3. L. W. Gellerman, *J. Genet. Psychol.* **42,** 207 (1933).
4. V. Denenberg, N. Talgo, L. Schrott, and G. Kenner, *Physiol. Behav.* **47,** 1031 (1990).
5. V. Denenberg, G. Sherman, G. Rosen, L. Morrison, P. Behan, and A. Galaburda, *Brain Behav. Immun.* **6,** 40 (1992).

Section II

Data Analysis

[11] Computer Program for Kinetic Modeling of Gene Expression in Neurons

James L. Hargrove, Martin G. Hulsey, and Diane K. Hartle

Introduction

Communication between neurons is rapid compared to the rate of signaling among most other cell types; responses to transsynaptic stimulation typically occur within milliseconds. This capacity for rapid communication is supported by biochemical and genetic adaptive mechanisms that are comparatively slow. Regulation of tyrosine hydroxylase (tyrosine monooxygenase) provides an excellent example; activity of this enzyme increases in response to elevated cyclic AMP as a result of acute and long-term regulation. Acute regulation results from phosphorylation by protein kinase A and other regulatory molecules that affect the K_m of the enzyme for tyrosine or allosteric properties, and is maximal within 2–5 min of stimulation (1, 2). Longer term adaptation is provided by an increased rate of enzyme synthesis mediated by an increased rate of transcription (3–9) and possibly by posttranscriptional mechanisms (10). This chapter describes a computer program that can be used to compare experimental data with theoretical predictions regarding the induction of gene products.

Basis for Kinetic Model of Gene Expression

Gene expression is not complete until the encoded mRNA and protein attain levels that are proportional to the rate of transcription; failure to achieve this balance indicates that a steady state has not been attained (11). The relationship between rate of transcription and level of protein can be stated in quantitative terms using the kinetic model shown in Fig. 1 (12–14). The level of mRNA in the cytoplasm at any moment (designated R_t) is proportional to the rate of transcription (k_{s1}, a zero-order rate constant) and the rate of decay (k_{d1}, assumed to be first-order). Other nuclear processes are ignored because they occur rapidly relative to the time needed to effect a change in protein concentration and seldom serve as points of regulation. The concentration of the encoded protein (P_t) is proportional to the level of mRNA; the proportionality constant, k_{s2}, equals the number of molecules of protein produced per molecule of mRNA per unit of time. Decay of the protein is

Methods in Neurosciences, Volume 10

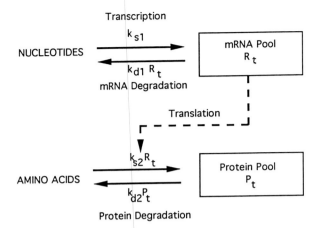

FIG. 1   The kinetic model is based on the assumption that transcription from any specific gene follows zero-order kinetics, meaning that under steady-state conditions a constant amount of mRNA is formed per unit of time. The rate constant for mRNA formation is designated k_{s1}. The mRNA is translated to form protein at a rate assumed to be first order, with a rate constant, k_{s2}. Degradation of mRNA and protein are both assumed to follow first-order kinetics, with rate constants designated k_{d1} and k_{d2}, respectively.

assumed to be first order; the decay constant equals k_{d2}. This simple kinetic model can predict experimentally determined induction kinetics, provided that the four parameters can be measured or estimated. This model gives rise to the following equations.

The rate of change in concentration of mRNA between two steady-state levels is a function of its rates of synthesis and decay, assuming zero-order synthesis and first-order decay:

$$dR/dt = k_s - k_d R_t \tag{1}$$

At steady state, the rate of synthesis equals the rate of decay (which is the product of the rate constant for decay and the concentration of product); the concentration of mRNA equals the ratio between these two constants:

$$k_s = k_d R_{ss} \quad \text{or} \quad R_{ss} = k_s/k_d \tag{2a}$$

The decay constant fixes the ratio between the rate of synthesis and the concentration of product at steady-state, and the half-life of a gene product is an inverse function of the rate constant for decay:

$$k_d = \frac{k_s}{R_{ss}} = \frac{\ln 2}{t_{1/2}} = \frac{0.693}{t_{1/2}} \tag{2b}$$

The time course required for the concentration of a gene product to attain a new steady state is related to the decay constant or half-life, no matter whether the concentration is increasing or decreasing:

$$R_t = R_0 + (R_{ss} - R_0)(1 - e^{-k_d t}) \tag{3a}$$

Equation (3a) can be rearranged to show that the fraction of the new steady state achieved at any moment after a stimulus can be predicted by a general, exponential function:

$$\text{Fractional response} = \frac{R_t - R_0}{R_{ss} - R_0} = 1 - e^{-k_d t} \tag{3b}$$

The concentration of a protein may be expressed in terms of the concentration of the corresponding mRNA:

$$P_t = P_0 + (P_{ss} - P_0)(1 - e^{-k_{d2} t}) - k_{s2}(R_{ss} - R_0)\left(\frac{e^{-k_{d2} t} - e^{-k_{d1} t}}{k_{d1} - k_{d2}}\right) \tag{4}$$

The computer program described below provides an interface to solve Eqs. (3a) and (4) using experimentally derived kinetic values. Equation (3a) describes a simple exponential function; the more formidable-appearing Eq. (4) describes a sigmoidal curve that depends on the solution to the simple exponential. The most important feature of this model is that it predicts the rate at which transitions will be made between initial and final levels of mRNA and protein, as determined by the rates of decay (half-lives) of these two populations of molecules. The periods predicted for induction of mRNA and protein do not depend on the degree of change [stimulation(-fold)] or actual concentration; indeed, the shapes of the induction curves will be identical whether concentrations are expressed as molar units, density units derived from autoradiograms, or relative levels derived by *in situ* hybridization or quantitative immunoassay. Although the model is intended to provide a quantitative link between absolute rates of transcription and level of mRNA and protein, it can be used to estimate half-lives that best fit experimentally measured induction curves by the method of approach to steady state (11). In this application, it is not essential that the true rates of transcription and translation be known. Instead, the levels of mRNA and protein may be assigned arbitrary initial values, and all subsequent values may be expressed as increase(-fold) or percentages of the final values. This is true because decay constants control the rate of change in concentration, and their units are expressed as t^{-1} ($k_{d} = 0.693/t_{1/2}$). Therefore, the program can be used to model time-dependent changes in gene products found in brain nuclei or glia without needing to know absolute rates of production or amounts of product.

Implementation of Kinetic Model

User Interface for HyperCard Stack

A program to solve the equations that result from the assumptions described above was written for the Macintosh computer using HyperTalk, the Hyper-Card programming language.*,† An alternative form using a spreadsheet was described earlier (15). The user interface for the HyperCard program is shown in Fig. 2. HyperCard permits a database to be assembled analogously to a stack of reference cards; in the present program, each card in the stack contains kinetic values for one gene. The program is operated by using a mouse to position the cursor within elements called buttons and "clicking" the mouse button. For instance, to add a new gene to the reference stack, one creates a new card by moving the cursor to the button labeled New; clicking the mouse causes a new card to be created that can be named and customized. Data are entered into the shadowed fields by highlighting the values and typing in new ones. Some features can be turned on and off by clicking the circles called radio buttons or the squares called check boxes marked by Xs. For instance, output is switched between numerical and graphical form by deleting the X in the box labeled "Record" on the right-hand side of Fig. 2. To prepare customized plots of the results, data can be saved using the Export button so that graphs may be prepared using professional graphics programs. HyperCard saves files according to the date and time of day and stores them in the HyperCard folder.

The screen depicts one card that is divided into an input field for data entry designed to follow the dual-compartment model shown in Fig. 1 and an output field into which results of calculations are entered. Results may be plotted as shown or entered in tab-delimited columns that represent the elapsed time, predicted mRNA concentration, and the protein concentration, respectively.

---

* Copies of the kinetic modeling programs written for HyperCard and for the Excel spreadsheet are available from Cosmic, 382 E. Broad St., Athens, GA 30602. Specify the operating system and diskette format, and part number UGA02503 for the MS-DOS version or UGA0254 for the Macintosh version.

† HyperCard and HyperTalk are registered trademarks of Apple Computer, Inc.

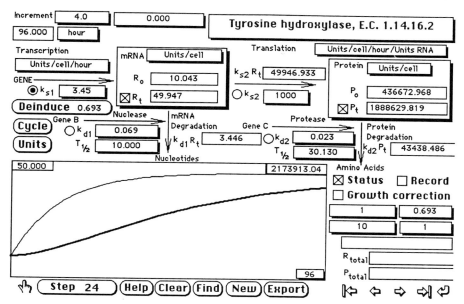

FIG. 2 The user interface for modeling gene expression using the HyperCard program is shown; it represents one file card in a database containing cards for numerous gene products. The large, open rectangle at lower left is the field in which results of calculations are displayed. The rest of the field represents the pools of mRNA and protein, which are connected by arrows similarly to the diagram in Fig. 1. Operation of the program requires that values for the kinetic parameters be entered into four input fields adjacent to the symbols k_{s1}, k_{s2}, k_{d1}, and k_{d2} as described in the text. The program is run by using the mouse to move the cursor to the Step button (lower left) and clicking it. Alternatively, clicking the Cycle button will automatically simulate a cycle of induction and deinduction.

Precision of the calculations is specified in the field on the left-hand side of the title bar. Cards with data for other genes are selected either by using the mouse to click the arrows at lower right or by entering the name of the gene after clicking the Find button at the bottom of the figure.

To use the interface, one first defines the units for the rate constants by clicking the Units button, which opens a dialog box into which the labels may be typed; the names in these fields do not affect the calculations. The period between data points, expressed in the time units used for the decay constants, is entered into the Increment field (top left, Fig. 2). The total number of data points to be calculated is entered into the button labeled Step (lower left, Fig. 2), and the total period spanned will equal the product of the increment size times the number of steps. In the simulation shown, the total period equals 96 hr.

Kinetic values are entered into the fields designated k_{s1}, k_{s2}, k_{d1}, and k_{d2}. These fields provide the information that is used by the program to solve the equations when the Step or Cycle buttons are clicked. Half-lives for mRNA and protein may be entered into the fields labeled $T_{1/2}$, and these will be converted to rate constants for degradation when the cursor is moved to the fields labeled k_d while holding down the option key on the keyboard. The program does not retrieve data directly from the half-life fields. Alternatively, one may calculate the rate constants for decay from the relationship $k_d = 0.693/t_{1/2}$. If the rate of transcription has not been measured, it may be defined as the value that produces the observed initial concentration of mRNA from the steady-state relationship [Eq. (2)]; likewise, the constant for translation (k_{s2}) is defined as the value that will produce the initial, steady-state concentration of protein (P_{ss}) once the other values have been defined:

$$P_{ss} = \frac{k_{s1}k_{s2}}{k_{d1}k_{d2}} \tag{5}$$

Running Program

To simulate the induction or deinduction of mRNA and protein, a new value is chosen for one or more of the four kinetic parameters. If theoretical curves are being fitted to experimental data, the ratio between the final and initial values should match the observed increase(-fold) unless a steady state has not been attained. Basal and induced values for one of the four parameters are entered into the field within the button labeled Induce/Deinduce, which toggles between the two levels. Clicking this button sends the new value of the kinetic constant to the selected field. In the example shown, the transcription field has been selected, as indicated by the filled circle in the "radio button" on the left-hand side of Fig. 2. Calculations begin when the Step button is clicked, and data are sent to the output field. If desired, the concentrations may be viewed as they are calculated by clicking the square next to Status (right-hand side of Fig. 2). The fields at lower right activate a correction for exponential cell growth that will not be described here.

The program may be used to simulate regulation of transcription, translation, or turnover of mRNA or protein in any combination. When first activated, it automatically calculates initial values for concentration of mRNA and protein (R_0 and P_0, respectively) assuming that steady-state conditions exist when $t = 0$. Any combination of the kinetic parameters can be changed thereafter by highlighting the parameters of interest and changing the values manually. For the next interval, the program will begin with the existing

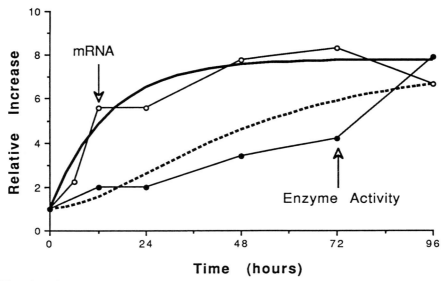

FIG. 3 The program was used to compare theoretical curves (heavy lines) with experimental results for induction of tyrosine hydroxylase and its mRNA by 8-bromo-cyclic AMP in PC12 cells. Half-lives used in the calculations were 10 hr for the mRNA and 30 hr for the enzyme, respectively. [Data were replotted from those of A. W. Tank, L. Ham, and P. Curella, *Mol. Pharmacol.* **30**, 486 (1986).]

values of R_t and P_t instead of recalculating steady-state concentrations. Results will continue to enter the output field in sequence until the Clear button is clicked. This permits changes to be made for parameters that increase gradually or that change after the initial stimulus has occurred, such as happens when stimulation is transient.

Modeling Induction of Tyrosine Hydroxylase

Transcription of the gene for tyrosine hydroxylase increases in the pheochromocytoma cell line designated PC12 after chemical depolarization, stimulation by neuropeptides, or activation of protein kinases A or C (1–10). Tank *et al.* (5, 6) have studied the induction of this enzyme and its mRNA after treatment with 8-bromo-cyclic AMP (8-Br-cAMP) and dexamethasone. They reported that the half-life of the protein was about 30 hr (5); other workers showed that the half-life of the mRNA in the presence of actinomycin D was about 10–16 hr (10). These half-lives correspond to decay constants of 0.023

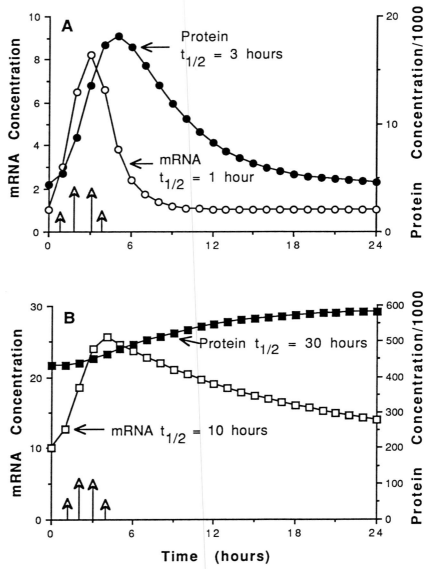

FIG. 4 Predicted effects of transient transcriptional activation are compared for short-lived and long-lived gene products. Transcription was increased for 4 hr (indicated by vertical arrows) and then returned to the basal level in both cases. (A) The short-lived mRNA increases nearly to a new, steady-state level and then decays back to the initial level to give a spike of protein synthesis. The protein does not attain the new steady-state level, but the increased activity is evident. (B) The long-lived mRNA

hr^{-1} for the protein and 0.0693 hr^{-1} for the mRNA [half-life of 10 hr; Eq. (2)].

Figure 3 compares the experimental results of treating the cells with 8-Br-cAMP (5) and predicted concentrations based on these initial decay constants. Because the absolute rate of synthesis and level of mRNA were not reported, the theoretical data were generated by setting the initial values for mRNA equal to 1 and plotting the relative increase with an arbitrary value for k_{s2} of 1000 (actual values range from 10^3 to 10^4). The theoretical curves, shown by lines without data points, generally lie within the standard error of the experimental data (not shown). Activation of preexisting enzyme was not observed here because the assay employed saturating levels of substrate and cofactor (5). The reported half-lives generate theoretical induction curves that are similar to measured values, suggesting that no major posttranscriptional controls are likely in this case. If discrepancies were observed, further studies of the mechanism would be merited.

Effects of Transient Activation of Transcription

Many stimuli do not result in new, steady-state levels of transcription; instead, transcription may increase briefly and then return to the original value despite the continued presence of the inducing agent. For instance, forskolin (3), nerve growth factor (7), and vasoactive intestinal peptide (9) cause transient activation of transcription from the gene for tyrosine hydroxylase in PC12 cells. The computer program can be used to model the consequences of transient activation because the kinetic parameters may be changed after any period.

Figure 4 shows predicted effects of transient stimulation on levels of short-lived and long-lived mRNAs and proteins. The long-lived mRNA and protein had half-lives identical to those for tyrosine hydroxylase and its mRNA (30 and 10 hr, respectively); the short-lived mRNA and protein had 10-fold shorter half-lives, equal to 1 and 3 hr, respectively. The rate of transcription was increased from the basal level by 5-fold for 1 hr, to 10-fold for the next 2 hr, back to 5-fold for 1 hr (indicated by vertical arrows in Fig. 4), and then

---

begins to increase but does not achieve a new steady state, instead slowly decaying back toward its initial level. The concentration of the long-lived protein is scarcely affected, increasing only 20% in the period shown. Notice the large difference in initial pool sizes for the two products, even though the rate of transcription is identical for both genes.

was maintained at the basal rate for the next 20 hr. This resulted in an 8-fold increase in the level of the short-lived mRNA, followed by a 4-fold increase in the short-lived protein (Fig. 4A). The concentration of mRNA returned to the basal level by about 9 hr after transcription was first increased, and the level of protein decayed to the basal level by the end of the day [Eq. (3b) indicates that a period equal to five half-lives must elapse before induction is complete]. In contrast, the concentration of the long-lived mRNA increased less than 3-fold and then slowly decayed toward the initial level. It did not return to the basal value during the period of simulation (Fig. 4B). The concentration of the long-lived protein increased by less than 20% during the 24 hr period, and did not begin to decline because the level of its mRNA did not return to the basal value.

These results demonstrate that the response of stable mRNAs and proteins to transient activation of transcription is attenuated because their initial concentrations are high relative to their rates of synthesis. In Fig. 4, note that the initial concentration of the proteins is more than 1000-fold greater than the concentration of the mRNAs. This accurately reflects the relative concentrations of mRNAs and proteins in living cells.

Uses of Computer Program for Simulating Gene Expression

Our original purpose in writing the program was to integrate quantitative data concerning rates of synthesis and decay for gene products. These values may be obtained using run-on assays for nuclear transcription with appropriate corrections (16, 17), quantitative assays for mRNA by solution hybridization (17, 18), and immunological or enzymatic assays for peptides or proteins (19). As demonstrated here, semiquantitative data can be used, so the technique could be applied to studies using *in situ* hybridization or immunological detection by light or electron microscopy (18, 19). A third use is didactic, because the program integrates data from molecular and cellular biology, indicating the enormous amplification involved in mammalian gene expression, the relative concentrations of mRNA and protein, and the predicted lag between induction of mRNA and protein. The lag is predicted partly because the mRNA and protein approach their new levels according to their independent rates of decay, and partly because the mRNA must increase before the rate of protein synthesis can increase. The results shown in Fig. 4B dramatize the need for control of enzyme activity by phosphorylation and other rapid mechanisms; proteins with slow half-lives cannot attain significant increases in concentration unless the rate of synthesis remains elevated for long periods. Finally, the program permits estimation of half-lives based on the concept of approach to steady state. This relies on following the time course of induction from an initial value to a new equilibrium or monitoring the course of deinduction to the original level.

Integrating Gene Expression with Metabolic Control Theory

The kinetic model described here predicts that the time required for induction of an mRNA or protein to 50% of a new, steady-state level cannot be shorter than the respective half-lives of these molecules (20). Many neuronal genes that are affected by transsynaptic stimulation probably never attain a steady state, because the time course of adaptation by this mechanism is much longer than the periods over which neuronal activity fluctuates.

The program represents an initial step toward making computer modeling of gene expression available to any laboratory that possesses a microcomputer. Limitations to the program derive from the untrue assumption that biological processes are instantaneous and the fact that enzyme induction may include control points that have not been modeled. These drawbacks are offset by the advantages of conceptual simplicity, use of kinetic parameters that can be measured by well-characterized experimental techniques, and the ability to use the program to make testable predictions about the effects of changing synthesis or stability of mRNAs and proteins.

Changes in gene expression at any level usually are adaptive in the sense that the organism is better able to make adjustments to events in the external environment. To determine the consequences of altered tyrosine hydroxylase gene expression for the availability of dopamine, it will be necessary to develop higher order programs that merge the kinetic ideas expressed here with metabolic control theory (21, 22), and to account for effects of phosphorylation, substrate availability, and rates of secretion and reuptake. General-purpose simulation programs such as STELLA* are already available for microcomputers at affordable cost. In the future, we anticipate that many laboratories will make use of these capabilities in testing the consequences of altered genetic regulation.

References

1. R. E. Zigmond, M. A. Schwarzschild, and A. R. Rittenhouse, *Annu. Rev. Neurosci.* **12,** 415 (1989).
2. R. Roskoski, Jr., L. White, R. Knowlton, and L. M. Roskoski, *Mol. Pharmacol.* **36,** 925 (1989).

* STELLA II is a general-purpose modeling program produced by High Performance Systems, 45 Lyme Road, Hanover, NH 03755.

3. E. J. Lewis, A. W. Tank, N. Weiner, and D. M. Chikaraishi, *J. Biol. Chem.* **258,** 14632 (1983).
4. E. J. Lewis, C. A. Harrington, and D. M. Chikaraishi, *Proc. Natl. Acad. Sci. U.S.A.* **84,** 3550 (1987).
5. A. W. Tank, L. Ham, and P. Curella, *Mol. Pharmacol.* **30,** 486 (1986).
6. A. W. Tank, P. Curella, and L. Ham, *Mol. Pharmacol.* **30,** 497 (1986).
7. E. Gizang-Ginsberg and E. B. Ziff, *Genes Dev.* **4,** 477 (1990).
8. E. J. Kilbourne and E. L. Sabban, *Mol. Brain Res.* **8,** 121 (1990).
9. M. Wessels-Reiker, J. W. Haycock, A. C. Howlett, and R. Strong, *J. Biol. Chem.* **266,** 9347 (1991).
10. S. Vyas, N. Faucon Biguet, and J. Mallet, *EMBO J.* **9,** 3707 (1990).
11. J. R. Rodgers, M. L. Johnson, and J. M. Rosen, *in* "Methods in Enzymology" (L. Birnbaumer and B. W. O'Malley, eds.), Vol. 109, p. 572. Academic Press, New York, 1985.
12. K.-L. Lee, J. R. Reel, and F. T. Kenney, *J. Biol. Chem.* **245,** 5806 (1970).
13. G. Yagil, *Curr. Top. Cell. Regul.* **9,** 183 (1975).
14. J. L. Hargrove and F. H. Schmidt, *FASEB J.* **3,** 2360 (1989).
15. J. L. Hargrove, M. G. Hulsey, F. H. Schmidt, and E. G. Beale, *BioTechniques* **8,** 654 (1990).
16. G. S. McKnight and R. D. Palmiter, *J. Biol. Chem.* **254,** 9050 (1979).
17. J. D. White and E. F. LaGamma, *in* "Methods in Enzymology" (P. M. Conn, ed.), Vol. 168, p. 681, Academic Press, New York, 1989.
18. G. R. Uhl, *in* "Methods in Enzymology" (P. M. Conn, ed.), Vol. 168, p. 741. Academic Press, New York, 1989.
19. J. W. Haycock, *Anal. Biochem.* **181,** 259 (1989).
20. J. L. Hargrove, M. G. Hulsey, and E. G. Beale, *BioEssays* **13,** 667 (1991).
21. J.-H. Hofmeyr, *Eur. J. Biochem.* **186,** 343 (1989).
22. C. I. Pogson, R. G. Knowles, and M. Salter, *Crit. Rev. Neurobiol.* **5,** 29 (1989).

[12] Computerized Optimization of Experimental Design for Estimating Binding Affinity and Binding Capacity in Ligand Binding Studies

Peter J. Munson and G. Enrico Rovati

Introduction

Ligand binding studies remain an important tool in physiology, pharmacology, biochemistry, and even molecular biology. Ideally, binding equilibrium measurements should allow for the description of the thermodynamic interactions between macromolecules, such as cell surface receptors, and their hormones or ligands. In fact, numerous technical difficulties (inability to establish true equilibrium, surface charge effects, etc.) have often hindered the measurement of the true free energy changes associated with such interactions. These difficulties notwithstanding, this experimental protocol has achieved wide popularity, and it has provided numerous useful insights into many biologically important receptor systems and subsystems. Indeed, the notion of the receptor itself would probably not exist were it not possible to demonstrate and measure the specific, saturable binding of its ligand.

On the other hand, the relative ease with which binding studies can now be performed has led some investigators to overlook important intrinsic limitations of this method in the rush to obtain publishable binding affinity (K_d) or binding capacity (B_{max}) estimates. For example, assaying binding to whole cells in an attempt to measure the affinity of a cell surface receptor incurs the possibility that internalized ligand will be included in the bound fraction. In this case, the estimated "affinity" may actually relate more to a (possibly passive) transport phenomenon than to a specific process. Many publications refer to estimated K_d values, even though rigorous demonstration of the saturability of binding has not been made. The fact that one can measure a slope on a Scatchard plot by itself does not establish the presence of a specific receptor. Moreover, inattention to the details of laboratory technique and to the data analysis methodology can lead to meaningless or misleading results.

Methods in Neurosciences, Volume 10

Good Studies Depend on Good Experimental Techniques and Good Designs

The techniques of "wet" chemistry of the binding procedure, preparation of receptor-bearing tissue, careful weighing and pipetting of ligand, standardization of conditions, and validation of the separation step are crucial to the success of a binding experiment. However, we do not consider such methodological problems here. The design of the experiment, namely, the choice of the concentrations in each tube in the rack or well on the plate, is a problem which is amenable to mathematical analysis, and it is the subject of this chapter. Too few tubes, too little ligand, or even too much ligand in all of the tubes will result in a suboptimal, sometimes useless assay run. In the case where biological tissue is scarce (e.g., when analyzing substructures of the rat brain), or when hundreds of assays are required (e.g., in a screening study), optimization of the experimental design can yield a dramatic improvement over conventional designs.

What we shall attempt to do here is build a computer model of the experimental system, simulate the experiment, collect the results, and analyze them—all before the "wet" experiment is actually performed. We essentially perform the experiment inside the computer using hundreds of different designs, to find out which one gives us the most accurate and precise estimate of K_d and B_{max}. To perform this modeling exercise, we shall need a good representation of the experimental system, including the nature and distribution of experimental errors. We also need a reasonably accurate model for the chemistry of the receptor system being studied, including preliminary estimates of the binding capacity and affinity of the system! Thus, there is an essential circularity to the design problem; we cannot optimize the design for estimating the binding affinity until we have a good idea of the binding affinity value.

To break this circle, one may begin with any reasonable design as a pilot study and gradually evolve toward a better design. In practice, it is found that two or three cycles of optimization are all that is required. For a long series of experiments, this is a perfectly acceptable route. However, even for short series or "one-shot" experiments, the idea of optimization can yield useful improvements even in the initial design. For example, one usually has some notion of the approximate magnitude of the binding affinity. If this number can be specified to within a factor of 10, a reasonably good initial design is possible. Still more useful is the protection against catastrophically bad designs: one cannot hope to measure three parameters if there are only effectively two concentrations used in the assay! Such indeterminate designs (which can be more subtle than simply having too few concentrations) show up immediately when one attempts to analyze the simulated experiment.

Good Binding Experiments Require Good Analysis Tools

If it were not for experimental error, many commonly used graphical techniques (Scatchard, Eadie–Hofstee, displacement curve) would all yield exact results, and the choice of analysis technique would be merely a matter of taste. In the presence of measurement "noise," such graphical methods are affected to a greater or lesser degree, becoming biased or inefficient to different degrees. Thus, we must consider carefully the choice of data analysis method.

Interpretation of binding studies must always take place in light of an assumed model. Common graphical techniques (such as the Scatchard plot), even common names or symbols (binding affinity, K_d, or dissociation constant) tacitly imply some sort of molecular or thermodynamic model. To proceed systematically, we must make these models explicit. Once the model is defined, we can specify a method for estimating the unknown parameters. A very general technique in this regard is nonlinear least-squares curve fitting, which is nothing more than adjusting the unknown parameter values so that the predicted curve reflects the experimental data as closely as possible. This technique has been widely studied statistically and it is generally agreed to be among the best available. We recommend appropriately weighted nonlinear least-squares curve fitting (WNLLS) as implemented in our computer program LIGAND (1), provided the assumptions of the model incorporated into LIGAND are met in the particular experimental system at hand.

Using the LIGAND program, we assume the measurement errors behave according to a Gaussian (bell-shaped) distribution with a variance that changes, and generally increases, with the value of the measurement. It is a fairly good approximation to assume that bound ligand concentration (B) can be measured with roughly constant relative precision (2). Mathematically, we say that the coefficient of variation, CV is equal to $\sigma_B/B = $ constant, or, equivalently, $\sigma_B^2 = $ constant $\times B^2$. Other, more complicated relationships are possible of course. The LIGAND program fits the exact (not approximate) mathematical expression for equilibrium binding to one or more classes of binding sites and has many options for more complex models such as cooperativity.

Most binding studies postulate the existence of one, two, or more independent classes of specific binding sites. However, it is important to remember that practically every ligand binding study is also influenced by nonspecific binding. Rather than "correct" the raw data for nonspecific binding, it is preferable to model such nonspecific binding explicitly using the parameter N. This will serve to give a more realistic picture of the complete experimental situation. Program LIGAND includes this feature.

Best Designs Give Lowest Standard Error for Estimated Parameters

How do we judge one experimental design against another? All other things being equal (reagents, operator, tissue preparation, number of replicates), the best experiment is one which gives the most precise values for the affinity and binding capacity, that is, the unknown parameters of the model. We measure precision in terms of the variance (squared standard error) of the parameters K_d, B_{max}, and N. Interestingly, if any one of these parameters is poorly estimated (say N), the large uncertainty tends to leak over into the other estimates. One could attempt to find designs which are optimal for estimating the affinity, K_d, if we were uninterested in the binding capacity. However, in general, we try to estimate all of the parameters with good precision. The mathematical criterion for comparing designs is called the D-optimality criterion (3) and is related to the geometric average of the variances of the parameters. Specifically, the criterion is the determinant of the variance–covariance matrix, a measure of the volume of the simultaneous confidence region for the parameters. Thus, a smaller volume means more precisely determined parameters and less overall uncertainty. To optimize an experiment, then, means manipulating the concentrations for each tube in the assay so that the overall confidence region for the parameters is smallest.

General Rules for Establishing Good Designs

Even without going more deeply into the mathematics of optimal design, we can offer some general rules for establishing good designs. The three most common protocols can be described as: (1) saturation studies, where only a single labeled ligand is present; (2) displacement studies, using a constant amount of labeled ligand displaced by the same ligand in its unlabeled form; and (3) heterologous displacement studies which use a distinct unlabeled ligand from the labeled ligand.

For saturation studies, space concentrations roughly logarithmically over as wide a range as possible. Center the range around the expected K_d value. Include excess "cold" tubes to estimate nonspecific concentration. A common shortcoming with these designs is that one cannot usually get more than a 100-fold concentration range, thus limiting the overall precision and reducing the ability to distinguish multiple binding sites. The affinity, K_d, also depends strongly on knowledge of the specific activity of the labeled species.

For homologous displacement studies, use the smallest feasible concentration of label to estimate the lowest possible K_d values. Use the widest possible range of unlabeled ligand, logarithmically spaced, to establish a complete binding isotherm. The range should be centered around an expected K_d value. Note that if labeled and cold compounds are not identical, then differences between the K_d of labeled and the K_i of unlabeled compounds can affect estimation of binding affinity and capacity.

Heterologous displacement studies should only be attempted if the K_d (and possibly the B_{max}) for the label is already established. Ideally, they should be run along side a "homologous" study. Use the same rules for design as for homologous displacement. Other general pitfalls which affect binding studies are degradation of ligands over the course of the incubation step (e.g., by peptidases), degradation or inactivation of receptor (e.g., by internalization or proteolysis), failure to reach equilibrium, failure to establish homogeneous solution of reagents (especially for membrane-bound receptors), incomplete or imprecise separation of bound ligand from free ligand (e.g., using filtration), and inadequate temperature control.

Systematic computerized optimization of binding studies may be valuable for assays which are to be repeated in a large number of samples, where the potential reduction of the amount of reagents can be significant. For all assays, the improvement of the precision of results is a desirable and sometimes essential goal. This is especially true in assays which are to be used to distinguish the effects of two or more treatment conditions, for example, a dietary pretreatment of animals prior to sacrifice and recovery of the receptor-bearing tissue. These goals are approached through the use of computerized optimization of experimental design. However, the benefits of this systematic approach can only be realized in an assay system which has first been subjected to a "common sense" appraisal of the overall quality of its design and experimental technique. Unless the assay results are somewhat reproducible from day to day, there is no sense in attempting a computerized design optimization.

Computerized Optimization of Experimental Design

To perform design optimization one must establish (1) the binding model, (2) rough values or ranges for the K_d affinity values, and (3) the relative precision with which the binding assay can be carried out. With these three facts in hand, one can then calculate a "minimal" design, namely, a set of experimental conditions which consume the fewest "tubes," yet result in the smallest overall error of the parameter estimates. One should not generally implement such a minimal design directly, but rather use it as a guide to setting up a practical design with more tubes and more conveniently obtained concentrations. One may also wish to build in some "robustness" to changes in the underlying assumptions. For example, a robust design would still give useful results if the actual K_d value were 10-fold different from the value assumed when setting up the minimal design.

For simple models, a few simple rules can often be used to set up such a robust design, whose performance will not be far from the mathematically determined optimal, minimal design. For one binding site with nonspecific binding, these rules are as follows. Include at least one point with a concentration as low as feasible (but no higher than 0.1 K_d). Include a point with a concentration between $1 \times K_d$ and $3 \times K_d$. Include a point with as high a concentration as feasible (at least $30 \times K_d$). Finally, include additional points with approximately a 5-fold differences in concentration from the middle point. Replicate each of these 5 points in duplicate or triplicate to give some measure of robustness against individual outliers.

For models with two or more binding sites, or two or more ligands, the computerized approach must be used. We illustrate a selection of optimal designs for the case where bound ligand concentration is assumed to be measured with constant relative precision.

Some Optimal Designs

Using the computer program DESIGN, we have calculated optimal designs for a variety of cases (more details are included in Refs. 4–6). The simplest case involves only a single ligand and a single class of binding sites, either in a saturation protocol or in a homologous displacement study. Here, we must estimate three parameters: K_d, the affinity; B_{max}, the binding capacity; and N, the dimensionless parameter for nonspecific binding. It stands to reason that we will need at least three points on the binding curve, since there are three parameters.

It turns out that the D-optimal design uses only three points, shown first on the Scatchard plot (Fig. 1A) and then on the displacement curve (Fig. 1B). Regardless of which protocol is used, saturation or displacement, the three points remain the same. Because this design is mathematically optimal, any additional tubes added to the assay should be placed at one of these three points, in roughly equal numbers. Of course, it is quite reasonable not to use just these three points. Using tubes with other concentrations can serve to validate the model (in this case a one-binding site model), to detect departures

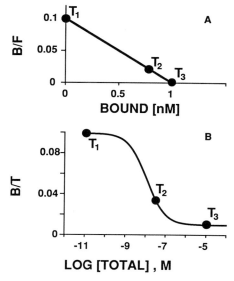

FIG. 1 D-optimal design for a single binding site, shown as a Scatchard plot (A) or as a displacement curve (B). The Scatchard plot has been corrected for the presence of nonspecific binding. Even though the Scatchard plot is linear, three points are required in the design owing to the presence of nonspecific binding. The parameter values for binding are $K_d = 10^{-8}\,M$, $B_{max} = 10^{-9}\,M$, $N = 0.01$. The variance model is based on a constant relative error in B ($a_4 = 2$), and the design will change somewhat for other variance models. [Adapted with permission from G. E. Rovati, D. Rodbard, and P. J. Munson, *Anal. Biochem.* **174,** 636 (1988).]

from the model, indicated, for example, by curvature in the Scatchard plot, and to provide a robust design which is not tightly linked to the assumed K_d value. The particular design shown here is not strongly dependent on the assumed B_{max} value, and it turns out to be generally true that optimal designs are not strongly dependent on binding capacity. The exact optimal design consists of three points: T_1, with the lowest feasible ligand concentration which still permits measurement of bound; T_3, with the highest feasible concentration; and T_2, a total concentration giving rise to the free concentration $F_2 = K_d(1 + B_{max}/K_d/N)^{1/2}$. The derivation for this expression is located in Ref. 4. One can assign "meaning" to each of the three points in Fig. 1: T_3 primarily determines the parameter N for nonspecific binding; T_1 determines the initial B/F ratio ($= B_{max}/K_d + N$); and T_2 in combination with the other points determines the slope on the Scatchard plot, and hence the affinity, K_d.

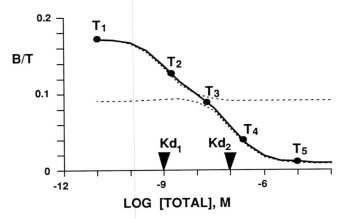

FIG. 2   *D*-optimal design for the two-binding sites model, shown as a displacement curve. The parameter values for the model are $K_{d_1} = 10^{-9}\ M$, $K_{d_2} = 10^{-7}\ M$, $B_{max_1} = 10^{-10}\ M$, $B_{max_2} = 10^{-8}\ M$, and $N = 0.01$. Constant relative error in B is assumed. The dashed lines indicate the separate displacement from each of the two binding sites which is summed to give the overall displacement. [Adapted with permission from G. E. Rovati, D. Rodbard, and P. J. Munson, *Anal. Biochem.* **174,** 636 (1988).]

For this and the following cases, we assumed a constant relative precision of measurement of bound concentration. We also specified a minimal and maximal ligand concentration, thereby preventing unrealistically small or "infinite" concentrations from arising.

Two classes of sites binding to the same ligand can be expected to give rise to a curved Scatchard plot or "gradual" displacement curve with a Hill slope less than 1.0. If the affinities of the two classes are well separated (say, by a factor of 5 or more), one can estimate K_d and B_{max} for each site. In this case, the optimal design becomes more complex, involving a minimum of five ligand concentrations (T_1 to T_5 on Fig. 2). Note that as in Fig. 1B, two of the points (T_2 and T_4 in Fig. 2) fall roughly at the midpoints of the curves associated with each individual site (dashed line, Fig. 2). As before, there are points (T_1, T_5) at the extremes of feasible concentration. The point T_3 serves to determine the proportion of the two sites present in the assayed tissue. From this analysis, one can see that if K_{d_1} and K_{d_2} are not sufficiently far apart, the design points T_2, T_3, and T_4 will be squeezed so closely together as to make separate estimation of two binding affinities impossible. Conversely, the farther K_{d_1} and K_{d_2} are separated, the more like two completely distinct and separate binding sites the system behaves. If a wide plateau exists around T_3, then we may even choose to treat the low affinity site

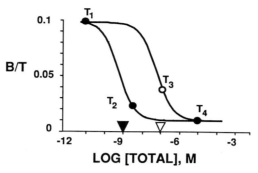

FIG. 3   *D*-optimal design for protocol including a homologous and a heterologous curve in a single experiment. The model has a single binding site with parameter values $K_d = 10^{-9}\,M$ (indicated by solid triangle), $K_i = 10^{-7}\,M$ (indicated by open triangle), $B_{max} = 10^{-10}\,M$, $N = 0.01$. Four points are required in this design, reflecting the four parameters in the model. [Adapted with permission from G. E. Rovati, D. Rodbard, and P. J. Munson, *Anal. Biochem.* **184,** 172 (1990).]

(K_{d_1}) simply as nonspecific binding. Dropping T_4 and T_5, we could then reduce the design to the one-site design shown in Fig. 1.

For a heterologous displacement, the problem gets still more complex. Now, each ligand (labeled and unlabeled) has a distinct binding affinity for the binding site. Call these affinities K_d and K_i, respectively. As before, there is a binding capacity, B_{max}, and nonspecific binding, N (there are actually two N's, one for the label and one for the cold, but we will disregard nonspecific binding for the unlabeled ligand). Although many investigators attempt to estimate the K_i from the ED_{50} or midpoint of a single displacement curve, the binding affinity of the label can also influence the estimate of the K_i [hence the need for the often used Cheng–Prussoff correction (7)]. Optimally, we would estimate all four parameters simultaneously, in a combination heterologous–homologous displacement curve. An optimal design for such curves is shown in Fig. 3. The results of such an experiment can hardly be called "curves" as measurements are made at only four points (T_1 through T_4). Of course, points could be included elsewhere on the curves. As before, the optimal experiment requires exactly the number of points as there are parameters in the model. The leftmost curve in Fig. 3 is the homologous curve and includes a point near the K_d concentration. The rightmost curve is the heterologous displacement and includes a point near the K_i. This simple arrangement of points is quite general regardless of the particular parameter values used in this case.

With two binding sites and two ligands (one labeled, one unlabeled), optimal designs start to get interesting. It is here that the payoff for using optimal designs is greatest, yet the simple, intuitive rules for generating designs for the one-ligand, one-binding site case seem to break down. Nevertheless, the computed mathematically optimal designs, after some thought, do make sense. There are several cases to be considered. If one of the ligands (labeled or unlabeled) cannot, by itself, distinguish two different binding sites (because $K_{d_1} = K_{d_2}$ or $K_{i_1} = K_{i_2}$), then it may be insufficient to run simply a heterologous and a homologous curve together. In such cases, one can resort to a "multiligand" design, described below.

We illustrate the case where both ligands can distinguish the two binding sites. Here, we must also consider the order of preference for the two ligands: if $K_{d_1} < K_{d_2}$ for the labeled ligand, the relationship of affinities may be the same or opposite for the unlabeled ligand. We consider the case when the two ligands have the same order of preference for the two binding sites. Finally, we must consider which ligand has greater selectivity for the two sites. Assembling all of this information (as values for K_{d_1}, K_{d_2} for label, K_{i_1}, K_{i_2} for unlabeled, B_{max_1}, B_{max_2}, and N) we can find a seven-point optimal design using two curves shown in Fig. 4. Here, the points T_2 and T_3 on the homologous curve and T_5, T_6, and T_7 on the heterologous curve fall near the respective binding affinity values, but the exact relationship is complex. As before, we need to locate one point (T_1) at the extreme low end and one point (T_4) at the extreme high end of the curves. The fact that three optimal points (T_5, T_6, and T_7) are located on the heterologous curve probably follows from the fact that the unlabeled ligand had greater selectivity for these sites than did the labeled. A complete description of the optimal designs for this and other possible ligand selectivity patterns is given in Ref. 5.

Finally, we illustrate four "multiligand" designs for the two-binding site case, for all four possible cases of ligand preference: same order of preference (Fig. 5A, also treated in Fig. 4), opposite order of preference (Fig. 5B), nonselective labeled ligand, L_1 (Fig. 5C), and nonselective unlabeled ligand, L_2 (Fig. 5D). A multiligand design potentially includes all possible combinations of concentrations of labeled and unlabeled ligand; that is, we may vary the "tracer" concentration as well as the "displacer" concentration. Therefore, we represent these designs on a two-dimensional field rather than on the one-dimensional axis used previously. The boundary of the feasible concentrations are indicated by the dashed box. Curiously, most of the optimal concentrations fall on the boundaries of this box, corresponding to points on either the homologous (unlabeled ligand $L_2 = 0$) or the heterologous (labeled ligand $L_1 = $ constant) curves. In all cases, new design points (M_1 and M_2) are observed within the interior of the box. These points give new information relative to the heterologous–homologous

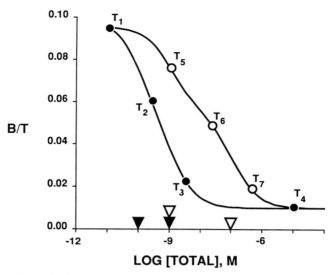

FIG. 4 D-optimal design for homologous and heterologous curves using a two-binding site model. The parameter values are $K_{d_1} = 10^{-10}\,M$, $K_{d_2} = 10^{-9}\,M$ (indicated by solid triangles) and $K_{i_1} = 10^{-9}\,M$, $K_{i_2} = 10^{-7}\,M$ (indicated by open triangles), $B_{max_1} = 5 \times 10^{-12}\,M$, $B_{max_2} = 5 \times 10^{-11}\,M$, $N = 0.01$. Again, the optimal design has seven points, as there are seven parameters. [Adapted with permission from G. E. Rovati, D. Rodbard, and P. J. Munson, *Anal. Biochem.* **184,** 172 (1990).]

curves alone. Although we have given an optimal design using only displacement curves for the same order of preference model (Fig. 4), identical experimental curves would be obtained if the values of K_{i_1} and K_{i_2} were switched (opposite selectivity pattern). This ambiguous result is avoided using the optimal multiligand design, which can easily distinguish these two selectivity patterns.

In the third case (Fig. 5C), we have previously shown (5) that these multiligand points provide essential new information, making the optimal multiligand design far superior to self- and cross-displacement which cannot even identify and estimate all of the parameters involved in the model without additional information. Experiments using nonselective labels, for example, use of [³H]dihydroalprenolol for the β-adrenergic receptor with selective cold ligands as a means of determining two binding affinities, would become much more informative with the use of an optimal "multiligand" design.

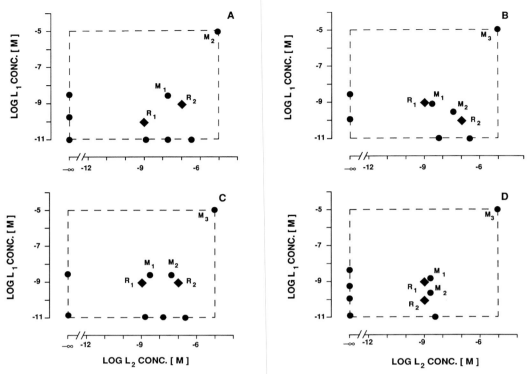

FIG. 5   Multiligand *D*-optimal designs for four patterns of ligand selectivity in the two-site binding model. In (A)–(D), the optimal concentrations (solid circles) of the labeled ligand (L_1) and the unlabeled ligand (L_2) are shown, selected from the feasible region bounded by the dashed lines. The new, interior "multiligand" design points are labeled M_1, M_2, etc. Points on the boundaries may be considered part of a homologous or heterologous displacement curve. The values for K_d and K_i are indicated by the location of the diamonds labeled R_1 and R_2. (A) Same order of ligand selectivity ($K_{d_1} < K_{d_2}$, $K_{i_1} < K_{i_2}$). (B) Opposite order of ligand selectivity ($K_{d_1} > K_{d_2}$, $K_{i_1} < K_{i_2}$). (C) Nonselective labeled ligand ($K_{d_1} = K_{d_2}$). (D) Nonselective unlabeled ligand ($K_{i_1} = K_{i_2}$). [Adapted with permission from G. E. Rovati, D. Rodbard, and P. J. Munson, *Anal. Biochem.* **184**, 172 (1990).]

Mathematical and Statistical Description

We will now outline the statistical and mathematical theory underlying computerized optimal design. A more detailed exposition of these ideas can be found in Ref. 1. We begin with a mathematical model of the binding experiment, the same model which is used in the LIGAND data analysis program.

We assume that the assay is governed by a simple equilibrium reaction whose equations are

$$B = \frac{B_{max} \times F}{K_d + F} + N \times F \tag{1a}$$

$$T = B + F \tag{1b}$$

or its generalization to multiple ligands or multiple binding sites. Thus, for any value of total ligand concentration, T, and binding affinity, K_d, capacity, B_{max}, and nonspecific level, N, there are values for bound and free ligand concentration (B and F, respectively) predicted by these equations. Adding a term to represent random experimental error, ε, the statistical model can be written:

$$B = f(T; K_d, B_{max}, N) + \varepsilon \tag{2}$$

where $f(.)$ is the function representing the binding reaction model. As stated earlier, we assume that ε has a Gaussian (normal) distribution with a variance σ^2 which may vary with B. In the data analysis step using weighted nonlinear least-squares (WNLLS), the sum of squared differences between the modeled and observed values of B,

$$\sum_{i=1}^{n} w_i[B_i - f(T_i)]^2 \tag{3}$$

is minimized by adjusting the unknown parameters, K_d, B_{max}, and N. The weights w_i should be set to $1/\sigma^2$, where σ^2 is determined according to a variance model. Assuming a fixed relative precision of measurement the variance model can be denoted

$$\sigma_B^2 = a_2 B^2 \tag{4}$$

where a_2 determines the magnitude of the relative error, or, more generally,

$$\sigma_B^2 = a_0 + a_3 B^{a_4} \tag{5}$$

where a_4 is generally between 0 and 2.

The values for a_0, a_3, and a_4 can be obtained by plotting the variance of replicate measurements versus the mean of each set of replicates. The general trend from a large number of sets of replicates can best be appreciated on a

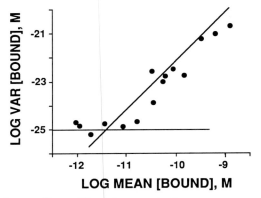

FIG. 6   Plot of variance of bound ligand concentration versus average bound concentration for 15 sets of replicates in a single binding assay. For each set of replicates, the sample variance and sample mean were computed and plotted on the log scale. The general trend is indicated with two line segments. The horizontal line gives a minimum value for the variance, independence of B. The increasing line has a slope of 2.0, corresponding to an exponent a_4 of 2 in the variance model. From the plot we may approximate a variance model for this case: $\sigma_B^2 = a_0 + a_3 B^{a_4} = 10^{-25} + 0.0025B^2$, corresponding to a 5% coefficient of variation in bound with a floor variance of 10^{-25}. [Adapted with permission from G. E. Rovati, D. Rabin, and P. J. Munson, in "Horizons in Endocrinology II" (M. Maggi and V. Geenen, eds.), p. 155. Raven, New York, 1991.]

log variance versus log mean plot (Fig. 6). The effort to refine the variance model has important implications for optimal designs. Optimal designs tend to concentrate points in the more reliably determined regions of the binding curve, namely, regions where σ^2 tends to be smallest.

The D-optimality criterion measures the overall standard errors of parameters as represented by the determinant of the variance–covariance matrix. Because we are dealing with a nonlinear binding model, we can only obtain an approximation of this matrix, after minimizing the sum of squares. We first form the design matrix X, where $X_{i,j} = \partial f_i/\partial \theta_j$, that is, the partial derivative of the prediction of bound for the ith data point in the design with respect to the jth parameter (one of K_d, $B_{\max}$, or N, in the simplest case), and its transpose, X^T. Using a diagonal matrix W of the statistical weights, w_i, the usual variance–covariance matrix for the parameter estimates is proportional to $(X^T W X)^{-1}$, and our optimality criterion is given by its determinant:

$$D = \det[(X^T W X)^{-1}] = 1/\det(X^T W X) \tag{6}$$

Optimization Algorithm

Using the foregoing notation, we can describe the computer algorithm which efficiently optimizes designs.

1. Initialize. Specify model, all parameter values, variance model parameters, and ranges of concentrations for all ligands.
2. Initial design. Specify an initial experimental design with sufficient points to span the allowable ranges of concentrations. Must have more points than there are parameters.
3. Calculate X, the design matrix.
4. Calculate $D = 1/\det(X^TWX)$
5. Adjust the initial design, incrementally changing each design point, and repeat Steps 3 and 4. Accept adjustments which reduce D. Terminate when no further reduction is possible.
6. Merge points in the current design which become close together. This reduces the number of points.
7. Adjust the number of replicates called for at each design point to further minimize D. Usually the replication factor remains at 1, but it may become fractional in more complex designs.
8. Print out the final design and the achievable standard errors on the parameters.

While not very complex, this algorithm has been adequate in our hands to find virtually all optimal designs for one- and two-site and one- and two-ligand binding models. Although there is no mathematical guarantee of convergence to a true optimum, an improved design results at every step of the procedure. Some cases of apparent nonconvergence have been observed when the final design contains many more points than the number of parameters. Overly restrictive concentration ranges may also force some cases to become indeterminate, so that no sufficient design exists. In such cases, D will become essentially infinite.

Computer Program DESIGN

Program DESIGN is a versatile computer program for optimization of ligand binding experiments. It is freely available from the authors and is implemented on IBM PC or compatible and Macintosh computers. DESIGN uses an exact mathematical model of the equilibrium ligand binding system, with up to two ligands binding to any number of classes of binding sites. The program permits the use of a variety of models to describe the magnitude of the measurement error, explicitly accounts for nonspecific binding, considers total rather than free concentration as the "independent variable" in the regression, allows designated parameters of the model to be fixed and thereby excluded from the optimization strategy, and allows a variety of a priori constraints on the experimental design. The program can be used to optimize various protocols, including the simple saturation or homologous displacement curve (1), the complete self- and cross-displacement for two ligands (four different curves simultaneously), and the "multiligand" experiment (5). A detailed description of the program DESIGN and of the characteristics of D-optimal designs can be found in our previous reports (4–6).

Practical application of D-optimal designs requires three components: (1) an accurate description of the (parametric) model of the biochemical system together with values for each of the parameters; (2) a description of the behavior of the statistical errors associated with measurement; and (3) a list of the minimum and maximum allowable concentrations for each reagent in the system. The program DESIGN allows for flexible specification of these three components.

The parametric binding model is taken from the family of models involving multiple ligand and multiple binding sites described for the LIGAND computer program. This family includes one- and two-binding site models with one or two ligands as special cases. Values for the binding capacities (B_{max}), binding affinities (K_d), and levels of nonspecific binding (N) must be specified by the user.

In the following example, we have assumed normally distributed errors with a constant 5% error in the bound ligand concentration. Although this may represent an unrealistically small value, the final designs we obtained do not depend on the absolute magnitude of the error. The parameter precision estimates ($\%CV$) in Table I may be inflated by an appropriate factor, for example, 2 for 10% error in B, for larger error rates. Other variance models may be used and are sometimes preferable (2).

Minimum and maximum concentrations are generally determined by experimental constraints such as solubility and availability of reagents.

The DESIGN system consists of two programs: the first, DESIGPRE, is a preprocessor program used to enter the assay protocol and the binding model, together with an initial design, and to convert them to the standardized input format required by DESIGN. The second, DESIGN, is the main optimization program which finds the D-optimal concentrations for the specified model. We will now give an example of optimizing a single homologous displacement curve for a two-site binding model.

The preliminary program, DESIGPRE, prompts the user to specify the type of protocol: (1) single homologous displacement curve (one ligand); (2)

TABLE I   Initial Design of Twelve Different
Concentrations Equally Spaced on a Log Scale

Total ligand (M)	Parameter names	Parameter values	Expected CV^a (%)
1.0×10^{-11}	K_{d_1}	$1 \times 10^{-9}\ M$	44.6
3.5×10^{-11}	K_{d_2}	$1 \times 10^{-7}\ M$	15.7
1.2×10^{-10}	B_{max_1}	$1 \times 10^{-10}\ M$	47.4
4.3×10^{-10}	B_{max_2}	$1 \times 10^{-8}\ M$	9.4
1.5×10^{-9}	N	0.01	5.1
5.3×10^{-9}			
1.8×10^{-8}			
6.5×10^{-8}			
2.3×10^{-7}			
8.1×10^{-7}			
2.8×10^{-6}			
1.0×10^{-5}			

^{a} Coefficient of variation for parameter based on 5% CV in measurement of bound ligand concentration.

single heterologous displacement curve (one radiolabeled ligand and a second unlabeled ligand); (3) homologous and heterologous displacement curves analyzed simultaneously; (4) multiligand experiment (one radiolabeled ligand); (5) two homologous displacement curves for two different labeled ligands analyzed simultaneously; (6) complete self- and cross-displacement experiment with two radiolabeled ligands and the corresponding two unlabeled ligands; and (7) multiligand experiment (two radiolabeled ligands).

We demonstrate protocol number one. DESIGPRE then asks for the number of classes of receptors (two in this case) and the parameter values ($K_{d_1} = 1 \times 10^{-9}$, $K_{d_2} = 1 \times 10^{-7}\ M$, $B_{max_1} = 1 \times 10^{-10}\ M$, $B_{max_2} = 1 \times 10^{-8}$ M, and $N = 0.01$, dimensionless). For each parameter DESIGPRE also asks whether that particular value should be considered constant, and therefore excluded by the optimization routine, or whether it is to be shared, that is, constrained to be equal to other parameters. Shared parameters will be "locked" together and treated as a single parameter. The user is also asked to enter the allowable concentration range for each ligand (from 10^{-11} to 10^{-5} M in our example).

Finally DESIGPRE asks for the number of ligand concentrations in the initial design. For cases 1, 2, 4, or 7 above, an initial design can be provided automatically based on equal logarithmic spacing of total ligand concentrations. Therefore, in our example, we enter the desired number of concentrations (12), thus obtaining an initial design with 12 different concentrations equally spaced on a log scale between 10^{-11} and $10^{-5}\ M$ (Table I).

After running DESIGPRE, we may start the main optimization program, DESIGN. First, we are prompted to specify the variance model parameters a_0, a_1, a_2, a_3, and a_4 which define the variance model $\sigma_B^2 = a_0 + a_1 + a_2 B^2 + a_3 B^{a_4}$. The choice of the variance model can potentially have a large influence on the results of the optimization process, even larger than on the analysis of the data. The default values for the parameters ($a_0 = a_1 = a_3 = a_4 = 0$ and $a_2 = 25 \times 10^{-4}$) are quite reasonable for many ligand binding experiments, since the correspond to a constant 5% error in B. This model, however, may underestimate the variance for very low values of B. Thus, one might wish to set a_0 to some small, non-zero value. Weighting models also affect the size of the expected precision of the parameters in both the original and optimal designs.

At this point DESIGN will compute the precision of the parameter estimates on the basis of the initial design and of the variance model specified (Table I). Next, we must specify the "merging factor," which determines when two nearly equal concentrations will be merged into a single design point (the default is 0.1 decimal log units). We must also specify whether the program should optimize the precision of all the parameters of the model (*D*-optimality) or of a single selected parameter (*G*-optimality), such as $K_{d_{21}}$, the affinity constant of ligand 2 for site 1.

Then we enter the number of iterations for adjusting the values for the total concentrations. After each iteration the program prints the value of the "objective function," that is, the log of the determinant of the information matrix in the case of *D*-optimality, or (variance)$^{-1}$ in the case of *G*-optimality (the higher the value, the lower will be the expected variance in both cases). Finally, we are prompted to enter the number of iterations for adjusting the replication factors.

After the specified number of iterations a final design is printed together

TABLE II   *D*-Optimal Design and Predicted %CV for
Parameter Estimates

Total ligand (M)	Parameter names	Parameter values	Expected CV^a (%)
1.0×10^{-11}	K_{d_1}	$1 \times 10^{-9}\ M$	34.8
1.4×10^{-9}	K_{d_2}	$1 \times 10^{-7}\ M$	12.1
2.0×10^{-8}	B_{max_1}	$1 \times 10^{-10}\ M$	36.8
3.8×10^{-7}	B_{max_2}	$1 \times 10^{-8}\ M$	7.4
1.0×10^{-5}	N	0.01	3.9

[a] Coefficient of variation for parameters based on 5% CV in measurement of bound ligand concentration.

TABLE III   Augmented Optimal Design for Five Parameters

Total ligand (M)	Parameter names	Parameter values	Expected CV^a (%)
1×10^{-11}	K_{d_1}	10^{-9} M	41.1
3×10^{-10}	K_{d_2}	10^{-7} M	15.1
1×10^{-9}	B_{max_1}	10^{-10} M	42.3
3×10^{-9}	B_{max_2}	10^{-8} M	11.1
3×10^{-8}	N	0.01	5.1
1×10^{-7}			
3×10^{-7}			
1×10^{-6}			
1×10^{-5}			

[a] Coefficient of variation for parameters based on 5% CV in measurement of bound ligand concentration. The CV has been multiplied by $(9/12)^{1/2}$ to make the 12 and 9 point (augmented) designs comparable.

with the expected coefficients of variation of the parameters (Table II). The user may now continue the iterative optimization process if the program has not yet converged at the true optimum (e.g., if the number of D-optimal concentrations remains greater than the number of parameters and/or if the value of the objective function does not appear to have reached a maximum).

Because there are five estimated parameters in our model, there are also five points in the D-optimal design (Table II). We next adjust the D-optimal concentrations slightly to make them easily obtainable by serial dilutions. To ensure robustness against errors in the parameter estimates, the design is also augmented with two additional concentrations, centered around the two original concentrations closest to K_{d_1} and K_{d_2} (8). Alternately, one could "split" all of the D-optimal concentrations into two nearby values. Of course, a modest penalty is paid in the expected precision (CV values), but the additional robustness and convenience is well worth the price. Table III summarizes the expected precision to be obtained using the augmented optimal designs.

Conclusions

It is almost axiomatic to say that the best experiments arise from thoughtful experimental designs. Yet it is equally true that investigators rarely utilize the statistical optimal design methodology. We have attempted to review the benefits to the investigator of the use of computerized optimization in the context of ligand binding studies. As the rules for optimal design for simple models are themselves quite simple and intuitive, it is quite possible to evolve nearly optimal designs for binding studies without the use of this systematic approach. However, for more complex models involving two or more ligands, and for newcomers to the practice of binding assays, the computerized approach described here has a great deal of value. In general, it is reasonble to expect around a 2- to 3-fold reduction in the variance of parameters of an optimized assay or, correspondingly, a 3-fold reduction in the number of tubes required to obtain the same variance, compared with a typical initial assay design.

Few published reports of the actual application of optimal design techniques to ligand binding studies are available. Our own experience (9) has shown that the payoffs from the investment in design optimization come less from the mathematical optimization per se, and more from simply attending to methodological problems identified in pilot studies, possibly related to stability of reagents, pipetting technique, or operator-to-operator variability. In fact, one of the most useful exercises is to collect the data as shown in Fig. 6, describing the shape of the variance model. Once the profile of measurement precision is established, it may become immediately apparent that initial designs are attempting to collect data in the wrong region of curve, and would thus be subject to large uncertainty. The remedy is obvious: move the concentration range to more precisely determined values.

Our experience has shown that only one or two rounds of pilot studies are required before the optimal design stabilizes. After this point, the long-term payoff of design optimization begins to accrue. This may take the form of more precise results than before, using roughly equivalent experimental effort, or a reduction in the total experimental effort, reagent costs, or tissue requirements yielding the same overall precision of results. Which form the payoff takes is up to the investigator and the particular needs of the study.

The trend toward increasing laboratory automation and robotics may make optimal design even more useful, as the tedious and error-prone steps of pipetting and separation are turned over to machines. The simple error model (Gaussian distribution) utilized in our work may more accurately describe the behavior of automated equipment than that of manual techniques, whose operating characteristics are less consistent. Full automation of a binding assay would also open up interesting new questions for the proper design of the experiment. While fully automated robots could in theory obtain hundreds of points along the binding isotherm, this might not be the best allocation of resources. Depending on the goals of the assay, increasing replication at fewer dilutions and randomizing the order of the assay may be more beneficial in terms of the ultimate information obtained. For large-scale screening assays, or for assays performed on a routine schedule, the benefits of formal computerized optimization will clearly be measurable.

Although we have addressed only one type of systematic experimental design optimization (D-optimal designs) for ligand binding studies, it is hoped that interested readers will attempt to apply these and related techniques to all forms of quantitative experimental assays. The availability of the computer program DESIGN should serve to remove one barrier to the application of these techniques. The thought process behind each experiment must still come from the investigator; only he or she knows the questions being posed and the suitability of the design of the experiment for answering them.

Acknowledgments

The authors wish to acknowledge assistance with manuscript preparation ably provided by Ms. Erlinda Inejosa-Ortañez.

References

1. P. J. Munson and D. Rodbard, *Anal. Biochem.* **107,** 220 (1980).
2. D. Rodbard, R. H. Lenox, H. L. Wray, and D. Ramseth, *Clin. Chem.* **22,** 350 (1976).
3. R. C. St. John and N. R. Draper, *Technometrics* **17,** 15 (1975).
4. G. E. Rovati, D. Rodbard, and P. J. Munson, *Anal. Biochem.* **174,** 636 (1988).
5. G. E. Rovati, D. Rodbard, and P. J. Munson, *Anal. Biochem.* **184,** 172 (1990).
6. G. E. Rovati, D. Rodbard, and P. J. Munson, *in* "Computers in Endocrinology: Recent Advances" (V. Guardabasso, D. Rodbard, and G. Forti, eds.), Vol. 72, p. 45. Raven, New York.
7. P. J. Munson and D. Rodbard, *J. Recept. Res.* **8,** 533 (1988).
8. E. M. Landaw, *in* "Mathematics and Computers in Biomedical Applications" (J. Eisenfeld and C. De Lisi, eds.), p. 181. Elsevier/North-Holland, New York, 1985.
9. G. E. Rovati, D. Rabin, and P. J. Munson, *in* "Horizons in Endocrinology II" (M. Maggi and V. Geenen, eds.), p. 155. Raven, New York, 1991.

[13] Computerized Analysis of Opioid Receptor Heterogeneity by Ligand Binding in Guinea Pig Brain

Mario Tiberi

Introduction

In molecular pharmacology, the study of receptor protein relies essentially on the binding of radiolabeled drugs to membrane preparations (crude or solubilized) or cytoplasmic preparations (e.g., steroid receptors) derived from specific target tissues. Generally, the radiolabeled compounds are neurotransmitters, hormones, or synthetic analogs (agonists or antagonists) that bind specifically to a given class of receptors. Saturation and competition binding studies have been mostly used to characterize the interaction of drugs with receptors. To get precise and accurate data from these two experimental approaches, investigators should consider two important factors. First, to adequately identify a receptor, one should use a highly selective, pure, stable radioligand with a high specific activity. Second, the choice of method for the data analysis is of great importance. Indeed, data obtained from binding studies are often analyzed using graphical methods. However, such graphical methods are sometimes subjective and may lead to erroneous estimates for the binding parameters. In the past, estimations of the binding parameters using this method was acceptable since no better and more suitable approach was available. In recent years, however, several computerized programs designed to facilitate analysis of the binding data have evolved (1–3). The different programs offer great advantages over the graphical or manual methods. Indeed, computerized curve-fitting programs allow investigators to use a more exact and complex mathematical model than the classic model describing the binding of one ligand to one class of sites.

De Léan *et al.* (4) have elegantly used the computer modeling method to demonstrate that the typical interaction of agonists and antagonists with β-adrenergic receptors was explained by the ternary complex model. The same approach was used to characterize the interaction of ligands with D_2-dopamine receptors in the pituitary gland (5). Computer analyses have also been useful to establish binding properties of ligands to different populations of adrenergic (6, 7), or opioid receptors (8–10). In this chapter, I will use the opioid receptor system as an example to illustrate the resolution of multiple

Methods in Neurosciences, Volume 10

opioid binding sites in the mammalian brain by applying a computerized approach and then compare this with a graphical/manual method.

Opioids mediate their physiological effects in the central nervous system (CNS) and in the periphery through the activation of multiple opioid receptors. Three major groups of opioid receptors have been extensively characterized: these are the mu (μ), delta (δ), and kappa (κ) receptors (11). Other opioid receptor types such as epsilon (ε) and sigma (σ) have been postulated, but their existence as direct opioid receptor entities remains controversial in the minds of several neuroscientists. In addition, several pharmacological and biochemical studies have suggested that within the three major groups of opioid receptors distinct receptor subtypes may exist. Wolozin and Pasternak (12) have proposed the existence of two isoforms for μ receptors. It has also been demonstrated that δ-selective ligands could interact with two different δ-receptor subtypes: δ-type I and δ-type II (13). Finally, several lines of evidence have suggested that opioids could interact with different κ-receptor subtypes (8–10, 14–17).

For several years, opioid research has focused on the synthesis of selective ligands for each of the three main receptor classes. The tritiated agonist [D-Ala^2,MePhe^4,Gly^5-ol]enkephalin (DAGO) binds selectively to the μ receptors (18). The δ receptors can be discriminated by the tritiated agonist [D-Pen^2,D-Pen^5]enkephalin (19), whereas the agonist [^{3}H]U69,593 interacts selectively with the κ_1-receptor subtype (8–10, 20, 21). Other radioligands such as [^{3}H]ethylketazocine (EKC) and [^{3}H]bremazocine can also bind to at least the three major opioid receptor classes. Before the availability of [^{3}H]U69,593 the latter two radioligands had often been used in the presence of unlabeled μ- and δ-selective ligands to discriminate specifically the κ-opioid receptor type (22, 23). Moreover, recent studies using [^{3}H]bremazocine in the presence of unlabeled μ-, δ-, and κ-selective compounds have shown that this nonselective radioligand could delineate an additional population of opioid binding sites that were referred to as the κ_2-receptor subtypes (8–10). In this chapter, I will attempt to illustrate by computer analysis the delineation of three of the major opioid receptor subtypes (μ, δ, and κ_1) in guinea pig brain using the nonselective radioligand [^{3}H]EKC and the unlabeled μ-selective agonist DAGO.

Methods

Guinea Pig Brain Membrane Preparation

Hartley male guinea pigs (300–450 g) were obtained from Charles River Labs (Wilmington, MA). The animals are sacrificed by cervical dislocation, and the whole brain minus the cerebellum is rapidly isolated. The brain is then cut into pieces and homogenized with a Polytron homogenizer (speed 4, 30 sec) in 50 mM Tris-HCl (pH 7.4 at 4°C). The homogenate is centrifuged at 49,000 g for 10 min in a Beckman RC-5B centrifuge. The supernatant is discarded, and the pellet is resuspended and centrifuged a second time under the same conditions. The final pellet is resuspended in 50 mM Tris-HCl (pH 7.4 at 25°C) at a final dilution of 1 g original wet weight per 100 ml of buffer.

Binding Assay

Binding assays are performed with 1.9 ml of membrane preparation incubated at 25°C for 40 min in a final volume of 2 ml. Under these conditions, equilibrium is reached. At the end of incubation period, membrane preparations are rapidly vacuum-filtered on glass fiber filters (GF/B; Whatman, Clifton, NJ) to separate bound and free radioligand. To reduce the nonspecific binding, filters are presoaked in 0.3% polyethyleneimine for 3 hr prior to use. Filters are then washed 3 times with 5 ml of ice-cold Tris-HCl buffer (pH 7.4 at 4°C) and immersed in 10 ml of scintillation fluid (Universol Cocktail, ICN Biochemicals, Costa Mesa, CA). Radioactivity retained on filters is counted at about 40% efficiency in a beta counter (Packard).

Competition experiments are performed by incubation of increasing concentrations of unlabeled $(-)$-EKC (0.2–100 nM) or unlabeled DAGO (0.5–64,000 nM) against a constant concentration ($\sim$0.94 nM) of [^{3}H]EKC (27 Ci/mmol; New England Nuclear, Boston, MA). Total and nonspecific binding are determined in the presence and absence of 10 μM $(-)$-cyclazocine. Under such experimental conditions, specific binding is 90%.

Theoretical Background

Graphical Method

The simplest binding model is defined as a bimolecular reaction according to the mass action law. At equilibrium, the interaction of a ligand with a single class of receptors is described by the following equation:

$$L + R \underset{k_{off}}{\overset{k_{on}}{\rightleftharpoons}} LR \tag{1}$$

where L is the free ligand, R is the free or unoccupied receptor, and LR is the ligand–receptor complex. k_{on} and k_{off} are association and dissociation rate constants, respectively. At equilibrium the association and dissociation reactions are occurring at the same rate, and thus the affinity constant (K_A) of a ligand for a receptor is defined as

$$K_A = [LR]/[L][R] \tag{2}$$

where square brackets indicate concentrations (the amount of ligand bound to the receptor is also referred to as [B]).

The affinity of an unlabeled ligand for a class of receptors or binding sites is determined by monitoring the displacement of the radioligand bound to the receptor. This is done by incubating a fixed concentration of receptors and radioligand with increasing concentrations of the unlabeled ligand. The binding data obtained are then plotted using a log scale for the x axis (which corresponds to concentrations of the unlabeled ligand) and a linear scale for the y axis (which corresponds to the amount of radioligand bound expressed either as a percentage of total or as an absolute value). Generally, this type of plot gives a sigmoidal (S-shaped) curve. From this plot, we can roughly estimate the affinity of the unlabeled ligand by measuring the concentration of ligand necessary to inhibit by 50% the binding of the radioligand to the receptor. This is referred to as the midpoint location or IC_{50} value.

The IC_{50} value can also be estimated more accurately using a linear transformation of the displacement curve. This method is also called a logit–log transformation, which is analogous to the graphical representation of the Hill equation. The method uses a log scale for the x axis (which corresponds to the concentrations of unlabeled ligand) and a logit scale for the y axis for which the points are calculated as follows:

$$\log\{[B_i] - [NS]/([B_0] - [NS]) - ([B_i] - [NS])\} \tag{3}$$

where $[B_i]$ is the concentration of radioligand bound at a given concentration of unlabeled ligand, [NS] is the amount of nonspecific binding, and $[B_0]$ is the total concentration of radioligand bound. These transformed binding data are then fitted by a simple linear regression. The slope (or pseudo-Hill coefficient) of such a plot gives an indication of the type of interaction that best describes the ligand–receptor binding model. A slope factor equal to 1 is indicative of a simple bimolecular reaction with a homogeneous class of receptors or binding sites. A slope factor smaller than 1 suggests an interaction with a heterogeneous class of receptors or binding sites (negative cooperativity). Finally, a slope value greater than 1 may indicate positive cooperativity.

As mentioned before, some investigators will use the IC_{50} value as an estimate of the affinity constant or inhibitory constant (K_i) of the unlabeled ligand for the receptor. This is a wrong assumption since the IC_{50} value relies on a complex relationship to the K_i but also to the K_D value (affinity constant of the radioligand), R value (binding site capacity), and the concentration of radioligand used. To integrate all the different parameters, some investigators have routinely applied the Cheng–Prusoff correction method (24) to adjust the IC_{50} value for the K_D value and the radioligand concentration as indicated:

$$K_i \cong IC_{50}/(1 + F/K_D) \tag{4}$$

where F is the amount of free radioligand. However, this correction method may also lead to erroneous estimations (25).

Computerized Analysis

ALLFIT and LIGAND programs (1, 2) offer a better alternative for the analysis of binding data. These programs analyze the binding curves using a general weighted nonlinear and least-squares curve-fitting method that uses the Marquardt–Levenberg modification of the Gauss–Newton algorithm (2). Prior to the analysis, a file should be created that contains the amount of radioligand bound ([B]) corresponding to the concentration of unlabeled ligand added. Values of [B] may be expressed either as concentrations or as counts per minute (cpm). In the latter case, the user should bear in mind that a conversion factor is essential to obtain the final parameters in concentration units.

Generally, the fitting strategy of competition curves involves first the use of ALLFIT, which analyzes the binding curves based on a four-parameter logistic equation (1):

$$Y = D + (A - D)/(1 + (X/C)^B) \tag{5}$$

where X and Y are the concentration of unlabeled ligand and radioligand bound, respectively, C is the IC_{50} value, B is the slope factor that will be used to assess the curve steepness, A is the upper limit of the curve when $X = 0$, and D is the lower limit of the curve when X is infinite. Then, the binding curves are analyzed using LIGAND (previously known as SCAFIT) according to a mathematical "n by m" model developed by Feldman (26). This mathematical model is based on the mass action law and describes the interaction of any number of ligands (n) with any number of independent classes of binding sites or receptors. The final parameters obtained with

ALLFIT may be employed as initial estimates when using LIGAND program and may provide insights for selecting the best fitting model.

The computerized curve-fitting programs are also designed to take into account the unequal precision obtained for the determination of bound radioligand on different points of the curve. Indeed, the deviations of the experimental points (for each bound radioligand determination) from their predicted values, that is, residuals, are weighted according to the reciprocal of their expected variance (Var) according to the formula previously described (1, 2):

$$\mathrm{Var}(B) = a_0 + a_1(B^{a_2}) \tag{6}$$

where a_0 corresponds to a small constant which is used to prevent the variance estimate from becoming too small when B is close to 0, a_2 is an exponent which ranges from 1 to 2, and a_1 is a proportionality constant about equal to the square of the relative error of B (4). The statistical difference between the weighted residual variance obtained with different n by m models (one-site, two-site, ...) is tested by comparing the different residuals using the so-called "extra sum of squares" principle (2):

$$F = [(SS_1 + SS_2)/(\mathrm{df}_1 - \mathrm{df}_2)]/(SS_2/df_2) \tag{7}$$

where SS_1 and SS_2 are referred to as the sum of squares of residuals for the fit with the simpler n by m model (less parameters) and the more complex n by m model (more parameters), respectively, and df_1 and df_2 are the number of degrees of freedom assigned for the different fits. The calculated F ratio is compared with a tabulated F-statistic value with ($\mathrm{df}_1 - \mathrm{df}_2$) degrees of freedom for the numerator and df_2 degrees of freedom for the denominator. The main goal of this statistical test is to evaluate the goodness of fit of a simpler model over a more complex model.

Once the fitting has been accomplished, it is very important to assess the residuals. Regardless of the fact that one may have obtained the best possible fit for a given n by m model, it is still possible that this corresponds to an inappropriate model based on the distribution of the residuals for randomness. LIGAND provides a test for randomness, the "sign runs test" which is based on the distribution of signs (+ or −) of the residuals. For instance, if too many points are found to be clustered above or below the fitted curve, the runs test would indicate that the fitting model is inadequate.

Results and Discussion

Figure 1 shows the graphical representation of competition for [³H]EKC binding by the nonselective agonist EKC and the μ-selective agonist DAGO, the competition curve of unlabeled EKC for the binding of [³H]EKC gives a S-shaped curve (Fig. 1A). However, the competition curve of unlabeled DAGO is represented by a shallow curve (Fig. 1B). The comparison of the two curves suggests that competition for [³H]EKC binding by DAGO is indicative of a complex interaction with opioid receptors in the guinea pig brain. Graphical determination of IC_{50} values indicates that EKC and DAGO have IC_{50} values of about 3 and 300 nM, respectively (Fig. 1).

A logit–log transformation was performed on both competition curves and then fitted by a simple linear regression to obtain with more accuracy the IC_{50} values and slope factors (Fig. 2). Using this mathematical transformation, we were able to calculate an IC_{50} value of about 4 nM for EKC with a slope factor of 0.83 (Fig. 2A). This slope factor value is slightly lower than 1 suggesting that unlabeled EKC interacts with more than one class of receptors or binding sites in guinea pig brain. Indeed, it has been shown that [³H]EKC binds to different classes of opioid receptors in guinea pig brain (8). The different binding affinities represent the interaction of EKC at three opioid receptors rather than different affinity states of the same receptor (8, 22). Logit–log transformation of the competition curve of DAGO for [³H]EKC binding gives an IC_{50} value of 296 nM and a slope factor of 0.3, which is lower than 1 (Fig. 2B). This suggests that DAGO interacts with more than one class of opioid receptors in the guinea pig brain.

This mathematical approach seems to offer one major advantage over the simple graphical method. Using this method, we are able to measure a slope factor that is important in defining the type of interaction of the unlabeled ligand with the bound receptor. However, these results indicate that the mathematical and graphical methods are equivalent for the determination of IC_{50} values. In fact, the respective IC_{50} values for EKC and DAGO are similar. K_i values have been calculated according to the Cheng–Prusoff correction method and are listed in Table I.

The analysis of these binding data by ALLFIT gives similar results (Figs. 3 and 4). However, the different steps involved in the determination of the binding parameters are less tedious and cumbersome when using ALLFIT. In addition to the fact that computerized programs provide more accurate estimates, the speed of processing the raw binding data is also faster than the graphical method. Computer modeling of the competition curves for EKC and DAGO by ALLFIT indicates that the slope factors are 0.83 and 0.25. Interestingly, the upper limit (parameter A) of the two curves are very close but the lower limit (parameter B) are different. Because the two curves

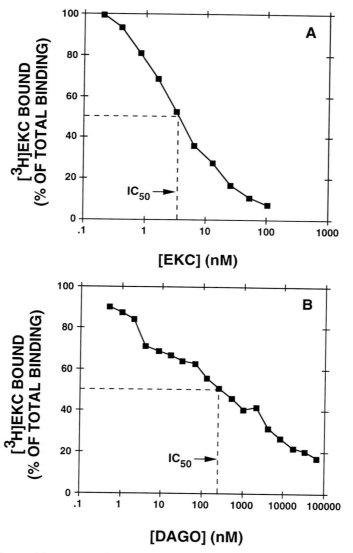

FIG. 1   Competition curve of unlabeled EKC (A) and unlabeled DAGO (B) for the binding of [³H]EKC. Competition curves were obtained with 10–18 points, each done in triplicate. The IC_{50} values measured graphically for EKC and DAGO correspond to about 4 and 300 nM, respectively.

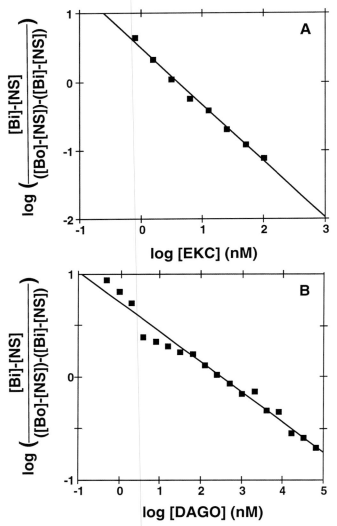

Fig. 2   Logit–log linear transformation of competition curves of unlabeled EKC (A) and unlabeled DAGO (B) for the binding of [³H]EKC. The simple linear regression of the logit–log transformation of the EKC and DAGO competition curves is described by $\log Y = -0.825 \log X + 0.496$ ($r^2 = 0.995$, $p < 0.05$) and $\log Y = -0.297 \log X + 0.734$ ($r^2 = 0.997$, $p < 0.05$), respectively.

TABLE I IC_{50} and K_i Values Determined by
Linear Logit–Log Transformation of
Competition Curves and Correction
of Cheng and Prusoff[a]

Ligand	IC_{50} values (nM)	K_i values (nM)	$1/K_i$ values (nM^{-1})
EKC	4	1.69	0.591
DAGO	296	126	0.008

[a] IC_{50} values were calculated from the equations described in the legend of Fig. 2. K_i values were determined by adjusting the IC_{50} values with the Cheng–Prusoff correction method [see Eq. (4)]. For the purpose of these calculations, the amount of free radioligand (F) was assumed to be roughly equal to the total concentration of radioligand used (~0.94 nM). The K_D value was established to 0.65 nM. This value has been shown to represent a rough estimate of the affinity of [^{3}H]EKC for μ-, δ-, and κ-opioid receptors (22, 23).

have been performed in the same membrane preparation using the same radioligand concentration, the two parameters should be equivalent.

If we look more carefully at the computer modeling of the competition of DAGO for [^{3}H]EKC binding, the runs test indicates that the residuals are randomly distributed along the fitted curve. This may be interpreted as an adequate model. However, the slope factor is inferior to 1, suggesting that DAGO interacts with more than one class of receptors or binding sites. Consequently, it becomes evident that a one-site model is not appropriate in this case, and this is supported by the fact that the IC_{50} value (parameter C) and the lower limit of the curve (D) have ±80% and ±400% errors, respectively Table II. To further analyze the binding curves, we can use simultaneous curve fitting of the two curves by unconstrained and constrained fitting. Unconstrained curve fitting for the unlabeled ligands did not modify significantly the final parameter estimates but slightly decreased the standard error (Table III). We can constrain both competition curves to share identical values for A and D parameters since they have been performed using the same membrane preparation and radioligand concentration. The upper and lower limits of these curves should then be identical. This constrained analysis did not result in a more suitable fit based on the goodness of fit. However, the use of these constraints reduced significantly the standard errors of the final parameters. Indeed, the IC_{50} and parameter C values for the DAGO curve now have ±24% and ±10% errors, respectively. Each of the two curves has a slope factor significantly lower than 1. We can also infer that the two curves are not parallel because if we constrain the curves to share

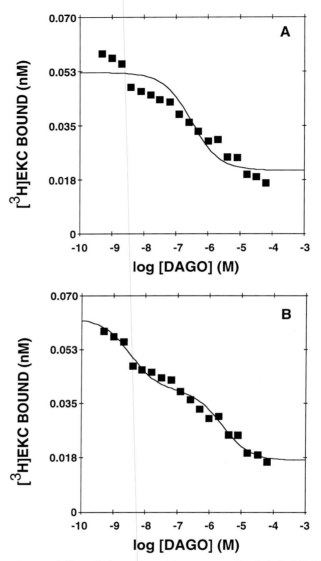

FIG. 3   Computer modeling of the competition curve of unlabeled DAGO for the binding of [^{3}H]EKC. The competition curve was fitted to a one-site model (A), a two-site model (B), and a three-site model (C) using LIGAND. The solid line represents the fitted curve, and squares correspond to the actual data points. A three-site model results in a significant improvement of the goodness of fit. Residuals were not randomly distributed in the case of the one-site model ($p < 0.01$) but were randomly

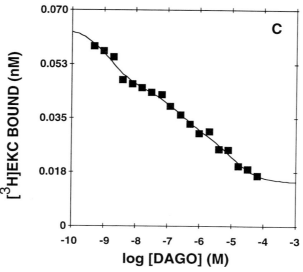

distributed for the two- and three-site models. Equilibrium binding affinity constants for [³H]EKC were constrained to remain constant at predetermined values of 5.26 nM^{-1} for the first site and 0.42 nM^{-1} for the second and third sites. The values of a_0, a_1, and a_2 weighting parameters were set to 1×10^{-6}, 1×10^{-3}, and 1.5, respectively.

identical values for the slope factor this results in a deleterious effect on the goodness of fit (data not shown). This may indicate that the two unlabeled ligands have distinct binding models.

To assess the best binding model further, the curves were analyzed independently and simultaneously using the LIGAND program. By fitting each curve independently, results indicate clearly that EKC and DAGO competition curves are fitted to a binding model describing an interaction of the unlabeled ligand with more than one class of receptors or binding sites. A binding model with two distinct classes of binding sites gives the best fit for the EKC competition curve (Table IV). Meanwhile, the competition of DAGO for [³H]EKC binding is fitted best to a three-site model (Table IV). To accomplish these different fits, the equilibrium binding affinity constants of [³H]EKC at the first site (K_1) and second site (K_2) were set constant to values of 5.26 and 0.42 nM^{-1}, respectively. These values were obtained from previous saturation studies (8). In the case of the DAGO experiment, K_2 and K_3 values of [³H]EKC were set constant and equal. By LIGAND analysis, the fitting of the DAGO competition curve to a one-site model did not result in a satisfactory determination of the binding parameters. As shown in Fig. 3A,

TABLE II   Binding Parameter Estimates Using ALLFIT Program[a]

| Ligand | Binding parameters ± standard errors | | | | Residuals | | Runs |
	A (nM)	B (nM)	C (nM)	D (nM)	+	–	
EKC	6.72E − 02 ± 1.71E − 03	8.29E − 01 ± 4.71E − 02	3.34E + 00 ± 3.45E − 01	6.90 E − 03 ± 5.80E − 04	8	4	5 ($p > 0.05$)
DAGO	6.54E − 02 ± 2.49E − 03	2.47E − 01 ± 3.68E − 02	5.06E + 02 ± 4.15E + 02	1.63E − 03 ± 6.21E − 03	6	13	9 ($p > 0.05$)

[a] Each competition curve was analyzed independently according to a four-parameter logistic equation [see Eq. (5)] using an unconstrained curve fitting (no shared parameters). A, upper limit of the curve (total binding); B, slope factor; C, IC_{50} value; D_X, lower limit of the curve (nonspecific binding). $E − X$ or $E + X$ refer to the antilogarithm (10^X).

TABLE III   Binding Parameter Estimates by Simultaneous Curve-Fitting Using ALLFIT Program[a]

| Ligand | Binding parameters ± standard errors | | | | Residuals | | Runs |
	A (nM)	B (nM)	C (nM)	D (nM)	+	−	
FIT 1							
EKC	6.60E − 02 ± 1.44E − 03	8.40E − 01 ± 5.16E − 02	3.56E + 00 ± 3.53E − 01	6.82E − 03 ± 6.61E − 04	7	5	5 ($p > 0.05$)
DAGO	6.60E − 02 ± 1.44E − 03	2.72E − 01 ± 1.36E − 02	2.15E + 02 ± 5.26E + 01	6.81E − 03 ± 6.62E − 04	9	10	9 ($p > 0.05$)
FIT 2							
EKC	6.72 E − 02 ± 1.98E − 03	8.29E − 01 ± 5.46E − 02	3.34E + 00 ± 4.01E − 01	6.90E − 03 ± 6.72E − 04	8	4	5 (p > 0.05)
DAGO	6.54E − 02 ± 2.33E − 03	2.47E − 01 ± 3.46E − 02	5.04E + 02 ± 3.86E + 02	1.70E − 03 ± 5.80E − 03	7	12	9 ($p > 0.05$)

[a] Competition curves were fitted simultaneously using a four-parameter logistic equation [see Eq. (5)]. FIT 1 describes the binding data derived from a constrained curve fitting for which A nd D parameters of each curves shared common values, respectively. FIT 2 corresponds to the results obtained for the inconstrained curve fitting (no shared parameters). The goodness of fit shows no significant improvement of FIT 2 over FIT 1 ($F = 1.09$; $p = 0.36$).

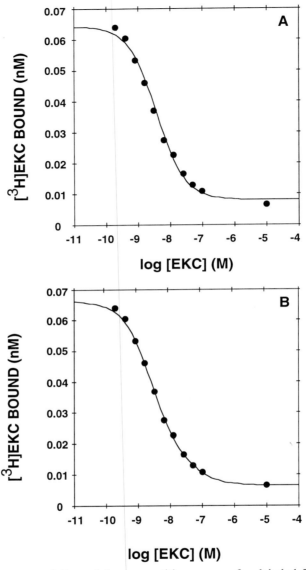

FIG. 4 Computer modeling of the competition curve of unlabeled EKC for the binding of [³H]EKC. The competition curve was fitted to a one-site model (A) and a two-site model (B) using LIGAND. The solid line represents the fitted curve, and circles correspond to the actual data points. A two-site model results in a significant improvement of the goodness of the fit. The residuals were randomly distributed

the fitted points are not randomly distributed along the curve. This is confirmed by the runs test which is statistically significant, suggesting that the final binding parameters are inadequate.

Figure 3B shows an improvement in the fit when we select a two-site model to explain the experimental data. The runs test is not statistically significant, suggesting that the residuals are randomly distributed along the fitted curve. However, when we look carefully at this curve, it is still possible to improve the curve fit. Indeed, as mentioned before [³H]HKC interacts with three distinct opioid receptors in guinea pig brain. Moreover, the R value obtained by saturation studies using this radioligand is higher than the total R value $(R_1 + R_2)$ obtained with the competition curve fitted to a two-site model (Table IV). By fitting the DAGO curve to a three-site model, it was possible to obtain a major improvement of the goodness of fit and a decrease of the root mean square (rms) error, suggesting that this binding model was the most appropriate (Table IV; Fig. 3C).

Independent analysis of the EKC competition curve modeled to a two-site model results in an improvement of the goodness of fit over the one-site model (Fig. 4). This confirms previous results showing that [³H]EKC could bind to high- and low-affinity sites in guinea pig brain (8). However, these results have established that [³H]EKC has binding affinity constants of 5.26 and 0.42 nM^{-1} at the high- and low-affinity sites, respectively. In the present study, those values are 2.29 and 0.025 nM^{-1}, respectively. The discrepancies obtained between the two values for the low-affinity sites are likely to be explained by the fact that the EKC competition is done with fewer points than the saturation studies. Furthermore, the total R value obtained by fitting the EKC competition curve to a two-site model is lower than the R value obtained by saturation studies.

To improve the assessment of the final parameters for both competition curves, we have used simultaneous curve-fitting analysis. In fact, by pooling the two curves we increase the number of points used for the computerized fitting. In this specific example, the number of points increased from 12 to 31. Using this strategy, the EKC competition curve is modeled preferentially to a two-site model with binding affinity constants that are in good agreement with the ones obtained previously by saturation studies (8, 22, 23). Moreover, the DAGO competition curve is still fitted to a three-site model which is

---

around the fitted curve in both fits. Equilibrium binding affinity constants for [³H]EKC were constrained to remain constant at 5.26 nM^{-1} for the first site and 0.42 nM^{-1} for the second site. The values of a_0, a_1, and a_2 weighting parameters were set to 1×10^{-6}, 1×10^{-3}, and 1.5, respectively.

TABLE IV   Binding Parameter Estimates Using LIGAND Program[a]

Ligand	Binding model	K_1 (nM^{-1})	K_2 (nM^{-1})	K_3 (nM^{-1})	R_1 (nM)	R_2 (nM)	R_3 (nM)	N	rms (%)
					Binding parameters ± standard errors				
EKC	One-site	1.56E + 00 ± 1.70E − 01	N.D.[b]	N.D.	6.91E − 02 ± 2.08E − 03	N.D.	N.D.	8.65E − 03 ± 6.85E − 04	0.7
EKC	Two-site	2.29E + 00 ± 3.62E − 01	2.48E − 02 ± 1.84E − 02	N.D.	6.42E − 02 ± 3.51E − 03	2.74E − 02 ± 1.17E − 02	N.D.	6.98E − 03 ± 5.05E − 04	0.4
DAGO	One-site	1.95E − 02 ± 8.09E − 03	N.D.	N.D.	3.92E − 02 ± 3.26E − 03	N.D.	N.D.	2.23E − 02 ± 1.86E − 03	1.7
DAGO	Two-site	4.22E − 01 ± 1.69E − 01	2.36E − 03 ± 6.76E − 04	N.D.	8.81E − 02 ± 9.11E − 03	2.80E − 02 ± 1.82E − 03	N.D.	1.80E − 02 ± 1.16E − 03	0.7
DAGO	Three-site	9.53E − 01 ± 4.16E − 01	2.83E − 02 ± 2.06E − 02	1.45E − 04 ± 8.48E − 05	7.62E − 02 ± 9.22E − 03	1.76E − 02 ± 3.35E − 03	5.83E − 02 ± 9.27E − 03	1.54E − 02 ± 1.71E − 03	0.5

[a] Each curve was fitted independently using a general and weighted nonlinear curve-fitting method (2). The interaction of the unlabeled ligand with bound receptors was described best by binding models involving more than one class of receptors. In the case of EKC curve, a two-site model was significantly better than a one-site model ($F = 12.79$, $p = 0.00$). On the other hand, the interaction of DAGO with bound receptors was characterized by a three-site model which was found to be significantly better than a two-site model ($F = 8.53$, $p = 0.00$) or a one-site model ($F = 45.42$, $p = 0.00$). K_1, K_2, and K_3 denote the equilibrium binding affinity constant to site 1, site 2, and site 3, respectively. R_1, R_2, and R_3 represent the binding capacity of each site. N denotes the bound/free ratio for nonspecific binding of each ligand. Root mean square (rms) error is the square root of the sum of squares divided by the number of degrees of freedom, and it represents the overall deviation of the measured bound receptor to the fitted curve.

[b] N.D., Not determined.

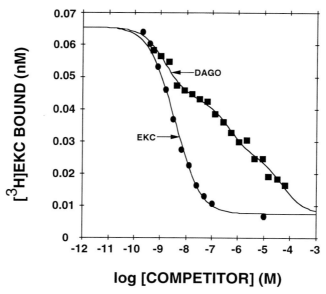

log [COMPETITOR] (M)

FIG. 5 Simultaneous analysis of unlabeled EKC and DAGO competition curves. EKC and DAGO competition curves were fitted to a two-site and three-site model, respectively. These binding models have increased significantly the goodness of fit over a one-site or two-site model. Solid lines represent the fitted curves, and circles (EKC) and squares (DAGO) are the actual data points. Equilibrium binding affinity constants for [^{3}H]EKC were constrained to remain constant at predetermined values of 5.26 nM^{-1} for the first site and 0.42 nM^{-1} for the second and third sites. The values of a_0, a_1, and a_2 weighting parameters were set to 1×10^{-6}, 1×10^{-3}, and 1.5, respectively.

statistically improved over the one- and two-site models (Fig. 5). The total binding capacity obtained using the simultaneous fitting is close to the binding capacity measured by saturation studies of [^{3}H]EKC (Table V). Thus, DAGO can bind to three populations of opioid receptor sites with different binding affinity constants. Based on this analysis, DAGO has an affinity of 3.76 nM^{-1} at the μ sites, 2.93×10^{-3} at the δ sites, and 2.73×10^{-5} nM^{-1} at the κ sites. These affinity binding constants agree well with values previously published for DAGO at these three opioid receptors (18, 19, 21–23). The total binding capacity obtained with the simultaneous fitting is 0.1590 nM. Consequently, the relative proportions of the μ-, δ-, and κ-opioid receptors are 18, 40, and 42%, respectively. Others studies have reported similar results (18, 19, 21–23, 27). It is worth mentioning that the κ-receptor subtype delineated by

TABLE V   Binding Parameter Estimates by Simultaneous Curve Fitting Using LIGAND Program[a]

| Binding model | Ligand | Binding parameters ± standard errors | | | | | | | |
		K_1 (nM^{-1})	K_2 (nM^{-1})	K_3 (nM^{-1})	R_1 (nM)	R_2 (nM)	R_3 (nM)	N	rms (%)
One-site	EKC	1.41E + 00 ± 5.44E − 01	N.D.	N.D.	4.98E − 02 ± 3.47E − 03	N.D.	N.D.	1.40E − 02 ± 1.97E − 03	2.4
	DAGO	5.18E − 03 ± 2.03E − 03	N.D.	N.D.	4.98E − 02 ± 3.47E − 03	N.D.	N.D.	1.40E − 02 ± 1.97E − 03	
Two-site	EKC	2.49E + 00 ± 1.71E + 00	2.44E − 01 ± 1.51E − 01	N.D.	3.52E − 02 ± 2.95E − 03	9.73E − 02 ± 6.04E − 03	N.D.	9.24E − 03 ± 1.11E − 03	1.0
	DAGO	9.69E − 01 ± 3.56E − 01	1.04E − 04 ± 2.50E − 05	N.D.	3.52E − 02 ± 2.95E − 03	9.73E − 02 ± 6.04E − 03	N.D.	9.24E − 02 ± 1.11E − 03	
Three-site	EKC	4.71E + 00 ± 1.67E + 00	2.15E − 01 ± 3.42E − 02	2.15E − 01 ± 3.42E − 02	2.84E − 02 ± 2.47E − 03	6.70E − 02 ± 7.01E − 03	6.36E − 02 ± 6.77E − 03	8.10E − 03 ± 6.36E − 04	0.6
	DAGO	3.76E + 00 ± 1.28E + 00	2.93E − 03 ± 1.27E − 03	2.73E − 05 ± 8.57E − 06	2.84E − 02 ± 2.47E − 03	6.70E − 02 ± 7.01E − 03	6.36E − 02 ± 6.77E − 03	8.10E − 03 ± 6.36E − 04	

[a] Competition curves were fitted simultaneously using a general and weighted nonlinear curve-fitting method (2). A three-site model was significantly better than a two-site model ($F = 88.55$, $p = 0.00$) or one-site model ($F = 26.17$, $p = 0.00$). For the three-site model, the binding affinity constant of EKC at sites 2 and 3 was constrained to share a common value. Abbreviations and symbols are explained in footnote a of Table IV.

[^{3}H]EKC in the present study is likely to be the κ_1-subtype as demonstrated elsewhere (8).

Identical results (data not shown) were also obtained using the curve-fitting programs EBDA/LIGAND that have been written (EBDA) or adapted (LIGAND) for utilization on microcomputers by McPherson (3). The binding data are first processed by the EBDA program, which is reminiscent of ALLFIT. However, EBDA does not offer the same flexibility and advantages. Indeed, the only parameter that can be constrained is the nonspecific binding (parameter D). Moreover, EBDA cannot perform simultaneous curve fitting. Despite this, EBDA remains a good program to get insights about the binding model and also to obtain binding parameters that can be used as initial estimates by LIGAND.

Conclusion

Computerized curve-fitting programs offer a better alternative over the graphical fitting of binding curves. As we have demonstrated with a specific example, the use of such computerized methods enables us to improve the estimates of the different parameters and consequently helps us to determine the best binding model applicable to the experimental data. This is supported by the results reported in Table I for the linear transformation of competition curves. Reciprocals of K_i values were calculated and compared with the affinity constants obtained by LIGAND (Tables IV and V). Differences between the two methods were at least of the order of 2- to 4-fold. In addition, determination of binding affinities for two or more sites by linear transformation of binding curves cannot be assessed easily and accurately.

Finally, for those working in the field of opioid receptors, computerized analysis of the binding curves will help to better characterize the plethora of receptor subtypes. Indeed the combined use of nonselective and selective ligands remains for many investigators the best and sometimes only approach to study the multiplicity of opioid receptors in the CNS. Undoubtedly, the development of highly selective agonists and antagonists for the different proposed opioid receptor subtypes (the κ receptors, for instance) is essential to better understand the exact mechanisms by which opioids interact at these different receptor subtypes.

Acknowledgments

The author acknowledges Dr. Jacques Magnan who kindly reviewed this chapter and provided helpful suggestions. The author also thanks Dr. Susanna Cotecchia for critical reading of the manuscript. Mario Tiberi is a fellow of the Medical Research Council of Canada.

References

1. A. De Léan, P. J. Munson, and D. Rodbard, *Am. J. Physiol.* **235,** E97 (1978).
2. P. J. Munson and D. Rodbard, *Anal. Biochem.* **107,** 220 (1980).
3. G. A. McPherson, *J. Pharmacol. Methods* **14,** 317 (1985).
4. A. De Léan, J. M. Stadel, and R. J. Lefkowitz, *J. Biol. Chem.* **255,** 7108 (1980).
5. K. A. Wreggett and A. De Léan, *Mol. Pharmacol.* **26,** 214 (1984).
6. A. De Léan, A. A. Hancock, and R. J. Lefkowitz, *Mol. Pharmacol.* **21,** 5 (1982).
7. R. S. Kent, A. De Léan, and R. J. Lefkowitz, *Mol. Pharmacol.* **17,** 14 (1980).
8. M. Tiberi and J. Magnan, *Can. J. Physiol. Pharmacol.* **67,** 1336 (1989).
9. M. Tiberi and J. Magnan, *Eur. J. Pharmacol.* **188,** 379 (1990).
10. M. Tiberi and J. Magnan, *Mol. Pharmacol.* **37,** 694 (1990).
11. S. J. Paterson, L. E. Robson, and H. W. Kosterlitz, *Br. Med. Bull.* **39,** 31 (1983).
12. B. L. Wolozin and G. W. Pasternak, *Proc. Natl. Acad. Sci. U.S.A.* **78,** 6181 (1981).
13. R. B. Rothman, W. D. Bowen, V. Bykov, U. K. Schumacher, C. B. Pert, A. E. Jacobson, T. R. Burke, and K. C. Rice, *Neuropeptides* **4,** 201 (1984).
14. B. Attali, C. Gouardères, H. Mazarguil, Y. Audigier, and J. Cros, *Neuropeptides* **3,** 53 (1982).
15. R. S. Zukin, M. Eghbali, D. Olive, E. M. Unterwald, and A. Tempel, *Proc. Natl. Acad. Sci. U.S.A.* **88,** 4061 (1988).
16. E. Castanas, P. Giraud, N. Bourhim, P. Cantau, and C. Oliver, *Neuropeptides* **5,** 133 (1984).
17. E. Castanas, N. Bourhim, P. Giraud, F. Boudouresque, P. Cantau, and C. Oliver, *J. Neurochem.* **45,** 688 (1985).
18. H. W. Kosterlitz and S. J. Paterson, *Br. J. Pharmacol.* **78,** 299P (1981).
19. R. Cotton, H. W. Kosterlitz, S. J. Paterson, M. J. Rance, and J. R. Traynor, *Br. J. Pharmacol.* **84,** 927 (1985).
20. R. A. Lahti, M. M. Mickelson, J. M. McCall, and P. F. von Voigtlander, *Eur. J. Pharmacol.* **109,** 281 (1985).
21. M. Tiberi, P. Payette, R. Mongeau, and J. Magnan, *Can. J. Physiol. Pharmacol.* **66,** 1368 (1988).
22. H. W. Kosterlitz, S. J. Paterson, and L. E. Robson, *Br. J. Pharmacol.* **73,** 939 (1981).
23. J. Magnan, S. J. Paterson, A. Tavani, and H. W. Kosterlitz, *Naunyn Schmiedeberg's Arch. Pharmacol.* **319,** 197 (1982).
24. Y.-C. Cheng and W. H. Prusoff, *Biochem. Pharmacol.* **22,** 3099 (1973).
25. P. J. Munson and D. Rodbard, *J. Recept. Res.* **8,** 533 (1988).
26. H. A. Feldman, *Anal. Biochem.* **48,** 317 (1972).
27. L. E. Robson, M. G. C. Gillan, and H. W. Kosterlitz, *Eur. J. Pharmacol.* **112,** 65 (1985).

[14]   Identification of Spikes by Computer

F. Delcomyn and J. H. Cocatre-Zilgien

Introduction

For many years, computer use in neurophysiological laboratories was rare and limited to general data processing, because the only available computers were slow, extremely expensive, and had limited memory. The astonishing increase in computing speed and memory during the 1980s, coupled with an equally astonishing drop in price, has changed this situation. The large amounts of digitized data created at the high sampling rates required for neurophysiological studies can now be processed on a microcomputer in real time for several channels simultaneously. This ready availability of computing power has resulted in a proliferation of computer methods, both custom and commercial, for the processing of neurophysiological data.

The main purpose of this chapter is to provide an overview of the factors that an investigator has to consider when selecting a commercial spike identification system or when designing a customized one. When applicable, we will illustrate the issues by describing features of our program Spike Finder (1). Discussion of more general principles of spike identification and sorting can be found in Glaser and Ruchkin (2), Abeles and Goldstein (3), and Schmidt (4).

Preliminary Choices

It is important to realize at the outset that, for several reasons, there is no universal method for spike analysis. First, data from different sources can be quite different from one another. For example, muscle potentials from electromyographic (EMG) recordings have rather different temporal characteristics than do action potentials from nerves. Hence, a program designed specifically to detect nerve spikes may not do so well with muscle potentials, or vice versa. Second, the aims of the detection may be different in different cases. For example, in electrocardiogram recordings, the focus of the analysis can be on the detection of the small P or T waves that frame the QRS complex, or on the rapid detection of QRS spike anomalies due to ectopic pacemakers. It is unlikely that a single system would work well for both purposes. Similarly, systems for the processing of electroencephalograms may be tailored to pick out "spike waves" for the detection of seizures, or to perform a long-term analysis of the frequency components of the recordings. The constraints imposed on a system are so different for each category of electrical data that it is critical to determine before anything else what the final use of the data will be, so as to select the best system to analyze them.

Several decisions need to be made before actual details of spike analysis are considered. First, one must choose between on-line and off-line analysis. The difference between these is that on-line systems work in real time, as data are being generated, whereas off-line systems do not, although the distinction between the two is somewhat artificial as far as the principles used for spike detection and classification are concerned. There are also hybrid systems that use a preprocessor or other hardware to make a preliminary analysis and save only the elements of interest. An on-line system may need a period of training or adjustment prior to the collection of the data itself. If data can change over time, such a system may require retraining, a drawback while the experiment is under way. There is also little control over what the system is doing in real time, especially when all attention may be absorbed by the conduct of the experiment itself. Real-time analysis is therefore best used for relatively stereotyped spikes. An off-line system will allow a more thorough analysis because data can be read for analysis more than once, and different features analyzed each pass. This is a definite advantage if data are complex.

A second decision that must be made is the resolution of the amplitude of the digitized signal. Properly speaking, this is a function of the hardware selected for digitization. If a signal is digitized into 12 bits, 4096 digital values are available to characterize voltages between the maximum and minimum allowable signal amplitude; if 16 bits are used, 65,536 values are available, hence providing 16 times the resolution. However, more bits of resolution also means slower digitization. In our experience, a resolution of 12 bits is adequate even for a mixture of large and small muscle potentials.

A third choice is that of sampling rate. This parameter can usually be adjusted by software just before data collection begins. The minimum sampling rate should be twice the high-frequency component of a spike, to ensure that at least one sampling point will occur during the spike itself, but at that rate spikes are defined by simple triangles. A rate of 5 kHz is acceptable for EMG recordings, but up to 30 kHz may be needed for the separation of different classes of nerve spikes. This represents a substantial throughput: 3.6 megabytes/min per channel. Saving data at that rate may severely limit the number of channels that can be sampled simultaneously. In some cases (see below), higher sampling rates are not always better.

A final consideration in selecting or designing a spike identification program is that it must be easy to use. At first glance, it may seem that a convenient and visually pleasing user interface may have only cosmetic value, but this is not at all the case. A well-designed user interface is critical not so much for preventing the user from crashing the program by pressing the wrong key, but for ensuring that the user will not avoid using features of the program because they are too complex, nonintuitive, or slow to use. By nature, spike data are very repetitive, one might even say boring. Specific program features usually exist to help users in their task, and if some are shunned for any reason it will likely be to the detriment of a better resolution of the discrimination between spikes.

Spike Detection

Spike analysis can be divided into two tasks: detection and classification (5). The task of spike detection is essentially that of separating spikes from noise, that is, from unwanted signals of any type. It is often based on the use of some sort of voltage threshold, a software equivalent of the window discriminators that for many years were the primary means of selecting specific spikes. If a threshold is the basis of spike detection, it follows that the larger the signal-to-noise ratio, the more reliable the ensuing spike detection will be. Even the best computer method will fail if the signals of interest are barely detectable.

There are several sources of noise that may have to be dealt with. There is first the irreducible noise inherent in the chain of amplification, recording, and digitization. To this can be added biological noise, the combined output of distant or small neurons that the investigator has deemed not to be of interest. Further noise can originate from movement artifacts and glitches. Some components of the noise can be reduced by careful positioning of the electrodes, proper shielding, and other methods particular to the experimental setting. It should be borne in mind that since the human brain is still far superior to a computer program in detecting certain types of patterns, some data interpretable by using conventional methods like filming may turn out still to be intractable to computer analysis.

The most common approach to spike detection is to use a voltage threshold. Any excursion of the signal over the threshold and then back below it is considered a spike. It is quite common even in automatic waveform analysis for the threshold to be selected manually (6). However, it is easier if an initial threshold is selected automatically. In our program Spike Finder, a default level for threshold is set based on the characteristics of a segment of data at the beginning of the data file. The program first finds the maximum and minimum voltages, then sets the threshold to 75% of the difference between the minimum and maximum. For our data (insect EMGs), this usually corresponds to an intersection at mid-spike level. The program allows this threshold level to be changed by the user at any time, an essential feature. A more complex but rather elegant alternative method of threshold selection is to position the threshold at a given number of standard deviations above the mean voltage (7).

Threshold-based methods of spike identification are complicated by shifts in the baseline of the recording. For example, movement artifacts can cause relatively large-scale and sometimes abrupt changes in baseline. These can lead to inclusion of noise as spikes or to missing spikes because an entire section of record is above or below the threshold. If a threshold-selection method is used to identify spikes, a stable baseline obtained through careful AC coupling and filtering (8) is important enough to sacrifice some detail of spike features that may be lost by the filtering. The baseline can also fluctuate at the line frequency of 50 or 60 Hz. It is possible to remove a substantial part of such a periodic component in the data through software by subtracting a sine or composite wave of appropriate amplitude in the spike-identifying program itself.

It is tempting to use as a threshold function some sort of running average, but then mean voltage increases with frequency, even if the spikes do not increase their amplitude as well (9). When the frequency is really high, the baseline often totally vanishes and the threshold reaches unpredictable or uncontrollable values.

Other methods exist to detect spikes without the use of a fixed threshold. One is to use the second (discrete) derivative to detect voltage peaks. This approach may be combined with data smoothing (1). Some authors use the software equivalent of a Schmitt trigger, with a particular signal-to-noise ratio, such as 2:1 (10). Any local high-frequency anomaly can be a putative spike. Nonthreshold spike detection methods in fact overlap greatly with the next task at hand, which is spike classification.

Spike Classification

Template Matching

The classification of spikes rests in great part on some form of template matching. Although the term template is usually reserved for a combination of several spike features or a group of similar ones (such as a given number of successive amplitudes), we will use a broader definition. Template matching is the matching of any spike feature, even something as simple as peak-to-peak amplitude, with some desired form to identify a typical spike of interest.

A primary spike feature is one that the investigator can easily grasp or visualize. Attributes to select for the classification of spikes should be chosen so that they enhance differences between the different classes of spikes. They can be used singly or in combination with one another. Just as for spike detection, amplitude is one of the most commonly used, either in reference to the baseline or peak to peak. A time component like spike width can also be used (11), or time from peak to the following zero-crossing or trough (12). These can be combined, like the amplitude at two adjustable epochs from the peak (13). Most models assume biphasic spikes, a high peak followed by a shallow trough, but in our experience a significant number of spikes can be triphasic (Fig. 1B); this fact must influence the choice of spike features. More geometrical aspects can be used, like the spike apex angle (10) or up- and down-slopes, which we use in Spike Finder (1).

More complex or less immediate features can also be used to define spike templates. Some can still have a physical meaning, like areas of the peak and the trough (14) or root mean square (rms) value (9). More abstract representations can be used, like principal component analysis, Fourier analysis, and other reduced feature sets (15, 16). Furthermore, nonparametric methods can be applied, like the branching study of slope changes (17). It is sometimes possible to isolate the simple features that make up a complex one (6). For example, in principal component analysis, one or both main vectors often contain a large percentage of amplitude information (18).

Other recent methods of spike classification employ neural networks to extract salient features of spikes (19), fractals for quantification of EEGs (20), or more than one recording site to increase resolution of spikes originating from different neurons (21). In the latter case, not only are the "points of view" different, which yield different amplitude ratios (10), but the differences in conduction velocity in different neurons can be put to advantage (22). Large neurons conduct action potentials faster than smaller ones. Cuff-like electrodes containing several parallel wires, as well as a matrix grid of electrodes that cover a large surface of neural tissue (23), have been used to record from several sites.

In the classification of spikes, there is usually the underlying assumption that spikes do not change shape over time, and that the information they carry lies solely in their time of occurrence. This assumption is not true in some cases (mollusks) where spikes do change width as a function of firing frequency, but even if they do not, there are numerous causes for spike shape variability. Few classification systems allow spike evolution in time (24, 25), although amplitude variations of 50% have been documented in recordings from a single source (10). Furthermore, in EMGs and our data, as the muscle

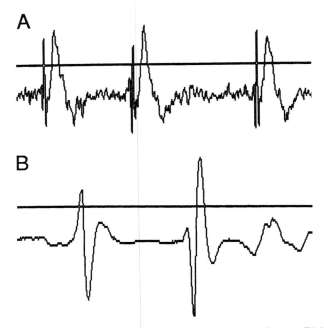

FIG. 1   Examples of muscle and nerve potentials from an insect EMG recording sampled at 8 kHz (A) and of recordings from a whole nerve cord sampled at 30 kHz (B). For each, 350 sampling points are displayed. The respective thresholds are represented by the horizontal lines. Up- and down-slopes can be visualized as lines joining the highest point of the spike to the two intersections with the threshold. Each muscle spike in the EMG is preceded by a nerve spike of similar amplitude. In the nerve cord recording, the two ''biphasic'' spikes are of opposite phase, and hence presumably travel in opposite directions on the nerve. If the small potential following the recovery phase of the first of these extracellularly recorded action potentials were larger or the threshold were lower, the first potential would be seen by the spike-detecting program as triphasic, and would require additional processing to be recognized properly.

contracts, the physical relationship of the electrodes to one another and to the muscle changes, which in turn changes the shapes of the spikes and increases the variability of the features used to define them. Superposition of spikes is also a common cause of episodic alteration of spike shape. Superposed spikes can be detected if the second falls in the refractory period of the first, or if more than two changes of slope are identified (9). Shifted subtractions (26) or an iterative removal of identified spikes from the data can be used if the density of spikes in the data leads to many superpositions.

Because of this problem of the lack of constancy of spike shape, it would be valuable for a user to be able to view the performance of the spike separation algorithm at any time, and to make adjustments to it based on the display. Plotting one spike feature against another at the time of spike analysis is a simple way of revealing clusters of spikes. Each feature itself can then be classified according to other spike features, and so on. This allows the selection of the features that are the most efficient for classification of a given set of spikes (13, 14). Cluster limits can be set manually (12) or automatically by clustering algorithms (6). This also helps to prevent the simultaneous use of features that actually have the same discriminatory power, an unnecessary load for the program and sometimes a source of confusion in the analysis (18).

Spike Finder

For our program we selected a template for spike identification that was relatively simple, so that it would be easy for the user to modify as desired. This was in line with our aim to automate but keep easy control of as many settings as possible (e.g., threshold, gain, or polarity). We wanted to analyze EMG records that contained both relatively broad muscle potentials and sharp nerve potentials (Fig. 1A) and to be able to pick one group of potentials and reject the other. We therefore selected the up- and down-slopes of a potential as the features used to define our templates. We define a slope as the ratio of the vertical excursion between the highest point above the threshold and the first point below it, to the number of sampling time intervals between these two points. The vertical excursion is expressed in "amplitude units," which can range from 0 to 4095 at 12-bit resolution. The calculated slopes are compared with numerical values that give the acceptable upper and lower limits of the up-slope and the down-slope. These slope limits may be changed by the user directly through the use of a pop-up menu or via program training that involves indicating which spikes are acceptable and which are unacceptable. Potentials for which both slopes fall within the limits are accepted, whereas those with one or both slopes outside the limits are rejected, either because they are too sharp or because they are not sharp enough.

The possibility of overriding any of the automated decisions of the detecting algorithm must be an integral part of the program. To this end, users of Spike Finder can employ interactive graphics to edit particular spikes in or out at will. Furthermore, any on-screen modification can iteratively teach the program what particular spikes the user is targeting for classification. In practice it corresponds to resetting of the upper and lower limits of the up-

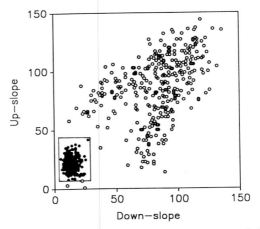

Fig. 2   Plot showing the relationship between the steepness of the down-slope for each electrical potential as a function of the steepness of the up-slope. This type of plot, relating one spike feature to another, should be available to the user at the time of spike analysis. Up-slope is plotted against down-slope for all spikes that cross a given threshold in EMG data. Filled circles represent slopes of muscle potentials; open ones represent nerve potentials or artifacts. The latter are arranged in a clawlike fashion because the sampling frequency (5 kHz) was close to the high-frequency component of the nerve spikes, so that the single sample usually taken from each spike can fall either on the up-slope or the down-slope of the spike. The small rectangle represents the slope limits for the acceptable muscle spikes. In Spike Finder, the slope limits are adjustable directly or through a graphic display (see text). However, it would also be useful if they were manually positionable on this plot as well, by adjustment of the four lines that define the rectangle of acceptable spikes. Slope units 12 V/sec. (Modified from Ref. 1.)

and down-slopes that define the spikes to be accepted. We did not implement interaction with a feature versus feature display as described above (Figs. 2 and 3) because we did not fully appreciate its value at the time the program was written, but we may do so for a future release.

Spike Validation

Spike validation is a key operation that is often neglected. The times of occurrence of the spikes are usually to be used in some analysis subsequent to the spike identification itself. It is therefore indispensible to have some measurement of the errors involved in spike identification, so as not to reach

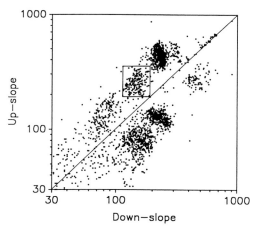

F<small>IG</small>. 3 Plot of up- against down-slope for nerve cord data sampled at 30 kHz. Clusters of spikes are apparent in the plot. Those on opposite sides of the diagonal, the locus of "symmetrical" spikes, correspond to spikes with opposite phases (initial positive or initial negative), since the rising phase of spikes that are positive first is not as steep as the apparent rising phase of spikes that are negative first (see Fig. 1B). The few points on the high end of the diagonal are analog-to-digital (A/D) glitches. For data such as these, it would be useful for the user to be able to manipulate the size and position of a rectangle that defines acceptable spikes, such as that drawn on the plot. Slope units 73 V/sec. (Modified from Ref. 1.)

erroneous conclusions in subsequent analysis of the time structure of the data. The more detection and classification are automated, or use complex features, the more this precaution is important to minimize errors.

Two types of errors can occur in both spike detection and classification, namely, false negatives and false positives. False negatives occur, for instance, if a genuine spike fails to cross the threshold and be detected because of a momentary dip in the baseline, or if two spikes are superposed in such a way that one or both of them are classified outside the limits of any cluster, as noise. On the other hand, false positives occur if, for example, some baseline high-frequency noise crosses the threshold, is detected as a potential spike, and mistakenly classified in a group. These two errors are very similar to the Type I (α) and Type II (β) errors of statistical tests.

To test our spike identification algorithm for the occurrences of these errors, we varied in a systematic way the two parameters that can be modified by the user of our program, namely, the sampling frequency and the positioning of the threshold, and the combination of both. For our EMG data, the percentage of false positives is less than 2% at sampling rates from 3 to at least 8 kHz, except when the threshold is too low, in which case the minimum number of false positives occurs at a frequency of 5 kHz. At that threshold level, higher sampling rates result in 10% of the spikes being false positives, a contradiction of the notion that the higher the sampling rate, the higher the resolution of the spikes, at least for unsmoothed data. Furthermore, the exact positioning of the threshold did not have a great effect on the percentage of false positives; this is presumably due to the fact that, as a first approximation, spikes can indeed be represented by triangles. The results show that up-slope is strongly correlated with down-slope in most cases, and therefore redundant, except for mixed nerves where spikes travel in both directions. In the latter case, classification of the two groups may be enhanced by noting which of the two slopes is the steepest (Fig. 3).

Instead of measuring and comparing identified items that can influence the magnitude of those errors, a more general approach has been suggested by Heetderks (5) with the use of an "apparent separation matrix" to quantify the separation between clusters. Heetderks has also demonstrated that up to 10 units can be separated at the 0.05 confidence interval level for both α and β; for more units one needs more than a single pair of electrodes.

Comparison between spike sorters has been done by several investigators (5, 15, 18, 27). Since spike classification is data dependent, one should not be too surprised by eventual discrepancies. Not surprising either is the fact that humans used as referees often do not agree with one another (27).

The task of separating even a few types of spikes can quickly become complex when real-life phenomena, like an unstable baseline, movement, superposition, and noise are added. As a result, there are only a few totally automated systems; the human brain must still contribute to a significant extent. This is why the user interface is as important as the actual methods used for spike detection, classification, and validation in the design of a successful spike identification system.

References

1. J. H. Cocatre-Zilgien and F. Delcomyn, *J. Neurosci. Methods* **33,** 241 (1990).
2. E. M. Glaser and D. S. Ruchkin, "Principles of Neurobiological Signal Analysis." Academic Press, New York, 1976.
3. M. Abeles and M. H. Goldstein, Jr., Proc. IEEE **65,** 762 (1977).
4. E. M. Schmidt, *J. Neurosci. Methods* **12,** 95 (1984).
5. W. J. Heetderks, *Biol. Cybernet.* **29,** 215 (1978).
6. M. Salganicoff, M. Sarna, L. Sax, and G. L. Gerstein, *J. Neurosci. Methods* **25,** 181 (1988).
7. C. Camp and H. Pinsker, *Brain Res.* **169,** 455 (1979).

8. F. Marion-Poll and T. R. Tobin, *J. Neurosci. Methods* **37,** 1 (1991).
9. R. J. O'Connell, W. A. Kocsis, and R. L. Schoenfeld, *Proc. IEEE* **61,** 1615 (1973).
10. W. E. Faller and M. W. Luttges, *J. Neurosci. Methods* **37,** 55 (1991).
11. J. C. Dill, P. C. Lockemann, and K.-I. Naka, *Electroencephalogr. Clin. Neurophysiol.* **28,** 79 (1970).
12. D. J. Mischelevich, *IEEE Trans. Biomed. Eng.* **17,** 147 (1970).
13. W. Simon, *Electroencephalogr. Clin. Neurophysiol.* **18,** 192 (1965).
14. G. D. McCann, *IEEE Trans. Biomed. Eng.* **20,** 1 (1973).
15. B. C. Wheeler and W. J. Heetderks, *IEEE Trans. Biomed. Eng.* **29,** 752 (1982).
16. G. J. Dinning and A. C. Sanderson, *IEEE Trans. Biomed. Eng.* **28,** 804 (1981).
17. P. F. Lister and M. L. Bishop, *IEE Proc.-E* **135,** 241 (1988).
18. J. F. Vibert and J. Costa, *Electroencephalogr. Clin. Neurophysiol.* **47,** 172 (1979).
19. R. C. Eberhart and R. W. Dobbins, *in* "Neural Network PC Tools" (R. C. Eberhart and R. W. Dobbins, eds.), p. 215. Academic Press, San Diego, 1990.
20. M. J. Katz, *Comput. Biol. Med.* **18,** 145 (1988).
21. W. M. Roberts and D. K. Hartline, *Brain Res.* **94,** 141 (1975).
22. E. M. Schmidt and M. W. Stromberg, *Comput. Biomed. Res.* **2,** 446 (1969).
23. B. C. Wheeler and J. L. Novak, *IEEE Trans. Biomed. Eng.* **33,** 1204 (1986).
24. E. W. Kent, *Electroencephalogr. Clin. Neurophysiol.* **31,** 618 (1971).
25. W. H. Calvin, *Electroencephalogr. Clin. Neurophysiol.* **34,** 94 (1973).
26. V. J. Prochazka and H. H. Kornhuber, *Electroencephalogr. Clin. Neurophysiol.* **34,** 91 (1973).
27. M. F. Sarna, P. Gochin, J. Kaltenbach, M. Salganicoff, and G. L. Gerstein, *J. Neurosci. Methods* **25,** 189 (1988).

[15] Computer Method for Identifying Bursts in Trains of Spikes

F. Delcomyn and J. H. Cocatre-Zilgien

Introduction

Bursts may be defined as periods of neural activity during which spike frequency is relatively high, separated by periods during which it is relatively low or spikes are absent altogether. The presence of bursts is a common feature of neural activity in many parts of the nervous system [e.g., striate cortex (1), basal ganglia (2)], especially those involved in the motor control of behavior like walking (3), flying (4), swimming (5), and respiration (6). In many physiological studies, it is critical to identify burst boundaries as clearly and reliably as possible, because the bursts may represent the product of the physiological process under investigation. When the bursting pattern is clear, with long silent periods between bouts of high-frequency spiking, no one questions the identification of burst starts and ends (e.g., Fig. 1). However, if there is considerable fluctuation of spike frequency within each burst, and the intervals between bursts contain a significant number of spikes, different investigators may disagree not only as to where each burst begins and ends, but even as to whether bursts are present at all.

In such circumstances, a formal method for the detection and identification of bursts would obviously be very useful. There is considerable literature on the analysis of spike trains (7–9); unfortunately, however, the identification of bursts has received little attention except in passing (10–15) and there are only two empirical studies to assist in burst identification (1, 16). There are several likely reasons for this apparent lack of interest in what is actually an important subject: in those instances in which there is a complete absence of spikes between the bursts, burst identification is trivial; bursts as such may not be the subject of a study in which they appear; and there is no sound theoretical basis for burst identification.

The purpose of this chapter is twofold. First, we will describe our empirical method for the detection and identification of bursts (16). We will show that an algorithm based on the analysis of histograms of interspike intervals from spike train data can detect and locate bursts with good reliability and few assumptions about data distribution. Second, we will provide a foundation for empirical or theoretical studies by pointing out ineffective methods for the detection and identification of bursts.

Methods in Neurosciences, Volume 10

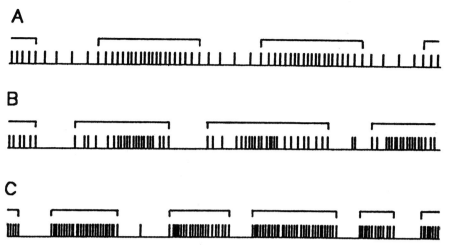

FIG. 1   Samples of spike trains analyzed by the burst-finding algorithm. The bracketed sections of the spike trains are the bursts identified by the program. The reader can test the effectiveness of the algorithm by covering the brackets with a sheet of paper and deciding where the burst limits ought to be. (A) Computer-generated sinusoidal pattern. (B, C) Motor patterns from the leg of a walking insect.

Burst Detection and Identification

There are two separate questions that must be answered about data in which one wishes to identify bursts: are bursts present in the data, and, if so, where do these bursts begin and end? We will refer to these as the questions of burst detection and burst identification, respectively. Although it seems logical to examine the data for the presence of bursts before trying to identify burst starts and ends, we have been unable to find any reliable method for the detection of bursts that does not first identify potential bursts.

Methods of Burst Detection That Are Unreliable

By definition, bursts constitute a nonrandom distribution of spikes in a set of data. Logically, a spike train that contains bursts ought to be recognizable as nonrandom by appropriate statistical tests. All approaches that we tried failed when tested against actual trains of spikes containing bursts.

Most tests for randomness check for stationarity. In other words, they examine arbitrarily chosen segments of the spike train to see if their statistical properties are invariant with time translations (7). For bursty spike data, only nonparametric tests should be used, because the underlying distribution of spikes is unknown. In many cases, typical tests [e.g., the Wald–Wolfowitz test and the inversion test (17)] sometimes yielded false-negative results; that is, because of interference between the segmentation necessary for the tests and the quasi-periodicity of the data, they failed to identify nonrandomness if the segments used happened to be ill-chosen in size and location. These tests were therefore unsuitable as methods for the detection of bursts.

Another approach is to compute serial correlations in interspike intervals before and after shuffling (18). The basis of the subsequent test is to compare the serial correlation of the original data with the serial correlation of the data after the intervals have been randomly reordered. If the data are nonrandom, there should be a significant difference between the correlations of the shuffled and unshuffled data sets. Here again, in some data sets containing a few large interburst intervals, the shuffling just moves the few large intervals about, and the correlations before and after shuffling do not turn out significantly different, hence again yielding false negatives. Serial correlation methods are therefore also unsuitable.

A third approach is to use autocorrelation methods for finding periodicities in the spike train directly. Autocorrelation studies are difficult for two reasons. First, data consist of discrete events, and a great number of spikes can occur in a burst before nonburst intervals are reached again; it is therefore possible to miss them if the autocorrelation is not carried out far enough. Typical autocorrelation studies rarely check beyond 50 events (9). Second, there are sometimes very few interburst intervals, and the data are often only quasi-periodic. In these circumstances, it is difficult to have clear periodicities emerge from the data.

Coefficient of Dispersion

One statistic that seems to have potential for being useful in detecting bursts in spike data is the coefficient of dispersion (CD, which is the variance/mean). This is an interesting statistic because it is a measure of aggregation (19, 20), and "bursty" data ought to show aggregation. A CD of 1.0 is observed in Poisson distributions (in which variance equals mean). A CD less than 1.0 means that the spikes tend to repel one another and that the distribution of their interspike intervals tends toward uniformity. This may occur at high firing frequencies because of the refractory period that follows each spike. A CD larger than 1.0, on the other hand, indicates a tendency for spikes to cluster together, which is one of the distinguishing characteristics of bursty data. Unfortunately, there is no easy means of testing whether a *CD* is significantly larger than 1.0.

One reason for thinking that the *CD* might help in the detection of bursts is that in our data (16) we observed that bursts were rarely present (by the criteria of the algorithm described below) when the *CD* was smaller than 1.0 and that the *CD* was always larger than 1.0 when bursts were present. The relationship was not perfect, but it suggests that the *CD* might be useful in association with another test or tests.

Methods of Burst Identification That Are Unreliable

Even if no good method exists to determine directly whether a spike train contains bursts, it is still possible to search for burst starts and ends in the data. One can then subsequently decide whether any starts and ends that have been identified actually represent valid bursts. This is the approach we adopted. As in the case of burst detection, there are several possible methods that might be used for such burst identification. Some of these methods have proved ineffective.

An automated method for burst identification must be able to identify aggregations of spikes without being misled by local variations in spike frequency within the aggregation, or by extra spikes between aggregations. We first attempted to use an averaging method to perform this task. The idea was to calculate a running average of interspike intervals and to compare each interval with this average. For some burst data, this method worked well, but in other data sets, bursts ended abruptly with a few high-frequency spikes. This often skewed the average enough that the next interval between bursts was taken to be part of the burst. Increasing the number of intervals included in the running average helped to some extent, but doing this meant that short bursts could not be detected, since averaging cannot detect bursts containing fewer spikes than the number used to calculate the average. For those reasons, we abandoned averaging techniques.

Burst-Finding Algorithm

The most successful method we have found for identifying bursts in a train of spikes is based on an analysis of the distribution of interspike intervals (ISIs) of the data set. Our reasoning is that in any spike train containing bursts, there will be two populations of ISIs. One population will consist of the relatively short intervals found between spikes within bursts, and the other population the relatively long intervals found between spikes that are not part of any burst, or between bursts. In our method, we analyze the distribution of ISIs in order to locate the transition between these two populations. In the ideal case (exclusively short intervals within bursts, exclusively long intervals between bursts), the two populations will be clearly separated and the transition point unambiguous. In other cases, the populations of intervals may overlap, making identification of the transition point between them more difficult. The method does not require any prior determination that bursts are in fact present. If there are no bursts in the data, then there will be only a single population of interspike intervals, and no transition point should be identifiable.

An overview of our method may help put the details that follow into context. We use three steps to identify burst starts and ends, then follow these by one further step to validate the existence of bursts statistically. The first step is to construct a histogram of interspike intervals. Next, an interval is selected that represents the "threshold" value that defines the transition from a burst to a nonburst state and vice versa. Then, each interval in the data set is compared with the "threshold" value, and the boundary spikes that represent the starts and ends of the bursts are identified. Finally, a chi-square (χ^2) test is conducted on the number of spikes in each burst and interburst period compared to the average expected number in the same periods of time.

We have implemented our method in a short computer program that uses the times of occurrence of the spikes as input and produces the times of occurrence of the boundary spikes that represent the starts and ends of bursts as output. The program requires three passes through the data and therefore will only work off-line. A listing of the program (in BASIC) and copies of sample data files may be obtained via electronic mail (address: delcomyn@ux1.cso.uiuc.edu) or by mail (send a blank, formatted, IBM-compatible diskette, 5.25 or 3.5 inch).

Step 1: Constructing the Interspike Interval Histogram

The heart of our method is a histogram of interspike intervals. The histogram is built by counting the number of intervals that fall into specified size categories (bins). Each bar of the histogram then represents the number of intervals of the specified size range. The first pass through the data extracts primary ISI statistics so that a histogram can be constructed. These statistics include the number of intervals, their maximum and minimum values, and range. The ISI sum and sum of squares are computed so that the mean and variance can be calculated, as are the coefficients of dispersion and of variation.

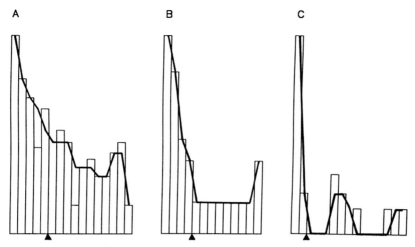

FIG. 2   Interspike interval histograms for spike trains partly shown in Fig. 1. The bars represent the histogram as generated by the program from the ISI data. The solid line in each represents the smoothed histogram. The threshold selected by the program is shown by the small triangle under each histogram.

The first step for creation of the ISI histogram is to compute the number of bins to be used in it. If there are too few, the shape of the distribution may be masked because of poor resolution, whereas if there are too many, analysis of the histogram may be rendered more difficult because of an excessive number of local minima. The number of bins is obtained using the formula $N\text{BIN} = 1.87(N - 1)^{0.4}$ (7). We used this formula so that we could apply tests of goodness-of-fit to the histogram should the need arise (7). The plain square root of the number of ISIs could also be used, since this does not change the number of bursts finally identified by the program (16). Bin width is set equal to the range of intervals divided by the number of bins. If necessary, the exact number of bins is then adjusted so that the minimum width for a bin is 1, and so that there are no empty bins on either end of the histogram.

Once bin number and bin width have been established, the histogram is filled by a second pass through the data. It is displayed using a logarithmic vertical scale, with high-frequency spikes (small ISIs, from bursts) represented on the left and low-frequency spikes (large ISIs, from periods between bursts) represented on the right. Figure 2 shows histograms derived from the data samples shown in Fig. 1.

Smoothing the Histogram

Smoothing the histogram makes subsequent analysis more reliable by reducing the possibility that small irregularities in the distribution of ISIs might cause errors in identification of the intersection between the two populations of intervals. We smoothed all histograms by using iterative running medians of 3, as described by Tukey (21). The original histogram is scanned so that each new value for a bin is the median of itself and those of the bins on either side. End values are simply copied on. This is of no consequence as most of the analysis will take place in the center of the histogram. The process is repeated as many times as necessary until no more changes occur. The program then identifies the bin with the greatest number of intervals in it. All subsequent operations are performed on the smoothed histogram.

Step 2: Analyzing the Histogram

Testing Assumptions

Our algorithm is based on two assumptions about the data, namely, that there is no trough in the distribution of short intervals to the left of the tallest mode of those intervals and that the number of short intervals in the data is greater than the number of long intervals. Because these assumptions are not always met, the program must test for them before proceeding with the analysis.

A trough in the short intervals may be due to potentials in the original spike train other than those of primary interest. For example, recordings taken from muscle can pick up spikes from the motor nerve innervating the muscle as well as potentials from the muscle itself. These nerve spikes precede the muscle potentials by only a short time, yielding a group of short intervals that are physiologically spurious (since the nerve and muscle potentials represent the same physiological event). When a trough is detected in the short-interval component of the histogram, the program is aborted, since analysis depends on accurate identification of a single mode in the short intervals. In cases like these, some form of filtering may have to be done prior to submitting the data for analysis.

If there are more long intervals than short ones, it means that much more time is spent in a nonbursting state than in a bursting one, and that there are then few bursts in the spike train. This phenomenon may be common in the brain [e.g., thalamic lemniscal neurons (10)]. Because our algorithm works from short to long intervals, as described in the next section, based on the assumption that the tallest mode is among the short intervals, it will stop if it encounters data in which the tallest mode is among the long intervals. To handle data containing only rare bursts, the program could be modified so that it worked from long to short intervals.

Selecting Threshold Interval

The objective of the histogram analysis is to find the boundary interval that represents the threshold, which is the transition point between intervals that are part of a burst and those that lie between bursts. This threshold interval is located at a local minimum between two modes, if there is more than one mode in the data, or at a "shoulder" of a unimodal histogram. The threshold is found by iterative, pairwise comparison of the value of a bin (the reference bin) to that of the bin to its right, until a bin is found that is equal to or higher than the current reference bin. The comparison is done using the smoothed bin values, and it starts from the tallest mode. A bin equal to or larger than the reference bin is taken to be the target bin, and the value of the interspike interval represented by that bin is saved as the threshold value for bursts. This procedure is analogous to taking the first derivative of a continuous line drawn through the tops of the bins of the smoothed histogram, with the threshold representing the point at which the derivative equals zero.

If no bin larger than any reference bin is found, or if bins of equal size continue to the end of the histogram, the program initiates a second analytical procedure, in which the slopes of the lines connecting the tops of adjacent bins are compared. In this procedure, slopes are computed for the lines connecting the top of a reference bin to the top of its leftmost and rightmost adjacent neighbors. If the second slope is equal to or steeper than the first, the right bin of the three involved in the comparison is taken as the new reference bin, and the process is repeated until the end of the histogram is reached or until a slope is found that is less steep than the one immediately to its left. This is analogous to analyzing the second derivative to find inflection points. A slope that is less steep represents an inflection point in the downward trend of bin heights, and is taken as the threshold for bursts. If no such slope is located, the program returns the message that no bursts were identified in the data. The thresholds identified by the program for the data in Fig. 1 are marked along the bottom of each histogram shown in Fig. 2.

One of the main reasons for developing this computer method was to have a consistent means of selecting burst starts and ends in data in which burst starts and ends were ambiguous to human observers. For this reason, the program will work entirely without human intervention. However, for some data, such as those representing the expression of some rhythmic behavior, there may be reasons extrinsic to the data themselves for thinking that bursts ought to be present in certain locations. In the case of small data sets of ambiguous data, it is possible for the algorithm to miss one or more groups of spikes that a knowledgeable human observer would classify as bursts. We have therefore made provision in the program for manual placement of the threshold by the user via interactive graphics. In our computing environment, the results of such placement can be made visible on screen, but this requires software beyond that being described here.

Step 3: *Identifying Bursts*

The actual location of the first and last spikes of each burst is determined in the third and final pass through the data. Each interspike interval of the data is compared with the threshold found by the analysis. If the interval is greater (longer) than the threshold value, it cannot be part of a burst; if the interval is smaller (shorter) than or equal to the threshold, the interval may be part of a burst. In our implementation, we have defined bursts as consisting of a minimum of three spikes (two intervals), but other users may require a smaller or larger number. The program checks that the requisite number of successive intervals has been found before recognizing a burst. When two successive interspike intervals are found to belong to different states (one within a burst, the other outside it), the boundary spike between them is noted as a burst start, or end, as appropriate, and its time of occurrence added to the output file.

Step 4: *Validating Bursts*

If the description of the spike train as consisting of bursts and intervals between bursts is statistically valid, the bursts must contain more spikes, and the intervals between bursts fewer spikes, than would be expected if there were no bursts in the data. To check this, a χ^2 test is performed with the null hypothesis that the expected frequency in any segment of data is the mean spike frequency of the entire train multiplied by the duration of the segment under consideration. The test is performed at the 0.05 confidence level. If the test does not reject the null hypothesis, the program reports that no valid bursts have been found and stops.

The statistical evaluation of the data for the presence of bursts is applied after tentative identification of burst starts and ends owing to the absence of usable statistical tests for the detection of bursts in trains of spikes. The presence of bursts is hence detected by a two-step process; the ISI histogram is analyzed for the presence of a bimodal distribution or an inflection point in a unimodal one, and if bursts are identified by this method, they are validated through the use of the χ^2 test. If the test rejects the presence of bursts that have been identified, it means that the histogram feature that allowed a threshold to be selected was a local anomaly in the distribution of what was a single population of interspike intervals.

Evaluation of Performance

Evaluation Based on Human Decisions

Any method must be evaluated on the basis of its performance. It is difficult, however, to evaluate a method for the automatic detection of bursts in spike trains because there is no objective standard against which it can be measured.

Investigators may well differ as to where the boundaries of a set of bursts should be placed. Readers can carry out simple trials for themselves on the sample data in Fig. 1 by masking the brackets with a sheet of paper to eliminate the strong bias they introduce. When faced with ambiguous data, humans often resort to strategems like squinting or viewing data from a distance that in effect reduce the detail of the train, since people detect global patterns more easily than local ones (22). This is equivalent to the overall or global view provided to the program by the histogram of interspike intervals.

The approach used by an algorithm based on a histogram of all the interspike intervals to find the location of bursts seems to employ some of the old principles of perceptual grouping of the Gestalt psychologists (23). Provided with the whole of the data, one can find bursts by the operation of various principles of grouping. Of those principles, that of proximity is used in the algorithm as a simple test of whether an interspike interval exceeds or does not exceed a critical value extracted from the histogram.

Evaluation Based on Comparison with Other Methods

Another approach to evaluating this method is to compare its performance to that of another existing method. This approach also has its weaknesses, mainly that different types of data may exhibit bursts with different temporal characteristics. A method of burst detection that works well with one type of data may not work well with another.

Comparison of the performance of our algorithm against that of the algorithm introduced by Legéndy and Salcman (1) illustrates the necessity of taking into account the type of data for which a burst detection algorithm was designed. Their method involves calculation and evaluation of the "surprise" S statistic. This is done as follows. The average of two successive

FIG. 3   Comparison of the performance of our histogram method and the "surprise" method of Legéndy and Salcman (1). The lower brackets represent bursts identified by the histogram method, and the upper brackets represent bursts identified by the surprise method. Note that the latter identifies shorter and fewer bursts than the former. See text for further details. Spike trains are from the legs of an intact walking insect (A) and one with a leg amputated (B).

ISIs is compared iteratively with the average ISI of the whole data set. When the pair average is one-half or less of the data set average, the three spikes thus defined are called a burst. The Poisson surprise S statistic for the burst so identified is then calculated. S is defined as the negative decimal logarithm of the probability of occurrence of the same number or more spikes in a random (Poisson) spike train of mean interval equal to the average interspike interval of the data. The surprise statistic is then maximized, first by adding more spikes to the right of the burst, then by removing spikes to its left and recalculating, until the largest value of S is obtained. We applied this procedure to our data, using a critical value for S of 1.301 (probability < 0.05).

Comparison of the approaches reveals two main differences (Fig. 3); the surprise method tends to find shorter bursts and fewer bursts than does our histogram method. The finding of shorter bursts can be explained in part by the sensitivity of the surprise method to the instantaneous firing frequency of a group of spikes. For long bursts that trail off in frequency in their extremities (Fig. 3A), the S statistic will not increase by the incorporation of more spikes, and hence only the central core will be identified as the burst. In comparison, the ISI histogram method is less dependent on the inner structure of the burst.

That fewer bursts are detected by the surprise method may be due to the reference ISI being the average ISI of the whole spike train. If large interburst intervals are interspersed in the spike train (Fig. 3B), the average ISI will reach relatively high values. As a result, the S value of a burst will often fail to reach the critical value before the iterative process leaves a particular concentration of spikes. This in turn leads to fewer bursts being detected by the method. In contrast, the ISI histogram is independent of the average ISI. In summary, one could say that the main difference of approach between the two methods is that the surprise method attempts to identify burst limits from within the bursts, whereas the ISI histogram attempts to do the same from outside the bursts.

Which approach works best? This question does not have a single answer. Our purpose in detailing these differences is to emphasize that each method, being designed to handle data with certain characteristics, does well with those data and not necessarily with other types. Because different bursts may have different properties, and different investigators have different notions of what constitutes a burst, no single algorithm will likely ever serve the needs of everyone.

Acknowledgments

Development of the method for burst detection described in this chapter was supported in part by a grant from the Whitehall Foundation. We thank George Gerstein for his help and encouragement during the course of the work described here.

References

1. C. R. Legéndy and M. Salcman, *J. Neurophysiol.* **53,** 926 (1985).
2. J. W. Aldridge and S. Gilman, *Brain Res.* **543,** 123 (1991).
3. F. Delcomyn and P. N. R. Usherwood, *J. Exp. Biol.* **59,** 629 (1973).
4. D. M. Wilson, *J. Exp. Biol.* **38,** 471 (1961).
5. S. Grillner, P. Wallen, A. McClellan, K. Sigvardt, T. Williams, and J. Feldman, *Symp. Soc. Exp. Biol.* **37,** 285 (1983).
6. C. M. Rovainen, *J. Comp. Physiol. A* **157,** 303 (1985).
7. J. B. Bendat and A. G. Piersol, "Measurement and Analysis of Random Data." Wiley, New York, 1966.
8. D. R. Cox and P. A. W. Lewis, "The Statistical Analysis of Series of Events." Wiley, New York, 1966.
9. J. W. de Kwaadsteniet, *Math. Biosci.* **60,** 17 (1982).
10. G. F. Poggio and L. J. Viernstein, *J. Neurophysiol.* **27,** 517 (1964).
11. A. Eckholm and J. Hyvärinen, *Biophys. J.* **10,** 773 (1970).
12. L. M. Mukhametov, G. Rizzolatti, and V. Tradardi, *J. Physiol. (London)* **210,** 651 (1970).
13. O. Benoit and C. Chataignier, *Exp. Brain Res.* **17,** 348 (1973).
14. G. Barrionuevo, O. Benoit, and P. Tempier, *Exp. Neurol.* **72,** 486 (1981).

15. G. J. A. Ramakera, M. A. Corner, and A. M. M. C. Habets, *Exp. Brain Res.* **79,** 157 (1990).
16. J. H. Cocatre-Zilgien and F. Delcomyn, *J. Neurosci. Methods* **41,** 19 (1992).
17. D. M. Himmelblau, "Process Analysis by Statistical Methods," pp. 71–76. Wiley, New York, 1970.
18. D. H. Perkel, G. L. Gerstein, and G. P. Moore, *Biophys. J.* **7,** 391 (1967).
19. R. R. Sokal and F. J. Rohlf, "Biometry," 2nd Ed., pp. 82–94. Freeman, San Francisco, California, 1981.
20. J. H. Zar, "Biostatistical Analysis," 2nd Ed., p. 410. Prentice-Hall, Englewood Cliffs, New Jersey, 1984.
21. J. W. Tukey, "Exploratory Data Analysis," pp. 205–264. Addison-Wesley, Reading, Massachusetts, 1977.
22. D. Navon, *Cognit. Psychol.* **9,** 353 (1977).
23. S. Coren and J. S. Girgus, *J. Exp. Psychol.* (*Human Perception and Performance*) **6,** 404 (1980).

[16] Computer Algorithms for Deconvolution-Based Assessment of *in Vivo* Neuroendocrine Secretory Events

Johannes D. Veldhuis, William S. Evans, James P. Butler, and Michael L. Johnson

I. Introduction and Definition of Deconvolution

A. Introduction

Many biological processes result from the combined effects of two or more components acting jointly or interacting over time. For example, the plasma concentration of a hormone, substrate, or metabolite is defined jointly by at least the rate of entry of that substance into the vascular compartment on the one hand, and its rate of exit from the blood compartment on the other. The mathematical process of recovering the values of the input function (e.g., secretion rates) and possibly other relevant functions (e.g., kinetics of metabolic disposal) from the observed overall outcome (e.g., plasma hormone concentrations over time) is referred to as deconvolution analysis (1–4). Deconvolution etymologically derives from the concept of "unraveling," "unrolling," or "disentangling." Conversely, the term convolution indicates that the two or more mathematical functions are convolved or interrelated with respect to the observed output; that is, they contribute jointly over time in a manner specified by each and their interactions.

B. Convolution Integral

A convolution integral represents the integral of the product of two or more functions, whose combined operation over time contributes to changes in the observed or measured commodity. In simple form, a typical convolution integral contains a secretion and an elimination function, both of which are related over time to changing hormone or neurotransmitter concentrations. This relationship can be given in general as follows:

$$C(t) = \int_0^t S(z)\, E(t - z)\, dz + \varepsilon \tag{1}$$

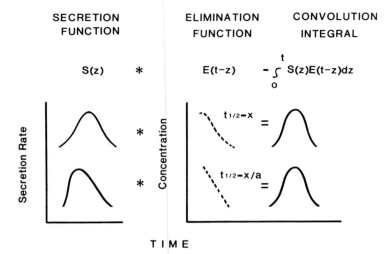

| SECRETION FUNCTION | | ELIMINATION FUNCTION | CONVOLUTION INTEGRAL |

$$S(z) \quad * \quad E(t-z) \quad = \int_{o}^{t} S(z)E(t-z)dz$$

FIG. 1   Concept of the convolution integral. A convolution integral is used to express the combined ("convolved, or intertwined") effects of two functions on an observed process. Thus, a convolution integral containing a secretion, $S(z)$, and clearance, $E(t-z)$, function can define resultant changes in hormone concentrations over time. The secretion function could assume one of several forms, for example, an instantaneous impulse (zero duration secretion burst, not shown), a square-wave time course of secretion (not shown), a Gaussian waveform (top curve), or a skewed distribution of secretion rates (bottom curve). The clearance function most often is represented as a mono- or biexponential decay curve, but could in certain physiological circumstances exhibit different properties (e.g., log–linear concentration-dependent decay constants). Convolution of a Gaussian or a skewed secretory function with a clearance function is illustrated. Either convolution integral (given an appropriate half-life, x or x/a) might provide a good description of the observed neuroendocrine concentration data (rightmost peaks).

where $S(t)$ is the secretion function of interest, $E(t-z)$ is the elimination or disposal function, and $C(t)$ is the concentration of the hormone at time t. The term ε denotes random experimental error. This concept is illustrated schematically in Fig. 1.

The secretion function may take a variety of possible forms (1–3, 5). For example, secretion might consist of an abrupt discharge of a finite number of molecules over a very short time span (δ function). Because this particular secretory impulse is probably uncommon, more extended episodes of hormone secretion over time can also be envisioned. Indeed, as discussed further below, there may be a constant rate of basal (time-invariant) hormone release combined with episodic or "burst" release (1–3, 5).

The elimination, or dissipation or disposal, function is used to denote the effects of one or more processes that attenuate in a time-dependent manner the magnitude of the observed signal. For example, metabolic clearance represents one such elimination function, which typically can be described by one or more exponential expressions of the form

$$E(t - z) = Ae^{-k_1(t-z)} + Be^{-k_2(t-z)} + Ce^{-k_3(t-z)} \tag{2}$$

where k_1, k_2, and k_3 are individual rate constants of elimination and t is time. These individual rate constants contribute to the overall decay process in amounts proportionate to their respective amplitudes A, B, and C. The term half-life, in the case of any given rate constant k, is given as $\ln 2/k$. More complex elimination functions consisting of nonzero values of A, B, and C describe two or more compartments, which have corresponding distribution volumes and rates of ingress and exit. Neuroendocrine systems in which a discharged compound is bound with high affinity to a receptor or plasma transport protein can also be visualized as multicompartment models.

II. Special Features of Neuroendocrine Data

A. Biological Constraints

Endocrine systems are virtually unique in their complexity of regulation, and in the challenges endowed by the form of data collection and quantitation. Specifically, in many neuroendocrine systems the total number of observations made over any given interval is remarkably limited compared to physical contexts. For example, in fluorescence spectroscopy, quantitative observations can be made at micro- and nanosecond intervals, whereas in living systems, and particularly in *in vivo* experiments, samples are typically collected every several seconds or minutes. In addition, *in vivo* neuroendocrine experiments are often extended over relatively short intervals consisting of hours. Moreover, biological data are susceptible to unpredictable variations introduced by differences among individual animals or subjects, uncontrolled or physiological variability in the system under observation, and experimental uncertainties inherent in the collection, processing, and assay of the samples.

A special problem in neuroendocrine data analysis is the uncertainty regarding the true theoretical form of the secretion function contained in Eq. (1) above. Specifically, in relation to pituitary hormone secretion, relatively few data exist to define in detail the temporal profile of episodic hormone secretion in conscious, unrestrained healthy animals. Some data collected,

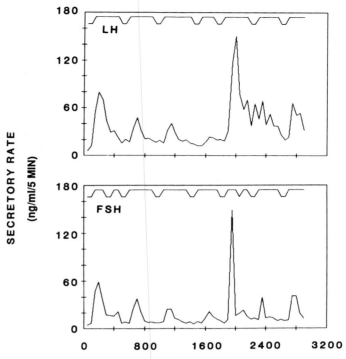

FIG. 2   Temporal profiles of pituitary LH and FSH secretion in a single conscious, freely moving, unanesthetized horse. Blood was sampled from the pituitary venous effluent at 5-min intervals in order to capture the patterns of *in vivo* gonadotropin secretion. Deflections above the data denote significant secretory peaks defined by CLUSTER. [Data from S. Alexander and C. Irvine (Christchurch, New Zealand). Adapted from J. D. Veldhuis and M. L. Johnson, *J. Neuroendocrinol.* **2**(6), 755 (1991), with permission.]

for example, in the horse (Fig. 2) indicate that the secretory event enacted at the level of the anterior pituitary gland at least for luteinizing hormone (LH) and follicle-stimulating hormone (FSH) is not instantaneous, but has a finite duration consisting of an increasing rate of secretion, a maximal value, and then a decline in secretory rate. The reproducibility of this time structure among animals, within different pathophysiological settings and in relation to different distinct hormones (or their isoforms), is not known. In addition, how the secretory signal is modified as it is transmitted throughout the systemic circulation, and altered in the course of observation and processing by the experimenter, has not been elaborated in any detail in neuroendocrine physiology.

B. Assay Characteristics: Precision, Sensitivity, Specificity, and Reliability

Although some of the sources of experimental variation noted above (Section II,A) are difficult to characterize in detail in various neuroendocrine experimental contexts, the dose-dependent precision profiles of the neuroendocrine assay can be submitted to more rigorous analysis and quantitation. Our experience with the human LH immunoradiometric assay (IRMA) indicates that the distribution of measurement error is approximately Gaussian (1, 2). To assess this, we measured the LH content of 10 replicates at each of 10 different concentrations of LH. The test samples were obtained by adding known amounts of highly purified pituitary LH (NIH I-3) to hypopituitary blood pooled from patients with undetectable LH concentrations as assessed in an *in vitro* bioassay (6). Only at the very limit of LH detectability was skew or evident departure from normality suggested. We believe that this experiment should be carried out for a wide range of specific neuroendocrine substances using appropriate conditions of assay, solvents, standards, and data reduction. Such information is important for the valid use of various mathematical algorithms to analyze neuroendocrine data, which commonly assume that experimental uncertainty is restricted to the dependent variable (Y axis) and that for any given value of the independent variable (X axis) the distribution of experimental uncertainty in the Y axis is Gaussian (1, 3, 7).

The precision of an assay can be described by the intrasample variance, standard deviation, or coefficient of variation, each of which can be calculated as a function of hormone dose or concentration (3). For example, the intrasample standard deviations of our growth hormone (GH) IRMA tend to increase initially as a power function (or quadratic function) of dose/concentration, and then over most of the assay range they vary as a linear function of GH concentration. We prefer to relate intrasample variance, or its square root, standard deviation, to hormone dose rather than coefficient of variation, which is a percentage expression. The latter values can be misleading at low and high ends of the working assay range (3).

In addition to experimental uncertainty in any assay system, the sensitivity and specificity of measurements must be defined unambiguously. For example, we and others have observed that certain antisera intended to measure LH concentrations specifically in blood detect variable amounts of uncombined α or β subunits, cross-react with other gonadotropic hormones [e.g., human chorionic gonadotropin (hCG)], and/or yield nonzero estimates even in hypopituitary serum (nonzero "blank" due to serum constituents or other components in the assay) (8). We would emphasize that deconvolution methods provide no more information than can be achieved by the specificity of the assay system employed.

The sensitivity of an assay refers to the minimal concentration of the substance, metabolite, or hormone that can be distinguished from a zero dose, that is, complete absence of the substance. Ideally, sensitivity should be determined using a reaction mixture that is identical to that of the experimental samples but contains zero analyte of interest. However, tissue fluids cannot always be depleted totally of a particular substance. In these circumstances, some modification of the buffer, solvent, or matrix may be required to create a ''zero-dose tube,'' but such modifications may cause an erroneous estimate of sensitivity.

An additional important feature in the analysis of experimental data is assay reliability, which is an index of the reproducibility of the measure. Reliability should be distinguished from validity, which refers to the ability of an assay system to reflect correctly the end point being measured; that is, the assay is as specific as supposed. When assays involve large numbers of experimental samples, care must also be taken to avoid systematic bias introduced into the assay because of drift or sample order-dependent trends in assay performance, for example, shifting baseline, decreasing or increasing sensitivity, and nonuniform experimental imprecision.

III. Specific Deconvolution Techniques

In general, the deconvolution techniques we use can be divided conveniently into two obvious categories. These categories include waveform-defined methods and waveform-independent techniques (1–5, 9–22). By waveform-defined algorithms, we imply that the secretion function in the convolution integral [Eq. (1)] is assigned a specific algebraic form, and that the parameters or constants have some presumptive biological implications. For example, our waveform-specific method assumes that a burst of neuroendocrine secretory activity results in a secretory event of finite amplitude, duration, and location in time. The values of these three parameters of the secretory waveform can be evaluated when the deconvolution technique is applied to experimental data (1–4, 11, 23). On the other hand, our waveform-independent methodology does not assume any particular function for the secretion event, but attempts to assign a numerical value of secretion to each sample observation. The assumptions underlying these two approaches, and the information they require and yield, are different in several respects (see below).

Both waveform-specific and waveform-independent deconvolution methods exist with multiple representatives, which have varying strengths and

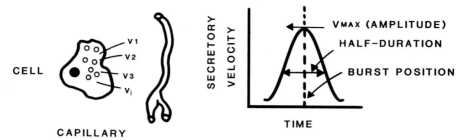

Fig. 3 Proposed model of burstlike neurohormone secretion. A "burst" of neurohormone release can be envisioned as a collection of individual molecular release rates, which are centered at some moment in time and due to the more or less abrupt discharge of molecules by a gland, neural network, or neuroendocrine ensemble. Each molecule of neurohormone is assumed to be secreted at its own individual velocity. The collection or set of such velocities dispersed over a finite time span can be considered a burst. The resultant secretory waveform (plot of secretion rate versus time) may be symmetric or asymmetric (skewed). Wherever possible, the secretory waveform should be assessed by direct observation and its nominal mean (and variance) given algebraically. The stability of the inferred waveform of release should also be evaluated in the various relevant experimental settings that may or may not alter it. [Adapted from J. D. Veldhuis and M. L. Johnson, *J. Neuroendocrinol.* **2**(6), 755 (1991), with permission.]

weaknesses. The full array of deconvolution methodologies in current use will not be reviewed here, but has been discussed elsewhere (1–5, 13). We have chosen an example of each specific deconvolution technique as used in our laboratory and show in detail its specific construction, application, limitations, strengths, and utility.

A. Waveform-Defined Method: A Specific Multiparameter Algorithm

Based in part on intuitive considerations in fluorometry and analytical spectroscopy (14), we have developed a waveform-specific method of deconvolution for neuroendocrine data that is based on the concept of a neuroendocrine secretory "burst" (4). By a secretory burst, we intend to designate a finite set of molecular velocities (secretory rates) of a neurotransmitter, hormone, or metabolite discharged more or less in temporal proximity, before and after which secretion falls to considerably lower (but not necessarily zero) levels (Fig. 3). This definition deliberately embraces an illimitable range of conceivable waveforms that may apply to neuroendocrine secretory activity.

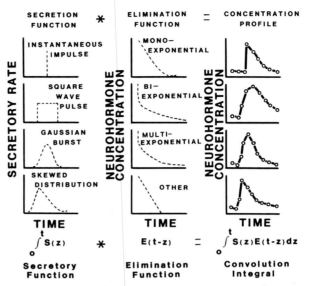

FIG. 4   Examples of several (of many) plausible secretory waveforms and elimination functions, which may be considered in deconvolution analysis. The secretory function, $S(z)$, could assume any of many conceivable forms, some of which are illustrated at left. Various elimination functions, $E(t - z)$, can also be envisioned (middle). Of note, the convolution integral (or mathematical combining of secretion and elimination functions) may therefore assume a variety of specific structures (right). We suggest that independent experiments be carried out to determine the most reasonable secretion and elimination functions to describe the behavior of any particular neurohormone. [Adapted from J. D. Veldhuis and M. L. Johnson, *J. Neuroendocrinol.* **2**(6), 755 (1991), with permission.]

As illustrated in Fig. 4, one extreme example of a secretory burst could involve a nearly instantaneous discharge of presynthesized molecules of neurohormone available for immediate and complete release. This mode of release would be represented mathematically by a δ function, which signifies a theoretically instantaneous increase in secretion rate, which is of zero duration, followed by a virtually instantaneous decrease in secretion rate. To our knowledge, physiological secretory events in general have not approximated a δ function very closely, since secretory episodes have a measurable (nonzero) time extent *in vivo* whether assessed remotely from or close to their source in various experimental animals (2, 6, 9, 16, 17, 20, 21, 24).

An alternative theoretical secretion burst could entail a rapid increase in secretory rate followed by a plateau or constant rate of secretion, which in turn is succeeded by a rapid decrease in secretion. This would constitute an approximately ''square-wave'' secretory event (Fig. 4). This presumptive secretory waveform has not yet been observed *in vivo* or *in vitro* to our knowledge. We would predict that, when this waveform propagates or moves away from its source under physiological conditions in body fluids, the leading and trailing boundaries undergo significant distortion due to admixture, diffusion, turbulence, convection, etc. Hence, a perfect square-wave secretory event translocated away from its source presumably would exhibit more smoothed ascending and descending limbs. Indeed, based on this consideration, we have suggested that this waveform might be approximated *in vivo* by a Gaussian distribution of release rates, at least when the waveform is thoroughly admixed in peripheral blood (1–4). Accordingly, as discussed further below, we have chosen either a Gaussian smoothly contoured symmetric waveform to approximate *in vivo* secretory bursts of anterior pituitary hormones or an asymmetric skewed but smoothly varying function with a more rapid increasing phase followed by a more slowly decreasing phase of secretion (1, 4, 5, 11, 13).

We wish to emphasize that the general form of Eq. (1) allows the use of any of a nearly infinite number of relevant mathematical functions to describe the secretory event, so long as the chosen function contains algebraic expressions whose values can be estimated by iterative or analytical procedures. By iterative, we mean a repetitive approximation by computerized numerical differentiation or integration, and by analytical we mean an explicit algebraic or closed-form solution, such as the general solution for a quadratic equation. Because of the wide choice of possible secretory functions to define neuroendocrine data, we recommend wherever possible that investigators design experiments suited to delineating the actual *in vivo* time course of the neurobiological signal of interest. For example, catheters may be placed in close proximity to the venous effluent of the secretory gland such as the pituitary, adrenal, ovary, or testis. Care must be taken not to disrupt normal blood flow to or from the endocrine gland, to collect blood that is truly representative of that which leaves the gland, and to make measurements of hormone release rates over time at frequent intervals and over a sufficiently prolonged duration to obtain representative information about the secretory behavior of the tissue. Moreover, the physiological state of the host (e.g., blood pressure, temperature, level of consciousness, presence of chemical anesthetic in the blood, and nutritional status) should be representative of the experimental conditions under which further studies will be performed.

A specific waveform-defined methodology that we have applied to the behavior of anterior pituitary hormone secretion *in vivo* is given below by $S(z)$:

$$S(z) = \sum_{i=1}^{n} A_i e^{-1/2(pp_i - z/SD)^2} \tag{3}$$

where *pp* is the center of the theoretical secretory burst in time (i.e., its location in time units), *t* is time, A_i is the theoretical amplitude or maximal rate of secretion attained within the *i*th (or *n*) inferred secretory event(s), and *SD* is the standard deviation, which is related to the half-duration (hd), or duration of the theoretical (calculated) secretory event at half-maximal amplitude, by HD = 2.354 × *SD*. We can add a constant to denote a basal rate of secretion, which may be zero (e.g., pure burst model of secretion). The value of the amplitude is expressed in units of mass of neurohormone secreted per unit distribution volume per unit time. The analytical integral of $S(z)$, or the total mass of neurohormone released in a burst, is given for a Gaussian function by the product of the individual secretory burst amplitude, its *SD*, and a constant (the square root of 2π) (4, 11, 23).

The above secretory function is an algebraic statement of a presumed Gaussian distribution of instantaneous molecular release rates. Intuitively, this representation of a secretory burst assumes that an endocrine gland, or ensemble of neuroendocrine cells, secretes an array of individual molecules, each of which is characterized by its own individual theoretical velocity. The collection of all such velocities within some relevant time interval constitutes a burst. In this model, a burst has specific quantitative features, which include a particular centered location or instant in time when its maximal value occurs (peak position, *pp*), a theoretical maximal secretory rate (the secretory burst amplitude, A_i), and a particular HD (see above). Moreover, any given series of observations would contain either no bursts or a finite number of secretory bursts, the summation of which describes the behavior of the overall secretory pulse train (4, 11, 23, 25). Basal (tonic) secretion can be added to the pulse train in Eq. (3), when desired.

We have recently introduced a distortion or asymmetry term to allow for an unlimited array of skewed secretory bursts (11). This concept is illustrated in Fig. 5. Accordingly, in our multiparameter model, a theoretical secretory burst can be estimated as an asymmetric or skewed event with any of an infinite range of possible slowly or rapidly increasing secretory rates before a maximal value is achieved, followed by a variable decline in secretion toward basal (1, 11, 23, 25). By casting the secretory burst in a particular algebraic form (11), we preserve for mathematical simplicity a constant or unit area of the secretory event during the fitting procedure that is used to estimate the skewness term and other secretory and clearance values.

The overall convolution integral assuming monoexponential disappearance can then be stated as follows [by combining Eqs. (1)–(3) above, with skewness also imparted, if desired]:

$$C(t) = \int_0^t \left[\sum_{i=1}^n A_i \, e^{-1/2(pp_i - z/SD)^2} \right] A \, e^{-k_1(t-z)} \, dz + \varepsilon \qquad (4)$$

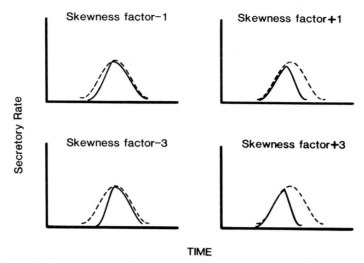

Skewness factor−1   Skewness factor+1

Skewness factor−3   Skewness factor+3

Secretory Rate

TIME

FIG. 5 Quantitative construct of skewed or variably asymmetric secretory bursts. A skewed (e.g., rather than Gaussian) release episode can be formulated mathematically to account for different rates of increasing (before the pulse maximum) and decreasing (after the pulse maximum) secretion within a theoretical secretory burst. All four of the arbitrarily skewed secretory bursts shown here have the same area (integral of secretion over time), center (location in time), amplitude (maximal secretory rate), and half-duration (time in minutes elapsing at half-maximal amplitude). This convenient feature simplifies the deconvolution algorithm, which given a particular half-life can then be used to solve for four particular secretory measures of interest in any given data set: (1) the locations, (2) the amplitudes, (3) the half-durations, and (4) the skewness of the underlying secretory events (see text). The interrupted curves depict a perfect Gaussian. [Adapted from J. D. Veldhuis, A. B. Lassiter, and M. L. Johnson, *Am. J. Physiol.* **259**, E311 (1990), with permission.]

To apply the above convolution integral to biological data, a family of equations must be stated. This family has as its members individual convolution integral equations that correspond to each measured serum hormone concentration. For example, if blood is sampled at 10-min intervals over 24 h, and 145 serum samples are assayed for hormone content, then 145 individual equations each containing a relevant convolution integral are written. Each equation describes the hormone concentration at a particular (sample) time point in relation to all prior secretion and ongoing metabolic clearance. Thus, the resulting family of convolution integrals takes the following form:

$$C(t_1) = \int_0^{t_1} \left[\sum_{i=1}^n A_i\, e^{-1/2(pp_i - z/SD)^2} \right] A\, e^{-k_1(t_1 - z)}\, dz\; +\; \varepsilon$$

$$C(t_2) = \int_0^{t_2} \left[\sum_{i=1}^n A_i\, e^{-1/2(pp_i - z/SD)^2} \right] A\, e^{-k_1(t_2 - z)}\, dz\; +\; \varepsilon \tag{5}$$

$$\vdots$$

$$C(t_j) = \int_0^{t_j} \left[\sum_{i=1}^n A_i\, e^{-1/2(pp_i - z/SD)^2} \right] A\, e^{-k_1(t_j - z)}\, dz\; +\; \varepsilon$$

Therefore, the hormone concentration measured at each given time (e.g., in each sample) is defined by a particular convolution integral that specifies the number, frequency, and duration of all prior secretory bursts and simultaneously designates the elimination rate constant(s). The resultant collection of integral equations is evaluated or solved simultaneously for all relevant secretion and clearance values given statistical assumptions inherent in nonlinear methods of curve fitting (7).

We use a robust iterative nonlinear least-squares method for simultaneously estimating all the individual parameter values of interest in the set of Eqs. (5) above. Such values include the locations of significant secretory bursts, their amplitudes, duration, and mass, as well as the individual half-life (or half-lives) of the hormone in the particular subject and under the particular conditions studied. The convolution integrals are solved analytically, which agrees well with numerical solutions (11). A quadratically convergent iterative methodology that represents a robust modified Gauss–Newton approach (1, 4, 7) is employed to evaluate the set of unknown secretion and clearance values using all the plasma hormone concentration measurements and their intrasample variances considered simultaneously (1, 4, 6, 7, 11). This essentially entails iterative nonlinear curve fitting to estimate multiple values of individual secretion (and clearance) parameters of interest. Such values can be estimated despite highly correlated (absolute value of $r \geq 0.95$) parameters and can be carried out to some chosen degree of precision (e.g., iterations continue until the estimated parameter values are stable within one part in 100,000) (1, 6, 7, 23, 25). Importantly, each parameter (e.g., the amplitude of individual hormone secretory bursts or the hormone half-life) is estimated with corresponding statistical confidence limits. The methodology for determining error propagation associated with each parameter estimate is discussed further below.

Whether an individual computed secretory burst represents a significant event statistically can be assessed from the mean area of the secretory event and the statistical confidence intervals for that area (1, 6, 7). If a proposed secretory event burst embraces an area whose lower bound at 95% confidence limits is not distinguishable from zero, then this peak is omitted and iteration continued with one fewer peaks. In addition, when the distribution of residuals (differences between the predicted curve and the observed serum hormone concentrations) is nonrandom as determined by the runs test, the reverse arrangements test, the Kolmogorov–Smirnov statistic, or autocorrelation, then we insert a putative peak in the region where two or more consecutive residuals are positive. This presumptive peak is then tested for statistical significance, after refitting all parameters, by evaluating whether its analytical area exceeds zero as judged by its appropriate statistical confidence limits.

Of considerable importance is the generality of the foregoing equations and approach. For example, the secretion function, while conveniently stated as a Gaussian, can also be formulated as a skewed asymmetric secretory event (11). Indeed, if the shape of the presumptive secretion waveform is known a priori, then the appropriate alternative secretion wavelet can be employed. Several plausible waveforms are illustrated in Fig. 4. Alternatively, if skew is suspected but not known quantitatively in advance, then it is helpful to know the hormone half-life so that one can iteratively estimate from the convolution integral equations the degree of skewness, as well as the secretory burst number, amplitude, duration, and mass (area). The choice of secretory waveform and elimination function should be motivated by corresponding experimental, clinical, or direct observations concerning the secretion time course and the removal kinetics (1, 4, 23, 25). The assumed waveform should lead to predicted hormone secretory rates and half-lives that agree with published literature values or other appropriate independent estimates. This appears to be true for a Gaussian model of randomly dispersed secretory events, at least as applied to pituitary hormones in healthy men (1, 4, 23, 25).

The appropriateness of the estimated secretion and clearance values in relation to the observed data can be assessed visually as well as mathematically. As shown in Fig. 6, the predicted "reconvolution curve" defined by the convolution integrals and the computed estimates of secretion and clearance should accurately describe the profile of observed serum hormone, metabolite, or neurotransmitter concentrations over time. Importantly, the distribution of residuals should be random normal around a zero mean (see residual plot in Fig. 6). The residuals denote the individual differences between the predicted (fitted, or reconvolved) neurohormone concentration curve over time and the measured neurohormone values (1). If the residuals are nonrandom (e.g., occur as a string of consecutively positive values), then

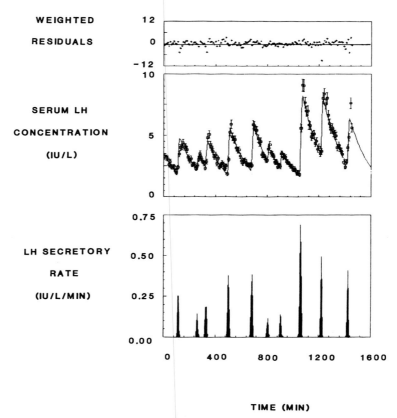

WEIGHTED RESIDUALS

SERUM LH CONCENTRATION (IU/L)

LH SECRETORY RATE (IU/L/MIN)

TIME (MIN)

FIG. 6   Example of multiple-parameter deconvolution analysis of a serum LH concentration profile. The smoothly varying continuous reconvolution (fitted) curve through the observed serum LH concentrations (middle) represents the predicted pulse profile based on the convolution concept. The secretion plot (bottom) consisting of punctuated episodes of LH release was calculated assuming homogeneous symmetric bursts of LH secretion and a single-exponential decay model. The residuals, or differences between the observed and predicted serum LH concentrations, are also shown at each time point (top). For statistical curve-fitting purposes, the sample residuals were weighted inversely as the dose-dependent intrasample variances.

some source of bias in the fit must be considered (e.g., the presence, location, or amplitude of the peak is incorrect and/or the hormone half-life is poorly estimated).

An advantage of the waveform-specific technique is enhanced precision of parameter estimation as data density increases. For example, in a computer model, sampling of blood at 5- to 10-min intervals for deconvolution analysis of serum "hormone" concentrations provides higher precision estimates (reduced confidence interval) of one or more measures of hormone secretory burst amplitude, duration, location in time, and/or simultaneous hormone half-life, compared to the results expected in analyzing data collected (for example) only every 30 min (1, 4). Hence, increased data density in this technique improves resolution. In contrast, techniques that rely on estimating hormone secretion in each sample (e.g., discrete deconvolution) based on concentration differences between consecutive samples show ill-conditioning, which represents the increased tendency of the estimated secretion values to ring, oscillate, or undergo sharp variations from sample to sample when the number of samples is increased (discussed below).

B. Waveform-Independent Methods

In some circumstances, the investigator has independent knowledge of the hormone half-life and would like to estimate the waveform of the underlying neuroendocrine secretory event. Here, a waveform-independent methodology would be suitable, since fewer assumptions about secretory characteristics would be necessary *a priori*. Among various waveform-independent methods [reviewed elsewhere (1, 2)], we have proposed the following general equation for deconvolution analysis:

$$C(t) = \sum_{i=1}^{n} S_i E(t - t_i) H(t - t_i)$$

$$H(t - t_i) = \begin{cases} 1, t - t_i \geq 0 \\ 0, t - t_i < 0 \end{cases}$$

(6)

where $H(t - t_i)$ is a so-called heavy-sided function, which is used to eliminate any effect of clearance on the hormone concentration before it is secreted (1, 2, 12).

In this formulation of waveform-independent deconvolution, a family of equations is written analogously to Eq. (5) above except that: (1) each sample is associated with some secretion rate, S_i, and (2) an investigator-specified hormone half-life must be given, for example, by a one- or two-exponential decay model (12). Iterative nonlinear methods are utilized to solve the family of equations, in which the unknown parameters are the individual sample secretion rates at each observation point. Thus, if blood is sampled at 5-min intervals for 12 h, then 145 individual sample secretion rates are estimated simultaneously from Eq. (6).

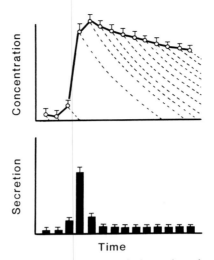

FIG. 7   Schematized model of our waveform-independent deconvolution technique. Observed sample hormone concentrations are considered to arise from the combined effects of individual sample secretion rates (bottom) and known clearance kinetics or decay (top). Error intrinsic in the hormone half-life estimate and dose-dependent experimental variance inherent in the hormone assay are both used to propagate conjoint error spaces and thereby estimate the statistical confidence limits for each of the calculated secretion values. In this model, some a priori knowledge or assumption regarding the half-life of hormone disappearance (decay) must be available.

Our waveform-independent approach is illustrated intuitively in Fig. 7. Note that to calculate an individual secretory rate that corresponds to each sample neurohormone concentration, the half-life of removal of the neurohormone from the sampling space must be assumed. Indeed, one cannot readily solve a matrix of unknown sample secretory rates (e.g., 145 individual secretory rates for 5-min sampling over 12 h) and simultaneously estimate one or more rate constants for an exponential decay function, since the number of degrees of freedom for the analysis would approach or fall below zero (1, 7, 12). Even when the hormone half-life is assumed from published estimates or independent measurements, few statistical degrees of freedom remain in the above waveform-independent deconvolution technique. Statistical power is gained in some measure, because many of the sample secretory rates fall to zero. Importantly, the amount of "ringing" or random oscillations in the secretory profiles is reduced compared to earlier methods that estimate each sample secretory rate from the difference in the concentrations of two consecutive samples. Consequently, an efficient iterative computer algorithm is required for convergent estimates of multiple sample hormone secretory rates. In this circumstance, we suggest the use of a linear convergence procedure, which although less rapid can deal with multiple highly correlated parameters (1, 7). In addition, we have adopted the use of a nonlinearly weighted smoothing function to minimize sharper transients in the secretory plot (1).

The applications of our waveform-specific and waveform-independent deconvolution method to a 16-h serum LH concentration profile are shown in Fig. 8. Note that the reconvolution curves (predicted profile of serum hormone concentrations over time) fit the observed data well. For comparison, we show predictions of a discrete deconvolution method from another laboratory. We note that, based on appropriate methods for error propagation (7, 12), our waveform-independent methodology estimates sample secretion rates and the associated standard error or statistical confidence limit for each estimate. To detect significant episodes of secretion, we then apply two criteria: (1) a peak of secretion must exhibit a significantly positive first derivative, followed by a significantly negative first derivative, of secretion (i.e., respectively, a significant rise in the rate of secretion over time followed by a significant fall) (1, 7, 12); and (2) the maximal rate of secretion attained within a secretory peak must exceed zero by some particular statistical boundary (e.g., two standard deviations). These requirements can be tested by evaluating the sample secretion rate and its derivative in relation to their individual statistical confidence limits. Although methods of error definition and propagation are discussed further below, in brief, our particular waveform-independent method evaluates the variance in each sample secretory rate in relation to dose-dependent intraassay variations and the experimental uncertainty inherent in the half-life estimate (1, 12).

Waveform-independent methods have the disadvantage that the hormone half-life must be known or assumed or constrained within a relatively narrow range a priori and/or that basal hormone secretion must be zero (1). However, such information is not necessarily available, may not be directly applicable to the experimental context under study, may not be derived correctly in relation to the particular isohormone being evaluated, or the specificity of particular assay, etc. In addition, a waveform-independent deconvolution methodology may have relatively few degrees of freedom and hence putatively less statistical power. Other waveform-independent methodologies may also exhibit similar or additional difficulties, including ill-conditioning (increasing variability in the secretion rate estimate at higher sample densities) (see Fig. 8 and below). In addition, in some algorithms, small variations in the alleged or computed half-life can produce significant changes in the number, amplitude, or duration of peaks detected and/or the amount of estimated basal secretion (1, 2, 14, 26, 27).

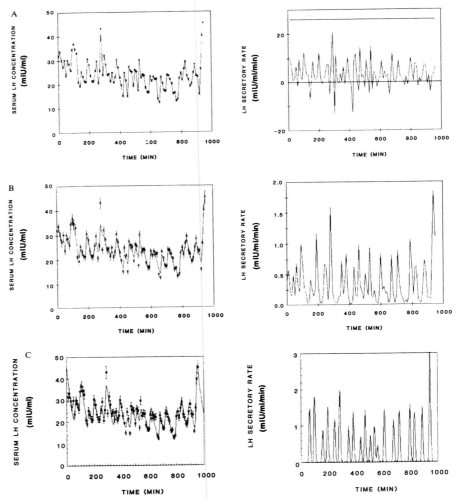

FIG. 8   Illustrative profiles of deconvolution-computed bioactive LH secretory events in a normal man. Resolution of LH secretory events by a discrete deconvolution algorithm of another laboratory (A), our waveform-independent technique (B), and our multiple-parameter method (C) is shown. Blood was sampled at 10-min intervals for 16 hr and submitted to bioassay using the rat interstitial-cell testosterone bioassay. The left plots show the data without (points simply connected but not fit) (A) or with computer-assisted fits (B and C). The right plots show the deconvolution estimates of LH secretory rates, which were obtained assuming a two-component LH disappearance model reported independently. In (A), secretion rates sometimes

C. Others

Other deconvolution methods have been considered and are reviewed in detail elsewhere (1–3, 5). One interesting recent approach suggested by the laboratory of Munson involves transforming the neurohormone concentration measurements into their corresponding natural logarithms, and then estimating the hormone half-life by the linear decline in the logarithm of hormone concentrations over time (27). This method assumes a single-exponential decay and zero basal secretion, and therefore cannot be applied in circumstances of known or presumed basal secretion which would falsely increase the half-life. Possible peaks are inserted individually so as to reduce the fitted variance of the data significantly while recalculating the half-life. A criterion for retaining peaks is the Akaike information content statistic. A stated difficulty with this approach is its vulnerability in estimating peak number to small changes in the fitted hormone half-life (27).

D. Ideal Methods

We believe that ideal novel deconvolution techniques would incorporate several specific features. The algorithms would allow for both waveform-defined and waveform-independent curve fitting of the neurohormone concentration profiles over time. The methodology should perform at an improved level with increased data density, rather than deteriorate as would be typical of an ill-conditioned problem. Moreover, statistically based estimates of individual parameter values (i.e., with corresponding statistical confidence intervals) should be carried out based on the degree of measurement error and other sources of experimental uncertainty that contribute to the data. As importantly, any methodology should be independently validated by assessing its performance on synthetic (computer-generated) data as well as *in vivo* biological data (28, 29). Where possible, the methodology should be capable of estimating both basal and pulsatile hormone secretion and assigning weights to their relative contributions. The convolution model should allow for one or more compartments of neurohormone binding, sequestration, metabolic degradation, and/or interconversion. A comparison of the predictions of the model and the observed data should reveal randomly

---

oscillate about zero ("ringing"). In contrast, methods (B) and (C) have positivity constraints. Formal comparisons have not yet been made among various deconvolution methodologies in the neuroendocrine field. [Adapted from J. D. Veldhuis and M. L. Johnson, *J. Neuroendocrinol.* **2**(6), 755 (1991), with permission.]

distributed residuals with no evidence of systematic bias in the fitting procedure and with a high level of goodness of fit assessed objectively. Finally, in
principle, the number of secretory bursts contained in the data should be
estimated with a corresponding standard error of the estimate, and information provided about the changes expected in the frequency estimate in relation
to different threshold stringencies and assumed or computed half-lives.

IV. Current Problems, Limitations, and Suggested Solutions in Deconvolution Approaches to Neuroendocrine Data

A. Ill-conditioning

In this section, we describe some of the mathematical features of deconvolution that can lead to high sensitivity to noise and certain methods to deal with
this difficulty. The convolution integral [Eq. (1)] can be written in matrix
notation as $c = ES$, where c is a vector of concentrations of any given
hormone, with components $c_n = c(t_n)$ measured at time t_n. S is a vector of
secretory rate, with components $S_n - S(t_n)$, and E is a matrix expressing
excretion or removal of the hormone, with components $E_{nm} = E(t_n - t_m)$.
In this notation, c is measured, S is unknown, and for simplicity let us assume
that E is also known. In the absence of noise, the solution S is easy to state:
$S = E^{-1}c$, where E^{-1} is the matrix inverse of E. While this is the formal
solution, it is often difficult to implement in practice, since the inverse of E
may be difficult to obtain.

It may appear at first glance that computing E^{-1} should pose no particular
difficulties, since e is a triangular matrix (since E is causal, $E_{nm} = 0$ for $n <
m$). Indeed this is true but only obscures the manner in which E^{-1} can
magnify the effects of any noise in the signal c. To address this question, one
needs to examine the so-called condition number γ of E (see, e.g., Ref. 30).
We may define γ as the ratio of the highest to lowest eigenvalue of the
(symmetric) matrix E^TE, where E^T denotes the transpose of E (some authors
define γ as the square root of the ratio). The important point is that the
magnitude of γ is a measure of how close E is to being singular (i.e., impossible
to invert) or, equivalently, the extent to which high-frequency variations in
the secretory rate S are significantly smoothed out by the excretion function
E. This implies that even small wiggles in the measured c can be magnified,
by multiplying by E^{-1}, to unacceptably high levels of variation in the estimated S.

Let us estimate γ in a particularly simple case of monoexponential excretion (i.e., removal of the given hormone by first-order kinetics). Let $a =$

$\exp(-k\,\Delta t)$, where k is the rate constant and Δt is sampling interval. Then E_{nm} is given by $a^{(n-m)}$ for $n \geq m$ and O for $n < m$. For a large data set, the maximum and minimum eigenvalues of $E^T E$ are approximately given by $(1 - a)^{-2}$ and $(1 + a)^{-2}$, respectively, so that γ equals $(1 + a)^2/(1 - a)^2$. As an example, for a hormone with $k^{-1} = 60$ min, and for which Δt is 5-min sampling, a $= 1 - 1/12$, and γ equals 600. The fact that γ is large indicates that variations in c on the order of 5 min will lead to large variations in the estimated secretory rate S. Notice the competing effects here. Lengthening the sampling interval Δt reduces a and improves the condition number γ, but at the expense of missing fine details and features in the very phenomenon of interest, namely, S. By contrast, more frequent sampling probes finer details of S, but at the cost of worsening the condition number and making the estimate of S more difficult.

A second way in which to appreciate this problem is in the frequency domain. Note that E_{nm} is a function only of the time difference $t_n - t_m$. In such displacement kernel problems, Fourier transforms can be a useful tool for estimating S, because no matrices need be inverted and because fast computational algorithms exist for computing the discrete Fourier transform (31). The first property arises from the transform of Eq. (1), which states that $\tilde{c}(w) = \tilde{E}(w)\tilde{S}(w)$, where the tilde denotes the Fourier transform, and the product is simply a scalar multiplication at each frequency w. The solution is given by a scalar division, $\tilde{S}(w) = \tilde{c}(w)/\tilde{E}(w)$. Causal excretion functions (e.g., single or multiple exponential decays) have the property that $|\tilde{E}(w)| \rightarrow 0$ like w^{-1} as $w \rightarrow \infty$. Furthermore, unavoidable contamination of $c(t)$ by white noise implies that $|\tilde{c}(w)| \rightarrow$ positive constant for $w \rightarrow \infty$. Thus the ratio $\tilde{c}(w)/\tilde{E}(w)$ becomes large at high frequencies, and indeed is dominated completely by the noise in that limit. A popular method of dealing with this is by Wiener filtering, which in effect smooths the data by multiplying $\tilde{c}/\tilde{E}$ by an appropriate (real) function of w which goes to zero for large w, thus canceling the high-frequency magnification of the noise.

Returning to the time domain, there are many methods to deal with ill-conditioning. These can be divided into two broad categories. We use both of these strategies in the deconvolution procedures outlined here. First, there are regularization methods (32–34), which in effect force a given level of smoothness to the recovered estimate of secretory rate $S(t)$. Reducing the space of possible functions S by convenient parameterization, as in Section III,A above, falls into this category. By allowing a slight departure of the fit from ideal, one gains a substantial improvement in stability and robustness of the resulting estimates of $S(t)$. The second category involves the use of mathematical constraints (30, 35). These are typically nonnegative constraints applied to the function S, since it is known a priori that secretion cannot be negative. Even by itself, the use of a nonnegativity constraint can often improve the behavior of a deconvolution method significantly. It must be emphasized that constraints are inherently nonlinear, and so analytical estimates of error propagation are correspondingly more difficult. Constrained solutions involve the Kuhn–Tucker conditions, which say that when S is positive it satisfies a least-squares criterion, whereas when S is zero the fit could only be improved by making S negative. It is the latter case wherein the constraint is binding, and prevents, for example, oscillations of recovered secretory rates about zero due to noise in $c(t)$, when the true S in fact is zero.

There is some ambiguity in recovering $S(t)$ from $c(t)$ when E is multiexponential, that is, the excretion process is more complex than first-order kinetics, which may arise when more than one compartment is involved or when intermediary metabolites contribute. In particular, for a monoexponential process, $c(t)$ is completely determined by $S(t)$, $E(t)$, and one initial condition, for example, the concentration at time zero, $c(0)$. For multiexponential processes, $c(t)$ is not determined by S, E, and $c(0)$. Rather, for nth order processes, in general n initial conditions must be specified for $c(t)$ to be determined. This has a practical consequence in the inverse, or deconvolution problem. That is, given $c(t)$ and E for a multicompartment model, it is not possible even in the absence of noise to reconstruct $S(t)$ without some additional assumptions about the initial conditions. Most commonly, it is assumed that all compartments are initially at concentrations of zero, and that, in consequence, the resulting $c(t)$ can be written as a single convolution integral or sum, with $c(0)$ being attained by an initial (virtual) spike at $t = 0$. But it must be emphasized that the original convolution equation itself requires these assumptions at $t = 0$. If different conditions obtain at $t = 0$, then caution is required in the interpretation of the recovered $S(t)$, at least for a time interval spanning several inverse time constants.

B. Error Definition and Propagation

An essential feature of statistically valid deconvolution approaches is the correct evaluation of experimental uncertainty. This is referred to as error propagation. Error assessment includes identifying all relevant sources of experimental variation in the data and evaluating their individual and joint influences on estimates of the calculated parameters (1, 7). We have used various individual statistical approaches in nonlinear systems to construct error profiles, which typically assume asymmetric highly correlated variance spaces when multiple parameters are involved (7). One class of techniques we use includes Monte Carlo procedures, in which random perturbations are made in the apparent parameter values and the effects on the resultant "fitted variance" are assessed. (The fitted variance is a measure of relative goodness of fit, and it represents the square root of the sum of the squares of the differences between the observed data and the fitted curve.) By varying two or more parameters simultaneously, one can create a joint variance or error space that describes the impact of small changes in several parameter values on the fit of the predicted curve. As reviewed elsewhere (1, 7), error propagation is a *sine qua non* of appropriate deconvolution techniques, since proper error propagation permits the investigator to evaluate statistical confidence intervals for the secretion and clearance terms of interest.

The absolute values of the statistical confidence limits for individual secretion and clearance estimates will depend on many factors, such as the quality of the data (the degree of experimental variation in the sample measurements, etc.), the applicability of the convolution model to the data, and the number and density of observations (which will control the number of degrees of freedom of the fitted model). Modeling studies using our multiparameter method indicate that, as anticipated, the spans of confidence intervals tend to vary inversely with sample number and directly with sample measurement uncertainty. We caution that deconvolution methods that do not provide (correct) error estimates in the secretion calculations are largely uninformative, since one cannot determine whether a putative secretory event or sample secretion rate is statistically significantly different from zero, or whether two estimates of half-lives are statistically different, etc.

C. Adequacy of Fit

A deconvolution solution that is adequate and also not overdetermined requires, as a minimum, that each added parameter significantly reduces the fitted variance, for example, as assessed by an F ratio test of the fitted variances or the Akaike information content index (1, 7). In addition, we require that the distribution of residuals, which are differences between the fitted curve and observed data, be random, for example, as evaluated by autocorrelation, the runs or reverse arrangements test, or the Kolmogorov–Smirnov statistic (1, 7). Importantly, any consistent discrepancies between the predicted curve and the observed data should be examined for possible underlying systematic bias. Finally, predictions of the deconvolution analysis and the validity of peak detection should be tested for accuracy by independent experimental means (see below, Section IV,E).

D. Weighting Functions Used in Reconvolution Curve Fitting

Neuroendocrine data often contain significant measurement variability (e.g., coefficients of variation of 3–20%). Such experimental variability is commonly dose-, assay-, hormone-, and condition-dependent. Accordingly, suit-

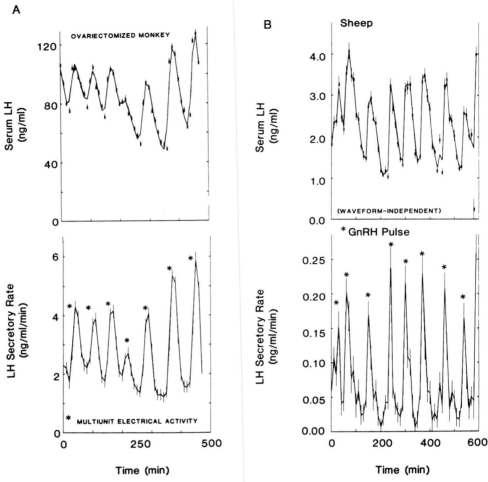

FIG. 9 (A) Correspondence between waveform-independent deconvolution-resolved LH secretory events (bottom) estimated from LH concentrations in the peripheral blood of rhesus monkeys (top, with computer-fitted curve) and simultaneously monitored electrophysiological activity in the mediobasal hypothalamus. Individual multiunit electrical discharges, which presumptively correlate with endogenous GnRH release episodes, are identified by asterisks and serve to mark the LH secretory events independently. [Data from J. D. Veldhuis and M. L. Johnson, *J. Neuroendocrinol.* **2**(6), 755 (1991), with permission.] (B) Concordance between LH secretory bursts, as resolved by waveform-independent deconvolution of LH concentrations measured in blood collected from the internal jugular vein of an ovariectomized ewe, and independently monitored GnRH release episodes in the hypophyseal–portal blood of

able and accurate estimates of measurement imprecision and other sources of experimental variations are necessary and useful. For example, in addition to their use in error propagation, the dose-dependent intrasample variances that we calculate can be used in an inverse weighting function in the iterative computation of the parameter values (1, 7). Measurements associated with larger experimental uncertainty (larger standard deviations) receive reduced weight in the fit compared to data with greater precision (smaller standard deviations).

E. Determining Peak Number (*Type I and Type II Statistical Errors*)

We believe that an important property of evaluating many kinds of neuroendocrine data is an accurate appraisal of the number of discrete release episodes contained within the observed time frame. Statistically, the estimation of a secretory burst number can be confounded by either a type I or type II error. A type I statistical error consists of falsely assigning a secretory peak when no event occurred. This can be equated with the concept of an α value in statistics, which denotes the probability of falsely rejecting the null hypothesis of no difference or no event. On the other hand, a type II statistical error consists of overlooking a significant peak in the data and is represented quantitatively by β. β denotes the probability of falsely accepting the null hypothesis of no pulse present in that region of the data. The power of the analysis is given by "1 minus β." Ideally, an investigator should minimize both type I (false-positive) and type II (false-negative) errors in the estimation of secretory peak number.

To determine how many false-positive or false-negative errors occur in the course of peak enumeration, the investigator requires independent and persuasive knowledge of the presence or absence of a peak within a particular data region. This information is often difficult to obtain *in vivo*. We have used several specific approaches to this end. Indeed, we prefer a combination of *in vivo* biological validation as well as mathematical (computer-synthesized) methods. For *in vivo* biological validation in the case of the LH pulse signal, we have utilized data collected by electrophysiological monitoring of the medial basal hypothalamus of the rhesus monkey, which was carried out in

the same animal. Measured serum LH concentrations are given at top with the predicted fit of the data. Deconvolution estimates of LH secretion are shown at bottom. Asterisks denote when GnRH peaks were found in the pituitary portal blood. [Data from I. J. Clarke (Melbourne, Australia).]

the laboratory of Ernst Knobil (28, 29). We have applied deconvolution analysis to serial measurements of serum LH concentrations in this *in vivo* animal model, and compared our enumeration of significant LH secretory bursts with the occurrence of multiunit electrical activity [a putative correlate of gonadotropin-releasing hormone (GnRH) neuronal activation] recorded independently in the hypothalamus. As shown in Fig. 9A, the deconvolution-estimated LH secretory profile in the rhesus monkey exhibited abrupt increases in LH secretory rates at episodic intervals, and these increases corresponded well with hypothalamic multiunit electrophysiological activity in the same animal (36).

As a second biological validating tool, we have utilized the data of Ian Clarke and co-workers, who have carried out frequent sampling of hypothalamopituitary portal venous blood in ovariectomized sheep (37). In this neuroendocrine model, episodes of GnRH secretion into portal blood can be considered putative independent markers of probable LH secretory bursts. As shown in Fig. 9B, deconvolution analysis of serum LH concentration profiles collected from the internal jugular vein of such sheep showed distinct LH secretory events scattered over the sampling session. The timing of the secretory events coincided significantly with the occurrence of peaks in portal blood GnRH concentrations.

Finally, as a third biological model for validating deconvolution detection of gonadotropin peaks, we have used data from William F. Crowley, who has studied men with idiopathic hypogonadotropic hypogonadism (e.g., Kallmann's syndrome) (38). These men were selected because of the absence of spontaneous LH pulsatility in the untreated state. However, LH peaks can be evoked by exogenous GnRH injections in these individuals. Thus, the correct identification of a GnRH-induced LH secretory burst can be denoted as a true-positive finding by deconvolution analysis. Overlooking such an event would be a false-negative or type II error, whereas falsely inserting an LH peak when no GnRH was injected would be a false-positive or type I error. Under these conditions, in 10 men we find a very high correspondence (>95%) between deconvolution-specified LH release episodes and exogenous GnRH injections.

In addition to the above *in vivo* biological validation, we have employed mathematical convolution models in which computer-synthesized hormone "pulse profiles" can be submitted to deconvolution analysis. As shown in Fig. 10, the convolution integral [Eqs. (1)–(4)] can be utilized to generate a synthetic time course of apparent serum hormone concentrations confounded by any desired degree of simulated experimental variation (noise) (11, 23, 25). Because the synthetic data are specified entirely by the convolution integral chosen, this approach permits one to generate pulse trains of various frequencies, amplitudes, burst duration, pulse mass, and burst asymmetry

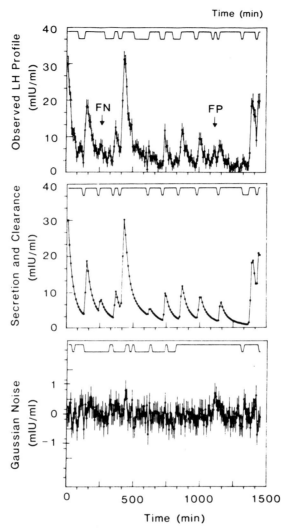

FIG. 10   Use of convolution modeling to test the performance accuracy of discrete peak-detection algorithms. The graphs illustrate the impact of random experimental variance on false-positive (FP) and false-negative (FN) errors exhibited by a particular discrete peak-detection algorithm applied to simulated plasma LH concentration profiles. Synthetic profiles were created by convolving constituent secretory and clearance functions, which were further perturbed with random biological and/or procedural noise. [Adapted from R. J. Urban, W. S. Evans, A. D. Rogol, D. L. Kaiser, M. L. Johnson, and J. D. Veldhuis, *Endocr. Rev.* **9,** 3 (1988), with permission.] Deflections above the graphs denote computer-identified peaks in the data.

(if desired) with or without baseline hormone secretion and with preferred half-lives. The deconvolution algorithm is then tested on the synthetic hormone profiles over time. Failure to detect a significant known synthetic hormone pulse is denoted as a false-negative error, whereas the incorrect detection of noise as an apparent peak is characterized as a false-positive error.

Using this construct of mathematically simulated neuroendocrine pulsatility, the following general findings have emerged for the two deconvolution methods described here: (1) the presence of both basal secretion and pulsatile hormone release makes the estimation of hormone half-life more difficult; specifically, if basal secretion were overlooked, the calculated half-life would be falsely prolonged; (2) increasing data density enhances the performance of the waveform-specific multiple-parameter deconvolution technique; and (3) low secretory pulse amplitude, high basal secretion, prolonged hormone half-life, increased experimental uncertainty (noise simulated in the data), and extended secretory burst duration for any given mass of hormone released all tend to attenuate the detection of discrete neuroendocrine secretory events. Although several reports have used synthetic pulse trains to test discrete peak-detection methods that are not deconvolution based, virtually no comparable observations (except those cited above) are available for specific deconvolution algorithms.

F. Estimating Combined Basal and Pulsatile Neurohormone Secretion

Although most efforts to analyze hormone concentration versus time series have focused on the identification and characterization of episodic fluctuations (pulses/secretory episodes) within such series, attention must also be directed to the so-called interpulse portions of the hormone profiles. In certain pathophysiologic situations, interpulse hormone concentrations are elevated (e.g., LH and FSH in postmenopausal women; GH in patients with acromegaly). Of the several mechanisms that could account for raised interpulse hormone concentrations, three of particular interest include (1) prolonged half-life of a secretory product such that a significant amount of the hormone secreted within a given burst is not cleared by the time a subsequent burst occurs, (2) true "tonic" or "basal" secretion occurring between the major secretory events, and (3) microbursts of hormone release.

Whether increased interpulse hormone concentrations reflect alterations in half-life and/or a tonic secretory process can be difficult to address unambiguously. Deconvolution analysis provides a potential tool with which to evaluate changes in half-life versus tonic secretion. However, some methods require a priori knowledge of half-life or tonic secretion (e.g., waveform-independent deconvolution). In contrast, waveform-dependent deconvolution can allow for the simultaneous estimation of both half-life and tonic secretion. A difficulty is that certain of the desired parameters, such as half-life, secretory burst amplitude, and basal secretion, are highly cross-correlated (i.e., nonorthogonal). The consequence of such correlations between fitted parameters is that uncertainties in the determination of one or more of these parameters can impart systematic errors to estimates of the other parameters. This problem becomes particularly significant when the cross-correlation coefficients between the fitting parameters approach ±1. Indeed as a cross-correlation coefficient approaches ±1, then a variation in one of the parameters can be nearly totally compensated for by a corresponding variation in another parameter; thus, unique values of the parameters cannot be determined without more information.

The cross-correlation between half-life and basal secretion rate is determined by several different variables, such as the time between successive data points and the time between successive secretion events. For half-life and basal secretion to be separable computationally, we suggest a minimum of three to five data points collected per half-life and three or more half-lives observed per secretion event. The former of these conditions can be met by altering the data collection protocol, but one has little external control over the number of half-lives per secretory event. If the data set contains fewer than these limits, the cross-correlation between the parameters can approach ±1 and so the parameter values may not be estimated adequately.

We have recently developed a method to improve the cross-correlation between the parameters when the secretion events occur more often than once every other half-life. This is done by stimulating a large secretory peak via an exogenously administered secretagogue, which is given several half-lives before the end of the data collection period. The large, generated peak can provide sufficient information about the decay process to reduce the cross-correlation between the parameters. An example of this approach is a computer-simulated 24-hr LH data set known to contain 17 secretory episodes (associated with an average half-life of 100 min) superimposed on a tonic secretory rate of 0.07 mIU/ml/min. When the basal component and half-life were estimated simultaneously using waveform-dependent deconvolution, 17 secretory episodes were correctly identified. However, the basal component was underestimated (0.041 mIU/ml/min) and the half-life overestimated (128 min). The correlation coefficient between the basal component and half-life was +0.996. When the same 24-hr data series was extended by 6 hr to include a large simulated secretagogue-evoked secretory episode, deconvolution analysis then yielded an estimated average half-life of 103 min, associated with a lowering of the basal component/half-life correlation coefficient to 0.987. The estimate of the basal component also improved considerably (0.063 mIU/ml/min). Thus, insertion of a stimulated hormone peak can provide additional statistical power and parameter information that may be adequate to decrease the cross-correlation between the half-life and basal secretion, and thereby allow unique values to be estimated.

G. Integrating a Systems View

Deconvolution analysis, as described above, evaluates only a limited aspect of a larger, more complex multidimensional neuroendocrine system, which has dynamic and interacting characteristics *in vivo*. For example, in the gonadotropic axis, release of LH and FSH triggers secretion of gonadal steroid and glycoprotein hormones, which in turn modify the release of hypothalamic signals that regulate the pituitary gland on the one hand, and influence the responsiveness of the anterior pituitary gland to these hypothalamic signals on the other. Such signals from the gonad can be either positive or negative. Consequently, a broader description of neuroendocrine systems as positive and negative feedback control loops must ultimately be incorporated into the high-resolution quantitative methodology of deconvolution analysis. Indeed, in the somatotropic axis, secretion of GH is controlled by both stimulatory and inhibitory hypothalamic factors [GH-releasing hormone (GRHR) and somatostatin], and the somatotroph is subject to modulating influences of sex steroid hormones, glucocorticoids, thyroid hormones, etc. In addition, GH itself and the product of GH action, insulin-like growth factor I (IGF-I), are capable of exerting significant negative feedback effects on GH secretion at several levels in the axis. A further complicating element in the systems view of the somatotropic axis is the presence of multiple high-affinity binding proteins in plasma, which variously bind not only GH but also the product of GH action, namely, IGF-I. A more comprehensive statement of the systems activity of a neuroendocrine axis will require incorporation of these additional regulated elements and their feedback controls.

V. Other Applications and Implications of Deconvolution Techniques

A. Model Synthesis

The proestrous LH surge in the rat consists of a rapid, acute increase in plasma LH concentrations. In principle, this massive rise in blood concentrations of LH could result from changes in one or more components of the neuroendocrine secretory and clearance system, namely, an increase in (1) the mass of LH secreted per burst, (2) the number of LH secretory bursts,

(3) the amplitude or duration of individual LH secretory bursts, (4) the half-life of secreted LH molecules, and (5) the amount of basal LH release. Modeling studies indicate that a combination of several mechanisms acting in concert would increase plasma LH concentrations in a multifold fashion (23, 25).

Deconvolution analysis can also be applied to experimental conditions in which the effect of hypothalamic-releasing factor is rapidly abolished. For example, injection of a synthetic antagonist of the decapeptide GnRH rapidly reduces plasma LH concentrations in postmenopausal individuals (39). Deconvolution analysis reveals that the GnRH antagonist at high doses essentially eliminates pulsatile LH secretion without influencing the apparent half-life of endogenous LH. Low or near-zero basal concentrations of LH that follow administration of a high dose of GnRH antagonist may be maintained by (1) low-amplitude LH secretory events not detected by 20-min blood sampling and/or (2) low levels of persistent or newly emerging basal (tonic, interpulse) LH secretion. More frequent blood sampling and increased assay precision would be required to distinguish between these two possibilities. Indeed, in young men treated with a GnRH antagonist, in whom blood is sampled at more frequent intervals and the plasma submitted to LH bioassay using the *in vitro* testosterone-secreting Leydig cells, residual low-amplitude LH release episodes can be recognized (40). Such episodes are of insufficient amplitude to trigger substantial testosterone secretion by the testes *in vivo*.

Other selected discoveries of the above methods of deconvolution analysis applied to evaluating neuroendocrine regulatory mechanisms include the following: (1) the amplitude-dependent decrease in postpartum hyperprolactinemia observed in healthy breastfeeding women (41); (2) the amplitude-specific increase in pulsatile prolactin secretion identified in healthy postmenopausal women treated with estradiol (42); (3) the ability of exogenous steroid hormones to modify the apparent half-life of FSH and the amplitude of spontaneous FSH secretory bursts in healthy men (43); (4) the finding that either amplitude and/or frequency control of otherwise randomly dispersed anterior pituitary hormone secretory bursts can give rise to the individual hormone-specific 24-hr rhythms in serum hormone concentrations (12); (5) a recent inference that the amplification by estrogen of GnRH action (induction of so-called self-priming by GnRH) can be accounted for by an increase in the duration of GnRH-stimulated LH secretory bursts with no significant change in their amplitude or in the half-life of secreted LH molecules (44); (6) the observations that the negative influences of obesity and aging on mean serum GH concentrations are endowed by a reduced GH half-life and decreased pulsatile GH secretion, the latter being mediated by suppression of amplitude and frequency of GH secretory bursts (45, 46); and (7) the conclusion that amplitude-dependent modulation occurs for multiple hormones, such as GH release in puberty (47), aldosterone and renin secretion in low-salt states (48), and adrenocorticotropin (ACTH) secretory bursts as a function of time of day (49). Accordingly, there is a diverse constellation of specific neuroendocrine regulatory mechanisms that can subserve pathophysiological variations in hormone secretion and clearance. Deconvolution techniques offer one (of many) tool(s) to explore these mechanisms *in vivo*.

B. Pure Secretion Profiles

The concept of a convolution integral can be simplified to evaluate pure secretory profiles, when the secretion pattern can be represented by the summation of discrete secretory events each of finite duration, amplitude, and location in time with or without basal release. For example, this idea can be applied to GnRH secretion measured in hypothalamopituitary portal blood. In this case, we can assume the complete absence of an elimination function, which is therefore assigned a value of unity in Eq. (1). The observed pattern of GnRH secretion over time might be accounted for by either (1) summating (overlapping) bursts of discrete GnRH release [Eq. (3)] or (2) individual bursts of GnRH secretion superimposed on a nonzero rate of basal GnRH secretion [Eq. (3) plus a basal (constant) term]. These two possibilities are difficult to distinguish purely on mathematical grounds.

Another example of possible summated secretion is the LH or FSH (or other) secretory profile obtained by collecting blood at frequent intervals from a conscious unrestrained animal via a catheter placed retrograde near the pituitary gland, for example, into the inferior petrosal sinus (Fig. 2). Such profiles can also be constructed as a summation of individual, partially overlapping secretory events or as punctuated release episodes superimposed on low levels of basal (tonic, time-invariant) hormone secretion. Independent additional information is required to distinguish between these two possibilities. However, either model illustrates the utility of nonsinusoidal waveforms to describe neuroendocrine events. Indeed, summated sinusoids (e.g., as generated by simple Fourier transformation) provide a very poor description of many physiological neuroendocrine variations. In contrast, the fit of neuroendocrine time series reconstructed by deconvolution analysis accounts for 90–95% of the variance inherent in the data. The remaining variability is presumptively due to intraassay measurement error and experimental uncertainty introduced by the biological preparation as well as the collection and processing of the sample for hormonal assay.

C. Multiple Pulse Generators

Of particular interest in some neuroendocrine investigations is the coordinate or time-linked secretion of neurotransmitters, hormones, and/or substrates and metabolites. For example, one necessary (but not sufficient) condition for the inferred dependence of pulsatile GH release on antecedent episodes of GHRH secretion is a high degree of concordance between hypothalamopituitary portal blood GHRH pulses and subsequent internal jugular pulses of GH release. To assess whether the degree of observed concordance between GHRH pulses and GH peaks is statistically significant (i.e., above that expected on the basis of chance associations alone), we need a specific quantitative hypothesis of the expected random concordance rate, its variance, and the corresponding probability distribution function. To this end, we initially utilized computer simulations based on the concept of bursts of hormone release. This was simplified to an "on–off" or binary model of randomly distributed hormone secretory peak maxima. These simulations provided estimates of expected random peak concordance rates for any particular specified combination of two or more pulsatile trains (37, 50). As anticipated intuitively, increased pulse frequency in paired pulsatile profiles leads to a higher number of randomly concordant (coincident) events.

Further computer simulations indicated that the particular number of expected randomly coincident events (e.g., peak maxima) between two pulse trains, with respective individual peak frequencies of m and n, is

$$x_e = \frac{mn}{z} \tag{7}$$

where z is the number of samples in each series (or, more accurately, the number of samples held in common by the two series). If peak maxima cannot be identified in the first and last samples, then the term z is replaced by $(z - 2)$. In addition, we found that the standard deviation or its square, the variance, of the expected mean number of randomly coincident events between two uncoupled pulsatile hormone series can be given as

$$x_{var} = \frac{mn}{z}\left(1 - \frac{m}{z}\right)\left(1 - \frac{n-1}{z-1}\right) \tag{8}$$

The probability of observing exactly x coincident events between two such pulsating and statistically independent series is given as the quotient in the following combinatorials, namely,

$$P(x) = \frac{\binom{m}{x}\binom{z - m}{n - x}}{\binom{z}{n}} \tag{9}$$

for $x = 0, 1, \ldots, \min(m, n)$. Equations (7), (8), and (9) match predictions of computer simulations and are implied by the hypergeometric probability density function (37, 50).

The above reasoning can be extended to three or more pulse generators firing independently, which would be expected to produce occasional synchrony between the various possible pairs of hormones as well as infrequent random coincidences among three or more hormones. Our modeling studies have identified algebraic descriptions that will approximate the expected behavior of three or more pulse generators activated independently, if we assume that secretory peaks can be considered as discrete events (e.g., a single-sample peak maximum can be used to define the location of any given pulse), and that such discrete event markers are distributed randomly over the sampling interval in a uniform noninteracting and nondiurnally varying manner (37, 50). Thus, P values can be calculated for the expectation of observing on the basis of chance alone any particular number of coincidences. However, an investigator will often wish to know the (summated) probability of finding at least the observed number of coincident peaks on the basis of chance alone. This value is the area of the right tail of the probability distribution, which can be calculated by subtracting from one the sum of all the individual P values for finding fewer than the observed number of coincidences on the basis of chance alone (37, 50).

If substantial diurnal variation in pulse frequency is introduced, or systematic interaction between a pulse generator and its feedback systems is allowed, then additional computer simulations should be carried out to estimate the expected number, variance, and probability distribution of randomly coincident pulses (37, 50). This is because the assumptions of the hypergeometric probability distribution and its corollary predictions (above) would not be satisfied. These assumptions most notably include the expectations that peaks can be discretized or marked as single-sample events distributed uniformly over the sampling interval, and that event locations are noninteracting.

VI. Summary and Conclusions

The concept of the convolution integral as developed originally in the physical and engineering sciences has substantial, interesting, informative, and novel implications to modern neuroendocrine research methods. Indeed, appropriate deconvolution techniques permit one to assess rates of hormone secretion and metabolic clearance simultaneously *in vivo* without the infusion of radiolabeled or cold hormone. Suitable deconvolution algorithms, such as those discussed in part here, are characterized by attention to statistical error propagation, a clear statement regarding assumptions inherent in the technology, a statistical evaluation of model adequacy, insight into type I and type II (false-positive and false-negative) statistical errors, the ability to distinguish basal from pulsatile hormone secretion, and ultimately an integrated systems view of the overall neuroendocrine axis.

Deconvolution methodology has provided new insights into mechanisms that subserve more complex neuroendocrine regulation *in vivo* (e.g., generation of the spontaneous proestrous LH surge, the compounding of partially overlapping secretory events such as GnRH release in portal blood during the LH surge, and the expected interaction on the basis of chance alone of multiple independent pulse generators). We believe that further developments in the analytical armamentarium of the neuroscientist will include enhanced methods of deconvolution analysis that incorporate some of the ideals stated here, including a systems view of the neuroendocrine axis.

Acknowledgments

We thank Patsy Craig for skillful preparation of the manuscript and Paula P. Azimi for the artwork. This work was supported in part by National Institutes of Health Grant RR 00847 to the Clinical Research Center of the University of Virginia, RCDA 1 KO4 HD00634 (J.D.V.), and RCDA KO4 HD00711 (W.S.E.), GM-28928 (M.L.J.), Diabetes and Endocrinology Research Center Grant NIH DK-38942, NIH-supported Clinfo Data Reduction Systems, the Pratt Foundation, the University of Virginia Academic Enhancement Fund, and the National Science Foundation Science Center for Biological Timing (J.D.V., W.S.E., M.L.J.).

References

1. J. D. Veldhuis and M. L. Johnson, *in* "Methods in Enzymology" (M. L. Johnson and M. Lakowitz, eds.), Vol. 210, p. 539. Academic Press, San Diego, 1992.

2. J. D. Veldhuis and M. L. Johnson, *J. Neuroendocrinol.* **2**(6), 755 (1991).

3. R. J. Urban, W. S. Evans, A. D. Rogol, D. L. Kaiser, M. L. Johnson, and J. D. Veldhuis, *Endocr. Rev.* **9**, 3 (1988).

4. J. D. Veldhuis, M. L. Carlson, and M. L. Johnson, *Proc. Natl. Acad. Sci. U.S.A.* **84**, 7686 (1987).

5. J. D. Veldhuis and M. L. Johnson, "Advances in Neuroendocrine Regulation of Reproduction" (S. S. C. Yen and W. W. Vale, ed.), p. 123. Plenum, Philadelphia, Pennsylvania, 1990.

6. J. D. Veldhuis, M. L. Johnson, and M. L. Dufau, *Am. J. Physiol.* **256**, E199 (1989).

7. M. L. Johnson and S. G. Frazier, *in* "Methods in Enzymology" (C. H. W. Hirs and S. N. Timasheff, eds.), Vol. 117, p. 301. Academic Press, New York, 1985.

8. J. D. Veldhuis, M. L. Johnson, and M. L. Dufau, *J. Clin. Endocrinol. Metab.* **64**, 1275 (1987).

9. J. D. Veldhuis, V. Guardabasso, A. D. Rogol, W. S. Evans, K. Oerter, M. L. Johnson, and D. Rodbard, *Am. J. Physiol.* **252**, E599 (1987).

10. J. D. Veldhuis, M. L. Johnson, A. Iranmanesh, and G. Lizarralde, *J. Biol. Rhythms* **5**, 247 (1990).

11. J. D. Veldhuis, A. B. Lassiter, and M. L. Johnson, *Am. J. Physiol.* **259**, E351 (1990).

12. J. D. Veldhuis, A. Iranmanesh, M. L. Johhson, and G. Lizarralde, *J. Clin. Endocrinol. Metab.* **71**, 1616 (1990).

13. J. D. Veldhuis, A. Faria, M. L. Vance, W. S. Evans, M. O. Thorner, and M. L. Johnson, *Acta Paediatr. Scand.* **347**, 63 (1988).

14. P. A. Jansson (ed.), "Deconvolution, With Applications in Spectroscopy," p. 99. Academic Press, New York, 1984.

15. R. P. McIntosh and J. E. A. McIntosh, *J. Endocrinol.* **107**, 231 (1985).

16. R. P. McIntosh, J. E. A. McIntosh, and L. Lazarus, *J. Endocrinol.* **118**, 339 (1988).

17. R. Rebar, D. Perlman, F. Naftolin, and S. S. C. Yen, *J. Clin. Endocrinol. Metab.* **37**, 917 (1973).

18. E. Van Cauter, *Am. J. Physiol.* **237**, E255 (1979).

19. W. J. Jusko, W. R. Slaunwhite, and T. Aceto, *J. Clin. Endocrinol. Metab.* **40**, 278 (1975).

20. F. O'Sullivan and J. O'Sullivan, *Biometrics* **44**, 339 (1988).

21. K. Albertsson-Wikland, S. Rosberg, E. Libre, L. O. Lundberg, and T. Groth, *Am. J. Physiol.* **257**, E809 (1989).

22. R. J. Henery, B. A. Turnbull, M. Kirkland, J. W. McArthur, I. Gilbert, G. M. Besser, L. H. Rees, and D. S. Tunstall Pedoe, *Chronobiol. Int.* **6**, 259 (1989).

23. J. D. Veldhuis and M. L. Johnson, *Am. J. Physiol.* **255**, E749 (1988).

24. C. M. Swartz, V. S. Wahby, and R. Vacha, *Acta Endocrinol.* **112**, 43 (1986).

25. R. J. Urban, M. L. Johnson, and J. D. Veldhuis, Am. J. Physiol. **257**, E88 (1989).

26. K. E. Oerter, V. Guardabasso, and D. Rodbard, *Comput. Biomed. Res.* **19**, 170 (1986).

27. P. J. Munson and D. Rodbard, "Proceedings of the Statistical Computing Section of the American Statistical Association," p. 295. Washington, D.C., 1989.

28. R. J. Urban, M. L. Johnson, and J. D. Veldhuis, *Endocrinology (Baltimore)* **124,** 2541 (1989).
29. R. J. Urban, M. L. Johnson, and J. D. Veldhuis, *Endocrinology (Baltimore)* **128,** 2008 (1991).
30. C. L. Lawson and R. J. Hanson, "Solving Least Squares Problems." Prentice-Hall, Englewood Cliffs, New Jersey, 1974.
31. W. H. Press, B. P. Flannery, S. A. Teukolsky, and W. T. Vetterling, "Numerical Recipes, The Art of Scientific Computing." Cambridge Univ. Press, Cambridge, 1986.
32. A. N. Tikhonov, *Sov. Math. Dokl.* **4,** 1624 (1963).
33. D. L. Phillips, *J. Assoc. Comput. Math.* **9,** 84 (1962).
34. S. Twomey, *J. Assoc. Comput. Math.* **10,** 97 (1963).
35. J. P. Butler, J. A. Reeds, and S. V. Dawson, *SIAM J. Numer. Anal.* **18**(3), 381 (1981).
36. E. Knobil, *Recent Prog. Horm. Res.* **365,** 53 (1980).
37. J. D. Veldhuis, A. Iranmanesh, I. Clarke, D. L. Kaiser, and M. L. Johnson, *J. Neuroendocrinol.* **1,** 185 (1989).
38. J. D. Veldhuis, L. St. L. O'Dea, and M. L. Johnson, *J. Clin. Endocrinol. Metab.* **68,** 661 (1989).
39. R. J. Urban, S. N. Pavlou, J. E. Rivier, W. W. Vale, M. L. Dufau, and J. D. Veldhuis, *Am. J. Obstet. Gynecol.* **162,** 1255 (1990).
40. S. N. Pavlou, J. D. Veldhuis, J. Lindner, K. H. Souza, R. J. Urban, J. E. Rivier, W. W. Vale, and D. J. Stallard, *J. Clin. Endocrinol. Metab.* **70,** 1472 (1990).
41. W. C. Nunley, R. J. Urban, J. D. Kitchen, B. G. Bateman, W. S. Evans, and J. D. Veldhuis, *J. Clin. Endocrinol. Metab.* **72,** 287 (1991).
42. J. D. Veldhuis, W. S. Evans, and P. Stumpf, *Am. J. Ob.-Gyn.* **161,** 1149 (1989).
43. R. J. Urban, K. D. Dahl, V. Padmanabhan, I. Z. Beitins, and J. D. Veldhuis, *J. Androl.* **12,** 27 (1991).
44. R. J. Urban, J. D. Veldhuis, and M. L. Dufau, *J. Clin. Endocrinol. Metab.* **72,** 660 (1991).
45. J. D. Veldhuis, A. Iranmanesh, K. Y. Ho, G. Lizarralde, M. J. Waters, and M. L. Johnson, *J. Clin. Endocrinol. Metab.* **72,** 51 (1991).
46. A. Iranmanesh, G. Lizarralde, and J. D. Veldhuis, *J. Clin. Endocrinol. Metab.* **73,** 1081 (1991).
47. P. M. Martha, Jr., K. M. Goorman, R. M. Blizzard, A. D. Rogol, and J. D. Veldhuis, *J. Clin. Endocrinol. Metab.* **74,** 336 (1992).
48. W. V. R. Vieweg, J. D. Veldhuis, and R. M. Carey, *Am. J. Physiol.* in press (1992).
49. J. D. Veldhuis, A. Iranmanesh, M. L. Johnson, and G. Lizarralde, *J. Clin. Endocrinol. Metab.* **71,** 452 (1990).
50. J. D. Veldhuis, M. L. Johnson, and E. Seneta, *J. Clin. Endocrinol. Metab.* **73,** 569 (1991).

[17] Graded Action Potentials in Small Hippocampal Neurons: A Computational Approach

Staffan Johansson and Peter Århem

Introduction

Computational methods are used in neuroscience to analyze problems at many different levels of complexity, from the behavior of individual molecules to cells and networks of cells and further to the behavior of whole organisms. The work presented here illustrates an approach used to understand electrical phenomena at the cellular level using small cultured hippocampal neurons. It was found that these cells generated action potentials with an amplitude and time course that systematically depended on the stimulus intensity (Fig. 1A) (1, 2). This contrasts to the "all-or-nothing" principle (see, e.g., Ref. 3) generally assumed to characterize the action potential of excitable membranes. To understand this phenomenon, we used a computational approach which was largely dependent on a voltage–clamp analysis of the currents underlying the action potentials (4). The main results of the analysis have been published (5).

Strategy of Analysis

Essentially a three-step strategy was adopted to analyze the amplitude variations of the action potentials. First, the underlying ion currents obtained from voltage–clamp measurements were characterized mathematically. Second, the equations obtained were then used to compute graded action potentials. Finally, the importance of the different parameters for causing the amplitude variability was investigated by modifying the parameters in the computations.

Characterization of Currents

The mathematical characterization of currents was based on voltage–clamp measurements with the tight-seal whole-cell method (6). Recording technique and equipment are described in detail by Johansson and Århem (4).

Methods in Neurosciences, Volume 10

Level of Analysis

We chose to analyze the currents on the "macroscopic," whole-cell level. It should be noted that different kinetics of single ion channels may result in similar kinetics of whole-cell currents (7, 8). The single-channel kinetics per se, however, was assumed not to be of importance for understanding the generation of graded action potentials by the whole cell membrane. [If phenomena depending on individual ion-channel events, e.g., impulses induced by single-channel openings (see Ref. 9) would be simulated, it would be necessary to consider the single-channel kinetics.]

Methodology of Current Analysis

When currents are to be used for computations it is essential to avoid artifacts introduced by the method of recording. Thus, no data from cells with "regenerative" inward currents indicative of poor space-clamp conditions were used. Further, for a correct description of potential-activated currents, leak and capacitative currents were subtracted. These aspects are described by Johansson and Århem (4).

Current Types

The voltage–clamp experiments (4) revealed three types of potential-activated currents activated by positive potential steps: (a) a classic, transient tetrodotoxin-sensitive Na^+ current (10), (b) a more sustained "delayed rectifier" K^+ current (10), and (c) a transient "A-type" K^+ current (11). The A current was only found in about 40% of the cells and was not required for the generation of graded action potentials.

Quantitative Description

A mathematical description of the time and potential dependence of the ion currents is necessary for the computation of action potentials. In the case illustrated here, the purpose of the computations was to understand the amplitude variability of the action potential. Thus, the Na^+ current and the delayed rectifier K^+ current were quantified but not the A current, since the latter was not required for the generation of graded action potentials.

As a basis for the description we used the equations developed by Hodgkin and Huxley (12) with the modifications introduced by Frankenhaeuser (13–15) for the nodal membrane in the myelinated vertebrate axon. One reason was that this is one of the best quantified nerve membranes available. Another was that this model uses the permeability concept rather than that of conductance. The former is more appropriate when applied to ion channels

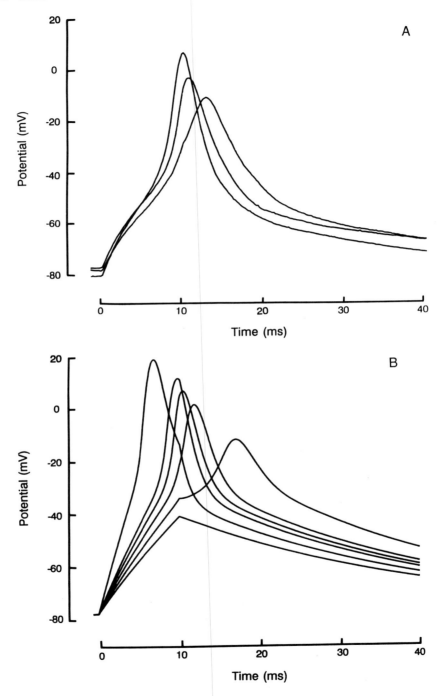

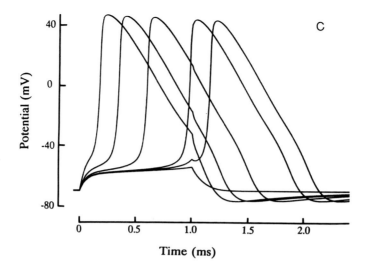

FIG. 1 Recorded and computed action potentials. (A) Experimentally recorded hippocampal action potentials (as described in Ref. 2). Stimulating current steps were 35, 40, and 45 pA, temperature 22°C (from the cell used for the standard empirical description of the currents; Ref. 4). (B) Computed responses of a hippocampal neuron to 10-msec current steps of 30, 35, 40, 45, 50, and 80 pA. (C) Computed action potentials for the nodal model. Parameters were from Frankenhaeuser and Huxley (15). Stimulating current steps were 0.350, 0.355, 0.360, 0.400, 0.500, and 0.800 mA cm^{-2}. Step duration was 1.0 msec to account for the faster time courses compared with those of the hippocampal model. [Reproduced with permission from S. Johansson and P. Århem, *J. Physiol. (London)* **445**, 157 (1992).]

which function as pores (see, e.g., Ref. 16). The quantification was based on the assumptions that the Na$^+$ and K$^+$ currents follow the constant field equation and that the Na$^+$ and K$^+$ permeabilities depend on membrane potential and time (13, 14, 17, 18).

Symbols used for Mathematical Descriptions

U	Membrane potential
U_R	Resting or holding potential
c	Capacitative current
L	Leak current
I_{Na}	Na$^+$ current
I_K	Delayed rectifier K$^+$ current
s	Stimulus current

P_{Na}, P_K	Permeabilities for I_{Na} and I_K
$\overline{P}_{Na}$, P'_K	Permeability constants (see equations)
m, h, n	Permeability variables for P_{Na} activation (m), inactivation (h), and P_K activation (n)
α, β	Rate constants for m, h, and n as indicated by subscript
A, B, C	Empirical constants in Eqs. (8)–(13) for α and β (given in Table I)
A_m	Membrane area
R	Gas constant
F	Faraday's constant
T	Absolute temperature
t	Time

TABLE I   Constants Used for Calculation of Rate Constants[a]

Constant		α_h	β_h	α_m	β_m	α_n	β_n
A	(msec^{-1} mV^{-1})	0.05	2.25[b]	0.06	0.07	0.016	0.04
B	(mV)	5	60	37	28	60	35
C	(mV)	6	10	3	20	10	10

[a] See text.
[b] msec^{-1}.

Equations and Computational Procedures

In analogy with the description of Frankenhaeuser and Huxley (15), the currents were calculated from Eqs. (1)–(7).

$$I_{Na} = A_m P_{Na} \frac{UF^2}{RT} \frac{[Na^+]_o - [Na^+]_i \exp(UF/RT)}{1 - \exp(UF/RT)} \tag{1}$$

$$P_{Na} = \overline{P}_{Na} h m^2 \tag{2}$$

$$dm/dt = \alpha_m(1 - m) - \beta_m m \tag{3}$$

$$dh/dt = \alpha_h(1 - h) - \beta_h h \tag{4}$$

$$I_K = A_m P_K \frac{UF^2}{RT} \frac{[K^+]_o - [K^+]_i \exp(UF/RT)}{1 - \exp(UF/RT)} \tag{5}$$

$$P_K = P'_K n^2 \tag{6}$$

$$dn/dt = \alpha_n(1 - n) - \beta_n n \tag{7}$$

α's and β's are described by Eqns. (8)–(13) with the values for A, B, and C (changed from those of Ref. 15) given in Table I.

$$\alpha_m = A[(U - U_R) - B]/(1 - \exp\{[B - (U - U_R)]/C\}) \qquad (8)$$

$$\beta_m = A[B - (U - U_R)]/(1 - \exp\{[(U - U_R) - B]/C\}) \qquad (9)$$

$$\alpha_h = A[B - (U - U_R)]/(1 - \exp\{[(U - U_R) - B]/C\}) \qquad (10)$$

$$\beta_h = A/(1 + \exp\{[B - (U - U_R)]/C\}) \qquad (11)$$

$$\alpha_n = A[(U - U_R) - B]/(1 - \exp\{[B - (U - U_R)]/C\}) \qquad (12)$$

$$\beta_n = A[B - (U - U_R)]/(1 - \exp\{[(U - U_R) - B]/C\}) \qquad (13)$$

For the integration method we used either the simple Euler method or an explicit exponential method (19). The choice of method is not very critical when a sufficiently small integration step can be used. For the computation of currents we used 10–50 μsec steps.

Equations (1)–(7), with constants from Frankenhaeuser and Huxley (15), describe currents which are qualitatively similar to those to be described here (4). To obtain quantitatively correct time courses, the following procedure is followed: (i) The exponents in Eqs. (2) and (6) are determined from double-logarithmic plots of the time course of the recorded currents. (ii) The rate constants are determined by a repeated comparative procedure in which the constants A, B, and C [Eqs. (8)–(13)] are adjusted to obtain a good fit between recorded and computed current time courses at different potentials. The constant A gives the magnitude of the rate constant. B and C relate to the curve describing the potential dependence of the rate constant: B gives the shift along the potential axis and C gives the slope of the curve. (iii) The permeability constants $\overline{P}_{Na}$ and P'_K are chosen to give the correct current magnitude.

Figure 2 shows recorded and computed current responses to various potential steps. The values for the constants A, B, and C are given in Table I. $\overline{P}_{Na}$ and P'_K were 1.3×10^{-4} and 2.4×10^{-5} cm sec^{-1}, respectively. It should be pointed out that if a kinetic scheme of the sequential type, as described for instance by Armstrong (7), is used as basis for the computations, considerably lower values for $\overline{P}_{Na}$ and P'_K have to be chosen. This is of principal interest since the permeability constants are generally assumed to reflect the density of ion channels in the membrane.

Simulation of Action Potentials

The next step in the analysis is to use the mathematical description above to compute action potentials. In this way it can be determined if the description accounts for the main features of the membrane potential under nonclamped

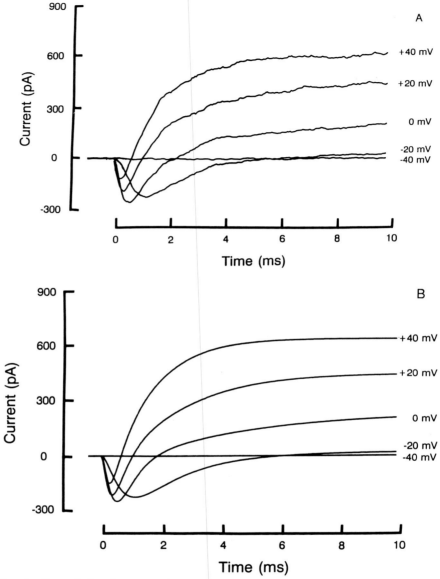

FIG. 2   Recorded and computed current responses to rectangular potential steps. (A) Recorded responses. Holding potential was -70 mV, and step potentials varied from -40 to $+40$ mV, as indicated. Leak and capacitive currents were subtracted. Remaining transients due to imperfect subtraction were truncated. (B) Computed responses. Parameter values were chosen to fit the observed currents in (A). Step potential values were as indicated. [Reproduced with permission from S. Johansson and P. Århem, *J. Physiol.* (*London*) **445,** 141 (1992).]

conditions. For the computations, the basic simplifying assumption of a uniformly activated membrane region with no current spread to neighboring regions was made. This assumption did not apply to the hippocampal neurons studied. In this case most neurons had neurites (2) to which current could spread. However, since electrophysiological data concerning the neurites are not available, the treatment described below serves as an approximation. Further, action potentials computed in this way were not expected to differ much from a situation in which current spread is possible (cf. membrane action potential and propagated action potential described by Hodgkin and Huxley, Ref. 12).

Equations (14)–(16) are used for the computation of action potentials.

$$dU/dt = I_C/C_m \tag{14}$$

where

$$I_C = I_S - (I_{Na} + I_K + I_L) \tag{15}$$

and

$$I_L = (U - U_R)/R_m \tag{16}$$

The ion currents I_{Na} and I_K are obtained from Eqs. (1)–(13), with standard values for the constants as described above for Fig. 2. The standard values for R_m and C_m are 4.3 GΩ and 7.0 pF, respectively. (These values are derived from the same cell as used for the description of the currents. This cell had an estimated membrane area of 100 μm^2, giving an estimated resistivity and capacitance per unit area of 4.3 kΩ cm^2 and 7.0 μF cm^{-2}, respectively.) Equation (16) is justified by the finding of a linear relation between leak and capacitative currents and the potential (4).

As for the computation of currents (see above), we used either of two integration methods, with similar results: the Euler method or an explicit exponential method (19). The integration step was 10–50 μsec for the hippocampal model and 5.0 μsec for the faster processes when data from myelinated nerve fibers were used (see below).

Computation of Potential Responses

First action potentials associated with rectangular current pulses of varying amplitude were computed. Some results are shown in Fig. 1B, together with experimentally recorded action potentials (Fig. 1A). Action potential amplitude increased with stimulus current in the computed case as well as in the experimental case. The computed peak versus stimulus curve was S-shaped in the suprathreshold stimulation range (cf. curve A in Fig. 5), as typically seen experimentally (2). Thus, the general agreement between computed and recorded action potentials was satisfactory, considering the uncertainties in the description of the currents and the individual variability of the experimentally recorded currents (4). Thus, we could conclude that the currents recorded under voltage–clamp conditions were sufficient to generate graded action potentials.

Analysis of Role of Membrane Parameters

From the above analysis it is clear that graded action potentials may be generated by cells with classic current types. Thus, modified Na^+ channel gating or persistent K^+ channel activation, proposed to principally explain graded potential responses in other preparations (20), could be excluded. Nevertheless, the difference from preparations generating all-or-nothing responses was not evident: the underlying current types were very similar. To understand the mechanisms for the amplitude variation, we computed (a) the time course of the time- and potential-dependent parameters for different stimulus strengths and (b) the effect of modified time- and potential-independent parameters on the action potential amplitude. Figure 3 shows parameters during responses to a weak and a strong suprathreshold current stimulus. It is clear from Fig. 3 that all computed parameters strongly depend on the stimulus intensity.

For case b above we adopted the strategy to focus on parameters with values that differ markedly in the present "hippocampal" model and in the model developed by Frankenhaeuser (15) for the nodal membrane of myelinated nerve fibers. The reason is that the latter model describes essentially all-or-nothing action potentials (see Fig. 1C). The effect of modifications of the membrane time constant and the permeability constants are discussed below.

Modification of Time Constant

The membrane time constant of a typical hippocampal neuron is much higher than that of the node. The standard values used here were 30 msec (resistivity 4.3 kΩ cm^2, capacitance per unit area 7.0 μF cm^{-2}) for the hippocampal model and 66 μsec (33 Ω cm^2 and 2 μF cm^{-2}) for the nodal model. We changed the values and made computations with the standard hippocampal parameters except for resistance and capacitance, which were set to the

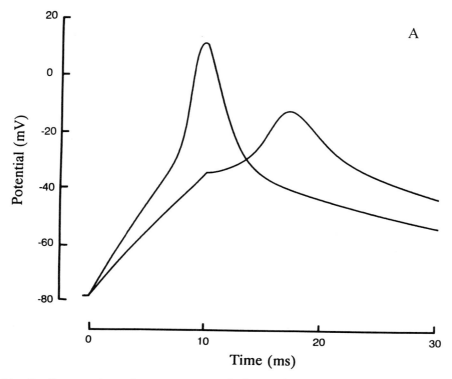

FIG. 3   Computed membrane parameters during two impulses of different amplitude. (A) Potential responses to 10-msec current steps of 35 (right) and 50 pA (left). (B) Na^+ and K^+ permeabilities during the impulses in (A) (left curves associated with the impulse elicited by 50 pA). (C, D) Variables m, h, and n during the impulses elicited by 35 (C) and 50 pA (D). (E, F) Na^+, K^+ and leak currents during the impulses (A) elicited by 35 (E) and 50 pA (F). [Reproduced with permission from S. Johansson and P. Århem, *J. Physiol. (London)* **445**, 157 (1992).]

nodal model values. Only when strong stimuli were used did the response peak clearly exceed the electrotonic response resulting from I_L and I_C (Fig. 4A). A small hump, corresponding to a regenerative response, was distinguished. This component and the total response were, however, graded in amplitude.

Conversely, computations were made with the standard nodal membrane parameters (Ref. 15; see above), except for resistance and capacitance, which were set to the hippocampal model values. Although a very small amplitude variation was detected, the regenerative potential responses were still essen-

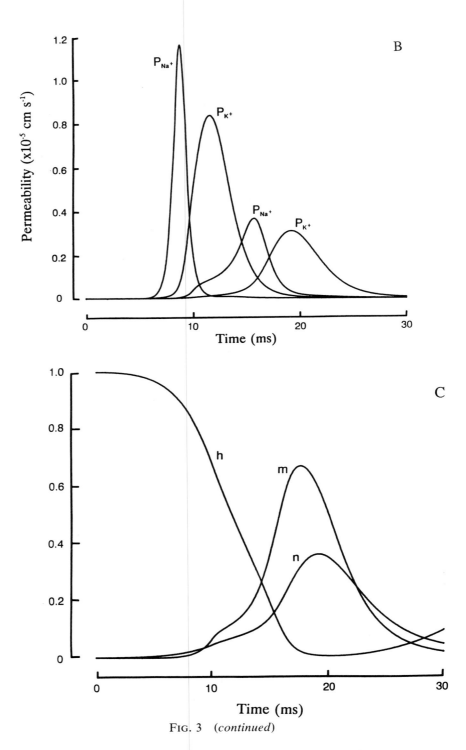

FIG. 3 *(continued)*

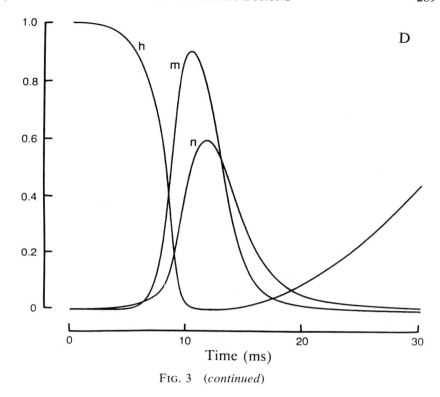

FIG. 3   (*continued*)

tially all-or-nothing (Fig. 4B). It was concluded that the amplitude variation of the hippocampal potential responses was not exclusively an effect of the time constant.

Modification of Permeability Constants

The ionic permeabilities of the hippocampal neurons are much smaller than those of the node of the myelinated nerve fibre. In the hippocampal model, the constant $\overline{P}_{Na}$ was 1.3×10^{-4} cm sec^{-1} and P'_K was 2.4×10^{-5} cm sec^{-1}, while in the nodal model the corresponding constants were 8.0×10^{-3} and 1.2×10^{-3} cm sec^{-1}, respectively. Effects of $\overline{P}_{Na}$ variations on peak membrane potential for the hippocampal model are shown in Fig. 5. When $\overline{P}_{Na}$ was reduced 5 times (curve F), almost no regenerative potential response was obtained in addition to the response resulting from I_L and I_C. In fact, responses smaller than those expected from I_L and I_C were obtained at strong stimulus currents, owing to the activation of outward currents. On the other

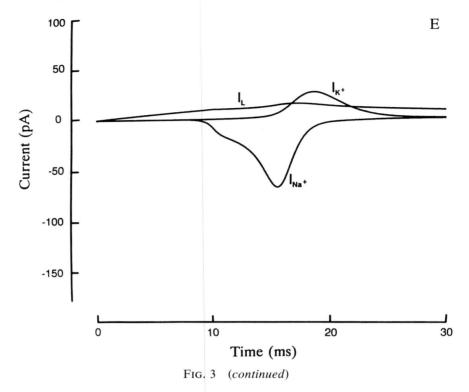

FIG. 3   (*continued*)

hand, when $\overline{P}_{Na}$ was increased, the potential responses grew larger; however, the amplitude variation gradually decreased (curves B–E in Fig. 5), giving an essentially all-or-nothing response at high $\overline{P}_{Na}$ values. Also when the constant P'_K was raised in proportion to the rise of $\overline{P}_{Na}$, higher values resulted in less graded potential responses (Fig. 5, curve G). Thus, the effect of raising $\overline{P}_{Na}$ was not exclusively due to a relative increase of the ratio $\overline{P}_{Na}/P'_K$ (see also below).

Many of the hippocampal neurons studied (4) showed a lower permeability for the delayed sustained current than that used in the mathematical model above. Therefore, it was of interest to compute action potentials with a reduced P'_K value. Such computations showed a clearly increased amplitude variability, with strong stimuli causing responses of higher amplitude than with the standard P'_K value (Fig. 6). When P'_K was zero, the action potential time course was drastically prolonged (Fig. 7).

From the analysis above we conclude that the existence of graded action potentials in the neurons studied critically depends on $\overline{P}_{Na}$, but not on the

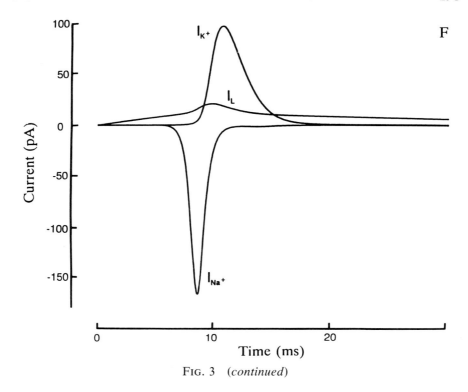

Fig. 3   (*continued*)

membrane time constant. The degree of amplitude variation depends on P'_K when standard hippocampal values are used for the other parameters. The $\overline{P}_{Na}$ dependence suggests that the amplitude variation is mainly determined by the density of Na^+ channels in the cell membrane.

Comparisons with Other Computational Studies

A few other computational studies that are directly relevant for the analysis of the amplitude variation of the action potential have been performed. It has been shown that the squid giant axon is expected to generate impulses of different amplitudes when near-threshold stimuli are given (21). However, the stimulus range for this is extremely narrow, and graded impulses are expected neither to be seen experimentally nor to play any physiological role (21). Further, it has been shown that theoretically this axon may propagate impulses of different amplitudes (22), although propagation of smaller im-

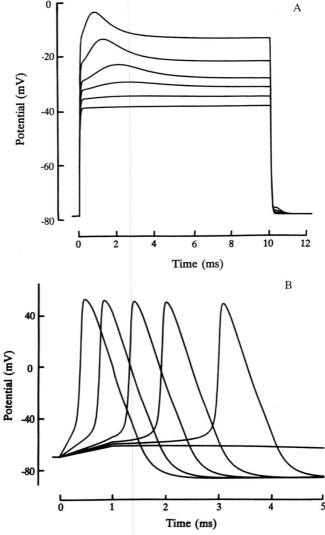

FIG. 4   Effects of the membrane time constant on computed potential responses. (A) Hippocampal model. Membrane resistance and capacitance were set to the standard values of the nodal model (see text). Stimulating rectangular, 10-msec current steps were 1.2, 1.3, 1.4, 1.5, 1.7, and 2.0 nA. Note that the lowest curve is almost indistinguishable from an electrotonic response. (B) Nodal model. Membrane resistance and capacitance were set to the standard values of the hippocampal model (see text). Stimulating 1.0-msec current steps were 60, 70, 80, 100, 200, and 500 μA cm^{-2}. [Reproduced with permission from S. Johansson and P. Århem, *J. Physiol.* (*London*) **445,** 157 (1992).]

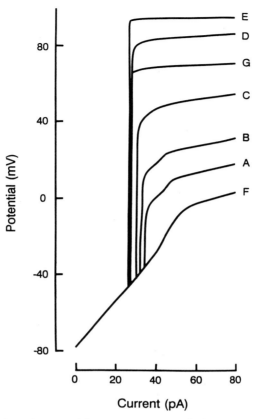

FIG. 5   Computed peak potential versus current relations at various $\overline{P}_{Na}$ values. Stimulating rectangular current pulses of 10-msec duration were used. (A) Standard hippocampal data. (B–E) $\overline{P}_{Na}$ increased 2 (B), 5 (C), 20 (D), and 50 (E) times. (F) $\overline{P}_{Na}$ reduced 5 times. (G) $\overline{P}_{Na}$ and P'_K increased 50 times. [Reproduced with permission from S. Johansson and P. Århem, *J. Physiol. (London)* **445,** 157 (1992).]

pulses may be unstable and not found experimentally. However, comp-utations by Cooley and Dodge (23) show that reduced Na$^+$ and K$^+$ conduc-tances theoretically can lead to responses that propagate decrementally over distances of several centimeters in the squid giant axon. Experimen-tal results supporting this view have been presented by Lorente de Nó and Condouris (24).

In the nodal model, the constant P'_K has only marginal effects on the action potential (15), and the function of the K$^+$ current remains obscure. In the present model P'_K was essential for the time course of the action potential.

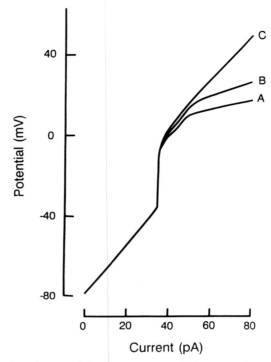

FIG. 6   Computed peak potential versus current relation at various P'_K values. Stimulating rectangular current pulses of 10-msec duration were used. (A) Standard hippocampal data. (B) P'_K reduced 5 times. (C) $P'_K = 0$. [Reproduced with permission from S. Johansson and P. Århem, *J. Physiol.* (*London*) **445**, 157 (1992).]

Conclusion

The present example of computational analysis revealed that the amplitude variation of the studied hippocampal neurons depended on classic Na^+ and K^+ channels and that the critical factor was most likely the Na^+ channel density. As other central neurons show membrane properties similar to those of the hippocampal neurons (25), we expect graded impulses to be more common in the nervous system than usually assumed. Thus a computational approach similar to that presented here may be of interest for the future understanding of information processing in the central nervous system.

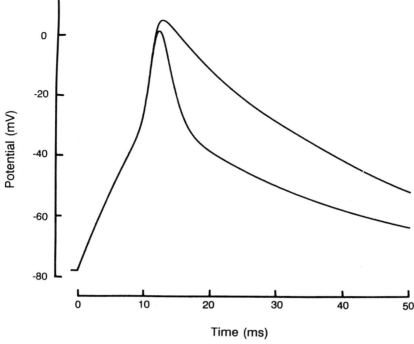

Fig. 7   Computed potential response at $P'_K = 2.4 \times 10^{-5}$ cm sec^{-1} (standard data; lower curve) and $P'_K = 0$ (upper curve). Rectangular stimulating current pulses of 40 pA and 10 msec duration were used. [Reproduced with permission from S. Johansson and P. Århem, *J. Physiol.* (*London*) **445,** 157 (1992).]

Acknowledgments

This work was supported by grants from the Swedish Medical Research Council (Project No. 6552) and Karolinska Institutets Fonder.

References

1. S. Johansson and P. Århem, *Neurosci. Lett.* **118,** 155 (1990).
2. S. Johansson, W. Friedman, and P. Århem, *J. Physiol.* (*London*) **445,** 129 (1992).
3. E. D. Adrian and A. Forbes, *J. Physiol.* (*London*) **56,** 301 (1922).
4. S. Johansson and P. Århem, *J. Physiol.* (*London*) **445,** 141 (1992).

5. S. Johansson and P. Århem, *J. Physiol.* (*London*) **445,** 157 (1992).

6. A. Marty and E. Neher, *in* "Single-Channel Recording" (B. Sakmann and E. Neher, eds.), p. 107. Plenum, New York, 1983.

7. C. M. Armstrong, *Physiol. Rev.* **61,** 644 (1981).

8. R. W. Aldrich, *Trends Neurosci.* **9,** 82 (1986).

9. S. Johansson, "Electrophysiology of Small Cultured Hippocampal Neurons." Academic Thesis from the Karolinska Institute, Stockholm (1991).

10. A. L. Hodgkin and A. F. Huxley, *J. Physiol.* (*London*) **116,** 449 (1952).

11. J. A. Connor and C. F. Stevens, *J. Physiol.* (*London*) **213,** 21 (1971).

12. A. L. Hodgkin and A. F. Huxley, *J. Physiol.* (*London*) **117,** 500 (1952).

13. B. Frankenhaeuser, *J. Physiol.* (*London*) **151,** 491 (1960).

14. B. Frankenhaeuser, *J. Physiol.* (*London*) **169,** 424 (1963).

15. B. Frankenhaeuser and A. F. Huxley, *J. Physiol.* (*London*) **171,** 302 (1964).

16. B. Hille, "Ionic Channels of Excitable Membranes." (2nd Ed.) Sinauer, Sunderland, Massachusetts, 1992.

17. D. E. Goldman, *J. Gen. Physiol.* **27,** 37 (1943).

18. A. L. Hodgkin and B. Katz, *J. Physiol.* (*London*) **108,** 37 (1949).

19. M. A. Wilson and J. M. Bower, *in* "Methods in Neuronal Modeling" (C. Koch and I. Segev, eds.), p. 291. MIT Press, Cambridge, Massachusetts, and London, 1989.

20. B. M. H. Bush, *in* "Neurons without Impulses" (A. Roberts and B. M. H. Bush, eds.), p. 147. Cambridge Univ. Press, Cambridge, 1981.

21. R. Fitzhugh and A. Antosiewicz, *J. Soc. Ind. Appl. Math.* **7,** 447 (1959).

22. A. F. Huxley, *J. Physiol.* (*London*) **148,** 80P (1959).

23. J. W. Cooley and F. A. Dodge, Jr., *Biophys. J.* **6,** 583 (1966).

24. R. Lorente de Nó and G. A. Condouris, *Proc. Natl. Acad. Sci. U.S.A.* **45,** 592 (1959).

25. S. G. Cull-Candy, C. G. Marshall, and D. Ogden, *J. Physiol.* (*London*) **414,** 179 (1989).

[18] Quantification of Fast Axonal Transport by Video Microscopy

Hugo Geerts, Rony Nuydens, Roger Nuyens, and Frans Cornelissen

Introduction

Neuronal cells are the most polarized cell type known. Their axons can reach lengths up to 1 m in humans. The transport of material over these long distances is very important for normal neuronal cell functioning. Therefore, an efficient system of mass transport has been developed to accommodate this distance. Only a few techniques are currently available to measure this form of transport. Little is known about the *in vivo* regulation of transport, and as a consequence its role in physiology or pathology has not always been appreciated.

Fast axonal transport (1–3 μm/sec) is used to transfer newly synthesized proteins from the cell body right to the peripheral nerve ending. In addition, mitochondria are transported by this mechanism. In the longest human nerve (1 m), this can take between 2 and 3 days. These proteins are transported in membrane-bound organelles and include receptors, glycoproteins, and the whole machinery for synaptic transmission at the presynaptic nerve ending. Retrograde transport (i.e., transport back to the cell body) carries down-regulated proteins, which head to the lysosomal compartment, and neurotrophic factors necessary for cell survival.

Slow axonal transport involves the motion of constitutive elements of the neuronal cytoskeleton such as microtubuli and neurofilaments. It is assumed to play a large role in plasticity changes of the neuronal cytoskeleton, for instance, in the neurite outgrowth response after axotomy.

Recently, however, deficiency of axonal transport has been suspected in a number of pathological situations. As a prime example, the aberrant cytoskeletal modifications in Alzheimer's disease are assumed to have a direct negative effect on this neuronal feature (1). This is expected to have implications on the NGF (nerve growth factor)-mediated cell survival of cholinergic neurons or on energy production along the neurites or the peripheral nerve ending.

Very little is known with regard to endogenous modulation of axonal transport or its changes during neuropathology. This is mostly due to technological problems associated with an accurate measurement of axonal transport. Classic techniques rely on the injection of radioactively labeled substances in the cell body of neurons *in vivo,* preferably the sciatic nerve. After many hours (for fast axonal transport) or even days (for slow axonal transport), the animal is sacrificed. The peripheral nerve is then cut into small pieces, and the radioactivity is measured in each portion. Alternatively, ligation techniques may be used, where accumulation of radioactive material is monitored. These techniques yield at best qualitative information with regard to changes in axonal transport. Specific proteins or compounds can be followed, and changes in protein content as a consequence of egression such as axotomy can be studied. However, rapid modifications of axonal transport or the direct effect of certain substances or conditions are difficut to study accurately.

Recent advances in neuronal cell culture together with technological achievements in high-resolution light microscopy and image processing make it possible to obtain a detailed quantification of fast axonal transport. We will describe two complementary techniques based on high-resolution video microscopy. The first technique uses Allen video-enhanced Nomarski interference contrast to visualize small vesicles (~100 nm) in neurites of living cultured neurons. An advanced image processing system then detects the individual moving vesicles and calculates parameters related to axonal transport. A major drawback, however, is the unknown content of these vesicles. We illustrate this technique by evaluating the effect of vanadate on axonal transport in embryonic hippocampal neurons.

The second technique, called nanovid microscopy, partially overcomes the problem of selectivity. By coupling small colloidal gold probes (30 nm) to antibodies against specific epitopes and detecting these probes individually, the motion of specific and discrete proteins can be followed. As an example, the transport of endocytosed 40 nm gold probes in neuroblastoma cells is studied.

Visualizing Fast Axonal Transport

In this section, we will briefly summarize the essential steps for good visualization of fast axonal transport. More elaborate details can be found in various references (2). The concept of video-enhanced contrast has been described in detail elsewhere (3, 4). (See Appendix at end of this chapter for addresses of suppliers.)

Microscope

Because the detection of the small endogenous vesicles (100 nm) and gold particles (down to 30 nm) relies most heavily on the extreme electronic contrast enhancement, any contrast produced by cell constituents through aberrations and incorrect adjustment of the microscope should be minimized. Differential interference contrast with monochromatic light (narrow-band green interference filter) is used. The high numerical aperture (N.A.) condenser and objective should both have immersion oil. Moreover, correct Koehler illumination is of the utmost importance. These requirements are available on most modern research microscopes. An important point to consider is the ease with which one can switch from one optical mode to another, as well as the availability of additional ports to introduce filters, polarizers, etc. For most purposes illumination can be provided with a simple halogen lamp. A mercury or xenon arc will provide crisper images, but the light intensity is less easily adjusted.

Camera

There are now an increasing number of suitable tube and CCD (charge-coupled device) video cameras and dedicated video microscopy systems available on the market. A simple high-resolution monochrome camera is generally sufficient to provide clear images. For both modes of quantitative axonal transport techniques (differential interference contrast and nanovid microscopy), contrast enhancement is crucial for correct visualization.

Analog contrast enhancement is done by gradually increasing the analog dc offset and the gain until an optimal image is obtained. The following characteristics are important or helpful. (1) Manual control of gain and offset should be possible. (2) The sensitivity of the camera must be high enough so that it can be saturated at high magnification. (3) The camera resolution has to be better than that of subsequent display and recording devices. For most purposes 500–700 lines in the horizontal direction are sufficient. (4) The dynamic range of the camera must be sufficiently large to cope with relatively large contrast differences. (5) The resolution and in particular dynamic range of the video monitor are at least as important as those of the camera. A simple black and white monitor (e.g., Panasonic WV5350) will most often be suitable. Most color monitors do not provide sufficient contrast. (6) For quantitative measurements the linearity of the detection system is important, and geometric distortion of the image must be minimal. (7) The dedicated image-processing system must be able to store a background image and to subtract this pixel by pixel in real time from subsequent images. This is needed to remove any imperfections in the optical path and to provide a clear uniform background. (8) The experiments are taped, and the analysis is performed off-line. To assure the highest quality, professional video equipment must be used. A broadcast quality video tape recorder (e.g., 1 inch, $\frac{3}{4}$ inch U-Matic or super VHS) usually is sufficient to record the dynamic video sequences. Alternatively, optical memory laser disks may be used. (9) From this video signal, further digital processing must be available.

The system we use consists of a Zeiss Axioplan microscope, usually with a $100\times$ Plan-Neofluar (1.3 N.A.) objective together with a high N.A. condenser. The temperature is kept at 37°C using a homemade air curtain system. A Chalnicon tube camera system (Hamamatsu) is coupled to a Hamamatsu C1966 video-enhancement system. Display is performed on a black and white Hamamatsu camera (Model C2130) or a Panasonic WV 5350 black and white monitor. The video signal is recorded on a Sony U-matic VTR (Model VO 5850P).

Quantifying Fast Axonal Transport of Endogenous Vesicles

Extracting Information from Video Signals

The amount of image information to handle is impressive. Consider a single sequence of 8 min (480 sec). At a rate of 25 frames/sec, this amounts to 12,000 video frames. One video frame usually can be digitized in 512×512 byte or 256 kbyte. The total mass of data therefore is about 3000 Mbyte or 3 Gbyte. Fortunately, only a small portion of the image (the axons) is of interest. The strategy therefore is to select a small area-of-interest (AOI) along a certain axon and to digitize only pixels from this small area.

In practice (5), this is performed by first indicating interactively the beginning and end point of a segment on the axon to be studied. A set of hurdles is then placed on the image (see Fig. 1a), and the gray values of these pixels are read out in real time. With our system, we currently can assure 180 randomly assigned pixels to be digitized in real time over a maximal time period of 24 min. With this system the memory needed is only $180 \times 25 \times 60 \times 8$ bytes or 2.16 Mbyte, a reduction with a factor of 1000. If necessary, the same procedure can be repeated for another axon. In our laboratory, this is achieved with a Imaging Technology Series 200 image processor, coupled to a MicroVax II system (Digital Equipment Corporation). The full 2K by 2K tileless organized image board can be randomly addressed, so that random pixels can be read out very fast. The program uses the ITEX software under VMS 5.3.

A second approach, currently under development in our laboratory, uses a DASM module (Analogic Devices), coupled to the SCSI port on a DECStation 5000/200 with 64 Mbyte of RAM memory. This system can digitize up to 12,000 pixels in real time and write them into the DEC memory. However, the pixels need to be contiguous so that only one large rectangular area can be followed. The available time span is then dependent on the available RAM

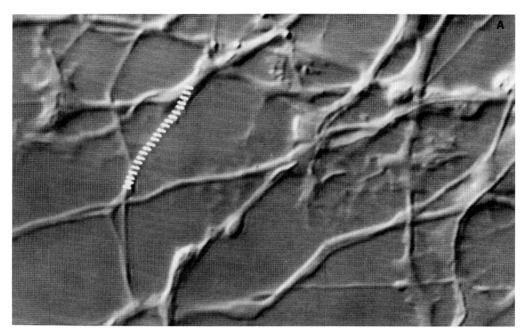

FIG. 1   (a) Typical microscopy image (Nomarski interference contrast) of embryonal hippocampal neurons, used for automatic axonal transport quantification of endogenous organelles. Seven days after plating, the cells have formed extensive neurites. Several axonal bundles are visible. The white pattern is an example of the hurdles placed over a specific axon. The gray values of all these points are digitized and read out at video rate. (b) Typical bright-field image in nanovid microscopy. Neurites of N4 neuroblastoma cells, differentiated for 48 hr with 1 mM dibutyryl-cAMP are visible. The small dark points are gold probes (40 nm), stabilized with bovine serum albumin–polyethylene glycol (BSA–PEG). They are taken up in endocytotic vesicles and display the typical features of fast axonal transport. The acquisition program digitizes a series of consecutive images of an interactively determined AOI.

memory and is currently only 150 sec. By carefully choosing the orientation of the axons, the total area to be considered in each image can be significantly reduced so that time periods of 5–8 min become possible.

The two approaches are different in that the digitization of the data actually needs the horizontal video signal as a trigger. In this way perfect synchronization is assured. In the latter method, the digitization program runs independently of the video timing. Therefore, in this case, careful adjustment in terms of size of transferred data and presence of other concurrent processes on the computer is needed.

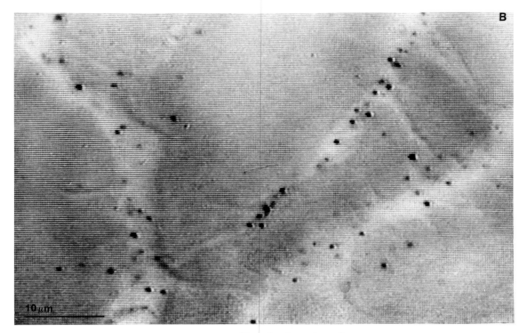

FIG. 1 (*continued*)

Alternative methods include the use of parallel disk transfer systems, where several hundreds of megabytes can be stored in real-time on Winchester disks (see, e.g., Recognition Concepts). Also laser-based optical Memory video disks systems (Sony) may be used, because they can yield perfect still images for analysis. Data reduction (e.g., the motion of vesicles along one neurite) is then performed off-line. Because the images can be retrieved at much lower speeds, cheaper digitizing boards and image processing hardware (such as Matrox, Datacube, Imaging Technology, or Data Translation) may be used. However, the price of the write-once media precludes a larger use.

Analyzing Dynamics; Software Considerations

First, all digitized gray values within the pattern are entered in a two-dimensional array, in which the columns represent time points and the rows the points (150–180) along the pattern (see Fig. 2). Drift in the illumination or background intensity is eliminated by subtracting the mean gray value of all

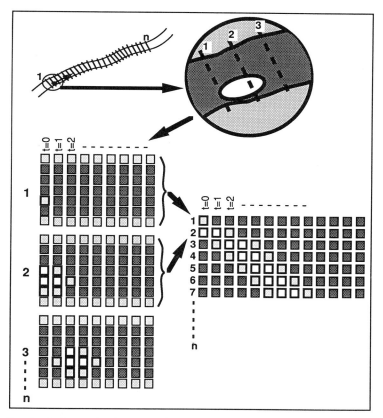

FIG. 2   Synthetic image in the analysis of axonal transport of endogenous vesicles. The gray values of all points along the lines (1...n) are digitized in real time. Each line is 5 to 9 points wide (in this example, 7). The gray values along the lines are represented in columns, and the rows indicate time. Appropriate segmentation techniques (see text) create a compressed image, in which each line at each time point is represented as a binary entry in the matrix. This can then be used to calculate further motion parameters. [With permission from F. Cornelissen *et al., J. Neurosci. Methods* **35,** 79 (1990).]

points in a certain time interval. Shading is reduced by performing the same operation in the space domain. Stretching of this "synthetic" image already reveals a lot of information.

The most important step is now the detection of individual vesicles in the synthetic image. In differential interference contrast microscopy, vesicles tend to display light and dark sides. This polarity is exploited to detect individual vesicles by the following algorithm:

Calculate absolute value of first-order derivative along the time axis.

Calculate histogram of this derivative image, and determine median and 90th percentile.

Threshold derivative image with a value equal to the median, augmented by the difference between median and 90th percentile. This procedure is further referred to as "median thresholding." The resulting binary image indicates the detection of moving vesicles.

Stationary vesicles are detected by a simple thresholding procedure in the horizontal (x) axis. The threshold value is calculated with the same median procedure as above. This second binary image yields the positions of the light sides of vesicles.

At each moment in time, vesicles consist of parts lighter or darker than the axonal background itself. A smoothed derivative over each column is likely to detect edges with relatively high edge values. Thresholding these values with the median method yields a third binary image.

A fourth binary image is calculated by taking the derivative in the x direction and thresholding the result in the y direction. This prevents the false detection of vesicles owing to a sudden change in illumination of the scene (e.g., as a consequence of manual focusing). In the vertical direction, edges arising from these artifacts tend to have more or less equal strength, so that thresholding in that direction prevent corruption of the binary image.

The four binary images are added and subjected to a "majority" operation. This procedure consist of shifting a 3*3 window over the whole synthetic image, calculate the sum of values within that window, and setting the resulting pixel to 1 if the sum exceeds a value of 12–15.

We are now left with a binary image. This image can contain gaps or other artifacts, owing to defocusing. We now proceed further by morphological operations.

A horizontal erosion of 1 cycle removes very small fragments ("speckles").

A horizontal dilation of 5 cycles joins together eventually split horizontal line fragments.

Two-dimensional closing of the binary image corrects for temporary translation of the axon during the motion of a vesicle.

These three binary images again are "added" by the "majority operation," described earlier.

We are now ready to start detection of "hits," or events of vesicles crossing the hurdles. Such an event is detected when more than half the points in the hurdle line are excited (having a value 1 in the resulting binary image). The original image is then compressed in the vertical direction such that each column now represents only the hurdles. The thickness of lines is normalized by horizontal thinning, followed by a continued two-dimensional thinning until a 1-point thick, 8-connected line pattern remains.

Motion Tracking

All lines and patterns in this final binary image are stored and labeled. A "fragment description array" is constructed, containing the coordinates of beginning and end points of each cluster, length, and average slope. This array is used to make the most rational connection between the end point of a previous fragment and the beginning point of another fragment. A similarity parameter is calculated for each candidate matching pair, based on four information units: distance between end point of first and beginning point of second fragments, orientation difference of the two fragments, length of the two fragments, and connectivity with other fragments.

Finally each resulting cluster is segmented into events representing "jumps" and "stop periods." First breakpoints separating jumps and pauses are found with a split-and-merge algorithm. A jump is defined as the track of a vesicle, moving with a speed exceeding 0.5 μm/sec. A segment is considered a pause when the speed over a period of 1 sec does not exceed 0.5 μm/sec. A jump time is defined as the time between two pauses. For the whole cluster, a cluster length and the overall transport speed are determined also. The vector scale ratio is calculated as the ratio of the total sum of absolute jump lengths divided by the cluster length. This ratio is related to the "effectiveness" of the axonal transport, that is, the number of switches between motion and stops. A flux parameter is calculated as the sum over all products of jump slopes with jump lengths over each cluster. The absolute value of this parameter is then summed over all clusters.

Illustration of Technique

The effect of vanadate, a known inhibitor of dynein- and (to a lesser extent) kinesin-mediated axonal transport, was investigated in embryonic hippocampal neurons. The technique allows the study of matched fields, that is, the same neurite is examined before and after vanadate perfusion during a time period of 10 min. This allows specific and acute effects of interventions to be studied. Three different doses of vanadate were added to the incubation

TABLE I   Effect of Vanadate Concentrations on Motion Parameters in Embryonic Hippocampal Neurons[a]

Vandate concentration (μM)	Number of tracks/8 min	Cluster length (μm)	Number of jumps/8 min	Jump slope (μm/sec)	Pause time (sec)
0	19	5.4	43	1.31	2.15
10	15	5.12	23	1.29	2.24
50	14	4.76	19	1.41	2.87
100	8	3.94	17	1.32	3.81

[a] The number of tracks and jumps clearly indicates a decrease in motion activity proportional to the vanadate concentration. The absolute value of the jump slope is not affected dramatically, but the pause time within clusters is significantly enhanced.

medium: 10, 50, and 100 μM. Table I illustrates the most relevant findings. Only the highest vanadate doses markedly depressed the axonal transport. It is clear that neither the jump velocity nor the jump length was affected, but the number of movements was reduced. In addition, the pause length was significantly increased. These data suggest that the initiation of a jump was impaired, rather than the jump itself.

Nanovid Microscopy in Axonal Transport

Nanovid Microscopy

The previous microscopic technique was based on the detection of 100 nm vesicles and mitochondria. However, the content of these vesicles is unknown. The obvious question then arises of how to follow the motion of specific, well-defined proteins. Such an approach has been pioneered in our laboratory (6–8). By coupling colloidal gold technology to high-resolution video microscopy, it became possible to follow individual gold probes 30 nm in size in living cells. When these gold probes were directly labeled with antibodies directed against certain proteins, in principle the dynamics of individual proteins could be followed.

The image formation in the nanovid microscope is illustrated in Fig. 3. Briefly, the colloidal gold particles observed in bright-field transmission light microscopy scatter light out of the aperture of the objective. Because the size of the gold particles (30 nm) is much less than the wavelength of light, we can apply the Rayleigh approximation to calculate the amount of scattered light. This intensity is proportional to the sixth power of the radius of the

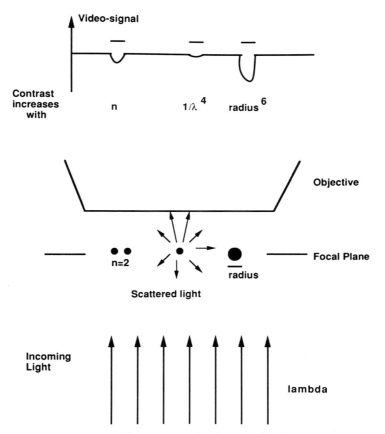

FIG. 3   Physical principles of image formation in nanovid microscopy. The transmitted light impinges on the colloidal gold particles and is scattered out of the objective aperture. The amount of scattered light is proportional to (1) the number of gold probes lying within the resolution of the light microscope (usually 300–400 nm), (2) the radius of the gold particles, and (3) the inverse of the wavelength. The tiny contrast difference in the video signal is then further amplified by a dc offset and analog contrast enhancement. Imperfections in the images are finally cleaned up by digital subtraction of a "mottle" image (an image carrying essentially out-of-focus information, such as uneven illumination or dust particles along the optical pathway). [With permission from H. Geerts *et al., Nature (London)* **351,** 765 (1991).]

particles. This feature virtually excludes the possibility of using double-labeling techniques (i.e., 20 nm for antibody *A* and 40 nm for antibody *B*) in this setup. The intensity is also proportional to the number of scatters, which can be illustrated as follows. The real resolving power of the light microscope of course does not exceed the limit imposed by the Rayleigh theorem, that is, the order of the wavelength used. Therefore, if two or more particles are within a distance less than or equal to the wavelength, they are not discerned as two individual points, but rather they are smeared and appear as a cluster with twice the contrast.

Once the image is picked up, the resulting (small) contrast variations in the video signal are amplified by analog contrast enhancement. This is performed by first subtracting an analog dc signal from the incoming video signal and amplifying the resulting difference. This usually is performed interactively by the observer. However, analog contrast enhancement tends to amplify lighter and darker zones in the image and dust particles along the optical pathway. Therefore, the sample is defocused, and a "mottle" image is digitized and stored in a separate image buffer. The mottle image is subtracted in real time pixel by pixel from the incoming video image, and the resultant image usually gets an offset of 100. This "mottle subtraction mode" yields a clean image in which the gold particles are clearly visible (see, e.g., Fig. 1b). This same procedure can be used for video-enhanced differential interference contrast, the image modality used for the quantification of axonal transport of endogenous vesicles.

This technique can be applied to a variety of problems in cell biology. The colloidal gold probes themselves bear a negative charge, which can be reversed by coating them with polylysine molecules. These probes then bind to negative sialic acid residues of glycoproteins present on the surface of cell membranes. In this way the dynamic organization of cell surfaces (two-dimensional Brownian motion with the creation of microdomains) or ATP-driven uniform motion of glycoproteins after cross-linking can be studied (2).

When probes are coupled to antibodies against specific proteins, the colloidal gold becomes a specific label for the protein. We have studied the endocytosis and intracellular fate of the transferrin receptor in A431 cells (6) and are currently studying the dynamics of the NGF receptor in septal neurons. Axonal transport can be quantified by studying the fate of endocytosed unlabeled 40 nm gold probes in neuroblastoma cells, such as the N4 or IMR32 lines.

Digitizing Images

The experiments are always taped so that a visual control can be performed afterward. The video sequences then have to be digitized in order to extract relevant information about the dynamic changes in the positions of the gold particles. Again the memory requirements are quite stringent. At this high magnification, an area of interest easily extends over 300 × 50 or 15,000 pixels. This means an information transfer of 15,000 × 25 frames, or 375,000 bytes per sec.

Several modalities exist as mentioned previously. Parallel-disk transfer is the best solution and can yield up to 1 gigabyte or 2730 sec of video sequences (45 min). In practice the amount of memory available largely depends on the budget. Optical memory laser disks can be taped in real time and read out at a much lower speed. This eliminates the need for large RAM memory and powerful image processing. However, the cost of this write-once media precludes the use of many hours of video sequencing. Alternatively, the RAM memory of modern computers can be stored at video rate. A 32 MByte RAM memory is then filled within 87 sec. These few examples illustrate the limits imposed on data storage.

In our setup, a 2048 × 2048 tileless frame buffer (Imaging technology S200) is used. We first interactively determine the area of interest (AOI) by use of a mouse-driven cursor on the S200 monitor. Depending on the size of the AOI, we then divide the remaining frame buffer in a number of areas, exactly equal to the AOI under study. This is done in such a way that the complete 2K × 2K frame buffer (including the high byte), is covered as closely as possible. The incoming video frame (512 × 512) is digitized, and the AOI is copied into subsequent areas. Depending on the size of this AOI, we can digitize between 60 and 200 sec of continuous video sequence at a rate of 5–6 frames per sec.

Detecting Gold Particles in Images

Once the images are digitized, the detection of colloidal gold can start. Because the gold particles are easily discernible in the visual image as black dots, the first approach is to simply threshold the image. The algorithm classifies the individual pixels by comparing the gray value to a predetermined value, into an object or background. However, even with the best mottle subtraction, there is still some slow drift in the background signal, which makes the detection very cumbersome. A better method is to include the information of adjacent pixels to decide whether a certain point belongs to an object or not. We routinely use a gray-scale morphology approach (9). This algorithm (available, e.g., in TCL or SCIL-TNO Delft or Multihouse) works in the following way.

By careful observation of the gray values we see that the image of a single colloidal gold particle is blurred to a spot of 200–400 nm diameter (2–3 pixels at the magnification used). Detailed analysis of the gray values further reveals a kind of two-dimensional Gaussian-shaped valley. The convolution or kernel $b(i,j)$ used in the gray-scale morphology has a profile resembling the Gaussian form. Gray-scale dilation is different from classic binary dilation, in that the dilated pixel is calculated from

$$d(x,y) = \text{Max}(i,j)[a(x + i,y + j) + b(i,j)] \qquad (1)$$

where $a(k,l)$ and $d(k,l)$ are the original and dilated image, respectively, and $b(i,j)$ is the kernel used. Conversely, erosion is defined as

$$e(x,y) = \text{Min}(i,j)[a(x + i,y + j) - b(i,j)] \qquad (2)$$

The closing operation is then a sequence of a dilation followed by an erosion. This operation pictorially can be compared to the movement of a ball, slightly larger than the greatest signal of a gold cluster, rolling over the two-dimensional array of gray values. This image formed ''by the rolling ball'' is then subtracted from the original image, yielding a much clearer contrast for the colloidal gold probes. An additional advantage of this method is that the threshold in this processed image is only dependent on the contrast difference between the gold and its local background and is relatively independent of global illumination. This obviates the need for sophisticated automatic threshold algorithms. However, the algorithm itself involves, for a 512×512 image, some 8 million linear and nonlinear operations. On dedicated image processors, this may take only a few seconds.

Calculation of Motion Parameters

Once the colloidal gold particles are detected in a series of time frames, we then proceed to the calculation of saltatory motion parameters. In a first analysis we have to associate a particle in each time frame to its position in the previous time frame. This is performed by assigning a number to each labeled particle in the first image. Only particles within a certain size window are included. In subsequent images, for each numbered ''template'' particle, we assign the same identity to the closest particle in its neighborhood. This distance, however, must lie within certain limits. An estimate for this distance can be calculated from the interframe time (160–200 msec), the maximal jump velocity (10 μm/sec), and the magnification (12 pixels/μm). at the end of this sweep, the limit is doubled and the remaining new particles are eventually assigned to solitary ''template'' particles. The remaining particles then get a new labeling number.

The second part of the analysis is the calculation of jump lengths and jump

TABLE II   Effect of Vanadate on Fast Axonal
Transport of Gold Probes in N4
Neuroblastoma Cells[a]

Vanadate concentration (μM)	Velocity (μm/sec)	Jump length (μm)
0	1.22/1.22 (29) (0.69–1.62)	2.05/1.61 (29) (1.03–4.56)
50	1.02/0.99 (4) (0.82–1.99)	1.5/1.24 (4) (1.03–4.43)

[a] Concentration of vanadate, 50 μM. Gold probe, 40 nm
BSA-PEG labeled colloid. Neuroblastoma cells were differentiated for 48 hr in EMEM (Earle's minimum essential medium) plus 10% FCS (fetal calf serum) containing 1 mM
dibutyryl-cAMP. Cells with clear neurite outgrowth were
chosen. Shown are the mean and median values, the number
of retained motions given in parentheses, and the minimum
and maximum of the parameters also in parentheses. On
application of vanadate, the number of motions in the neurite
is drastically reduced. In addition, a slight decrease in jump
length is noticed.

velocities for each set of particles. Only jumps exceeding a certain length are
taken into account. A histogram of these parameters is constructed, and the
relevant values are stored in an appropriate data file. It must be stated
however, that this analysis can also yield information on Brownian motion
that is important, for instance, in the study of the anti-NGF receptor antibodies (10). Close observation indicates that the particles are moving on the
extracellular surface of the growth cone before they are endocytosed.

Application of Nanovid Microscopy

We have used nanovid microscopy in the study of axonal transport in N4
neuroblastoma cells. These cells readily endocytose BSA (bovine serum
albumin)–PEG (polyethylene glycol)-stabilized colloidal gold (40 nm), so that
clear labeling of moving vesicles is obtained (see Fig. 1b). These vesicles
display the typical characteristics of saltatory motion. In addition, the absolute values of jump velocities and jump lengths are similar. We observe
mostly unidirectional jumps, and reversal of jump direction is rarely noticed.

Table II presents the results of the analysis on the motion of endogenous
40 nm colloidal gold probes in N4 neuroblastoma cells. Here the same neurite

is observed over the same time period (8 min) before and after vanadate application. Only jumps with a length exceeding 1 μm are retained. In the presence of 50 μM vanadate, the motion characteristics drastically decrease. As in the case of the embryonic hippocampal neurons, we do not observe a dramatic decrease in the jump velocity, but the number of motions is drastically reduced. Note, however, that only jumps exceeding a certain length are retained. In addition, a slight decrease in jump length is observed.

Conclusions

This chapter describes in detail two technical approaches to studying fast axonal transport in neuronal cell cultures. One technique monitors the movement of endogenous vesicles with a size exceeding about 100 nm. The other technique can calculate motion parameters of specific labeled proteins or protein complexes.

When comparing these techniques in neuronal cell cultures with the classic techniques *in vivo* in peripheral nerves, the following comments can be made. First, *in vitro* techniques allow for the direct study of certain interventions, notably pharmacological agents or simulations of ischemia or anoxia. These effects can only be studied indirectly *in vivo,* because so many factors may intervene (metabolism of active compounds, bioavailability, endogenous nonneuronal modulation mechanisms). Second, axonal transport may be studied at the mechanistic level in neuronal cell cultures, allowing subtle differences to be picked up. In addition, this may open the way for a detailed molecular biological understanding of the endogenous regulation of fast axonal transport (phosphorylation status of proteins, intracellular calcium levels). Finally, the *in vitro* technique allows the study of central nervous system neurons, like hippocampal neurons or septal neurons. This is obviously still a major problem in *in vivo* studies, owing to the small distances and the rather invasive labeling techniques.

Some comments can be made concerning the limitations of the techniques in neuronal cell cultures. First, the size of the colloidal gold probe can impose hydrodynamic restrictions on the motion characteristics. The results illustrated in Tables I and II indicate that the observed jump velocity is very similar in the two techniques. Notice, however, that the cells are different and the gold may be enclosed in a specific population of vesicles. We believe that the size of the colloidal gold is not important for the hydrodynamic resistance, since the gold is endocytosed in vesicles. Second, in the DIC method, between 10 and 20% of the visually detected motions are not "seen" by the automatic program. The is due to the small endogenous contrast within the image. The number of undetected motions may be reduced by carefully adjusting some parameters, at the risk of inducing false-positive detections. The conservative setting we use therefore accounts only for 80–90% of the motions.

Another (temporary) disadvantage is the limited size of the area of interest. In both techniques, the spatial extension of the motion is determined by technological limitations. The present state of computer hardware only allows covering a segment about 12 μm in length. Some saltations therefore may be truncated in length.

With regard to hardware requirements for analysis, the DIC method is registered on-line during the daytime. This means that, including the interactive session, a video sequence of 8 min can be digitized in 10 min. Analysis usually is performed overnight in batch mode and requires about 30–40 min per 8-min video sequence on a 1 Mips system. With the introduction of more powerful computer systems this time can drastically be reduced. The nanovid analysis, on the other hand, uses the gray-scale morphology technique and is best performed on a powerful general-purpose or dedicated image analysis system. In our laboratory, the S200 image processor can calculate gold positions over a 120-sec stretch in about 4 min. Subsequent reconstruction of tracks requires about 3 min on a 15 Mips Vaccelerator (Horizon Technologies) card, residing in the MicroVax system.

The use of such an automatic analysis system may facilitate considerably the gathering of information about the nature of axonal transport in cell biology. Such an approach may finally yield new insights into the role of axonal transport in neurodegenerative diseases.

Appendix: Addresses of Manufacturers Mentioned in Text

Analogic Corporation, 8 Centennial Drive, Peabody, MA 01960
Datacube, 4 Dearborn Road, Peabody, MA 01960
Data Translation, 100 Locke Drive, Marlboro, MA 01752
Digital Equipment Corporation, Maynard, MA 01754
Hamamatsu Photonics, 1126 Ichino-Cho, Hamamatsu City, Japan
Imaging Technology Inc., 600 West Cummings Park, Woburn, MA 01801
Matrox Electronics Systems, 1055 St. Regis Blvd., Dorval, Quebec H4T 1H4, Canada
Multihouse TSI, Osdorpplein 228, 1068 ER Amsterdam, The Netherlands
Panasonic, One Panasonic Way, Secaucus, NJ 07094
Recognition Concepts, 341 Ski Way, P.O. Box 8510, Incline Village, NV 89450
TNO Institute of Applied Physics, Delft, The Netherlands

For additional information on video microscopy, the reader is referred to the monograph by Inoue (11).

References

1. W. Burke, D. Park, H. Chung, G. Marshall, J. Haring, and T. Joh, *Brain Res.* **537,** 83 (1990).
2. M. De Brabander, R. Nuydens, R. Nuydens, A. Ishihara, B. Holifield, K. Jacobson, and H. Geerts, *J. Cell Biol.* **112,** 111 (1991).
3. R. Allen, N. Allen, and J. Travis, *Cell Motil.* **1,** 291 (1981).
4. S. Inoue, *J. Cell Biol.* **89,** 346 (1981).
5. F. Cornelissen, R. Nuyens, R. Nuydens, and H. Geerts, *J. Neurosci. Methods* **35,** 79 (1990).
6. M. De Brabander, H. Geerts, R. Nuydens, and R. Nuyens, *Am. J. Anat.* **185,** 282 (1989).
7. M. De Brabander, H. Geerts, R. Nuydens, R. Nuyens, and F. Cornelissen, *in* "Electronic Light Microscopy." (D. Shotton, ed.), in press.
8. H. Geerts, R. Nuydens, and M. de Brabander, *Nature (London)* **351,** 765 (1991).
9. S. Sternberg, *Comput. Vis. Graph Image Process* **35,** 335 (1986).
10. H. Geerts, M. De Brabander, R. Nuydens, S. Geuens, M. Moeremans, J. De Mey, and P. Hollenbeck, *Biophys. J.* **52,** 775 (1987).
11. S. Inoue, "Video Microscopy." Plenum Press, New York, 1986.

[19] Use of Computers for Quantitative, Three-Dimensional Analysis of Dendritic Trees

Brenda J. Claiborne

Introduction

Prior to the introduction of computers into the neuroanatomy laboratory in the mid-1960s and early 1970s (1–3), it was extremely difficult, if not impossible, to quantify the three-dimensional structure of individual cortical neurons. Previous attempts to quantify dendritic morphology were based almost entirely on two-dimensional measurements taken from photographs or camera lucida drawings of stained neurons. In the past decade, several computer–microscope systems have been devised specifically to incorporate the third dimension, or depth, into an analysis of dendritic structure (4–7, reviewed in Refs. 8–10). These systems allow investigators to quantify a number of dendritic parameters, including total dendritic length. Using such data, investigators have been able to classify populations of neurons into subgroups (11), to quantify dendritic growth and regression during development (12–15), to compare dendritic structure following various experimental treatments (16, 17), and, more recently, to create electrotonic models of dendritic trees (6, 18–20).

There are two basic approaches to the three-dimensional analysis of dendritic branching patterns using computer–microscope systems. In one approach, dendritic trees are reconstructed from serial sections (9, 10). This method is most commonly used to quantify Golgi-impregnated neurons from sections typically between 75 and 150 μm thick, but it can be time-consuming and error-prone. Not only is it difficult to follow dendrites from one section to another, but, given the common problems of differential tissue shrinkage and broken dendrites, it is often not possible to reconstruct an entire neuron using this procedure (21). A second approach allows the operator to analyze a dye-filled neuron directly from a thick brain slice, or "whole mount" (11, 22). The primary advantage of this method is that an entire dendritic tree is contained within the whole mount, thereby eliminating the problems associated with serial section reconstructions.

This chapter describes the use of a computer–microscope system to quantify the three-dimensional structure of single neurons from whole mounts of

Methods in Neurosciences, Volume 10

cortical slices and also details the labeling procedures that are necessary prerequisites to this type of analysis (11, 15, 22). The approach can be divided into three essential techniques. First, slices of brain tissue are maintained *in vitro* and neurons are labeled by intracellular dye injection. Second, the thick (400 μm) slices are processed as whole mounts. Finally, an entire dendritic tree is encoded in three dimensions directly from a whole mount. The resulting data file consists of a series of three-dimensional coordinates that specify the exact branching pattern of the tree. It includes information on dendrite diameters, branch points, and termination points, and can be used to calculate such parameters as individual branch lengths or the total dendritic length of the neuron. Alternatively, the data file can be entered directly into a simulation program and used to build an electrotonic model of the dendritic tree. The above approach is applicable to neurons from any region of the mammalian brain, although it was developed originally to analyze the dendritic trees of neurons in the hippocampal formation of young adult rats (11, 15, 22).

Neuronal Labeling

Intracellular Injection

Neurons in the *in vitro* brain slice preparation are labeled by intracellular injection. Preparation and maintenance of *in vitro* slices from various regions of the mammalian brain are now established laboratory procedures (23). In our laboratory, slices of the hippocampus are prepared by first removing the entire brain from a deeply anesthetized rat and immersing it in iced saline for approximately 30 sec (11, 15). The hippocampus is then dissected out, placed on a McIlwain tissue chopper (Brinkman Instruments, Westbury, NY), and sectioned into 400-μm-thick transverse slices. Alternatively, slices can be cut on a vibratome. The slices are immediately transferred to buffered saline (22) in an *in vitro* chamber at 32°C and supported on a membrane with 12-μm-diameter pores (Nuclepore Corp., Pleasanton, CA). Although slices can be supported on a nylon net, the indentations from the net often distort the structure of the labeled dendrites.

Neurons are filled with the enzyme horseradish peroxidase (HRP) using sharp micropipettes (11, 15, 22). Micropipettes with short shanks are made using a Brown–Flaming pipette puller (Sutter Instruments, Novato, CA). Their tips are enlarged slightly by reducing the heat setting on the puller or by beveling (Sutter Instruments Beveler) such that their final resistances are between 80 and 120 MΩ. The micropipettes are filled with a 2–3% (w/v) solution of HRP (Grade I, Boehringer-Mannheim, Indianapolis, IN) in 25 mM Tris buffer (pH 8) and 0.5 N KCl.

To reduce the probability of labeling neurons with cut dendrites, only neurons located in the middle of the slice are filled. After impaling a neuron with a suitable resting potential, -60 mV or better, the membrane is allowed to seal around the pipette for at least 5 min. Then HRP is injected using positive current pulses with a duration of 250 msec, an amplitude of 3–4 nA, and a frequency of 2 pulses/sec. During the injection, the current trace from the intracellular amplifier is monitored to ensure that the tip is not blocked. Experience has shown that the time for complete labeling depends on the size of the dendritic tree. Dentate granule neurons are filled completely after 5 to 8 min of dye injection, whereas CA1 pyramidal neurons require 20 to 30 min. After the injections, the slices remain in the chamber for 1 to 2 hr to allow for diffusion of the HRP.

Alternative Labels

The advantage of using HRP as a label is that it yields a dense product after a reaction with 3,3'-diaminobenzidine tetrahydrochloride (DAB). On the other hand, HRP is a large molecule, and hence it is difficult to eject from an intracellular pipette and diffuses slowly to distal dendritic tips. Biocytin (biotin-lysine, Molecular Probes, Eugene, OR) and neurobiotin (Vector Labs, Burlingame, CA) are both small molecules and are now used by some laboratories for intracellular labeling. They are made visible by reacting the tissue with avidin coupled to HRP (Sigma ABC kit), followed by a DAB incubation as described below. Fluorescent dyes, although advantageous for intracellular labeling because of their low molecular weights, are not useful for three-dimensional analyses because their fluorescence quenches quickly under illumination, even when antibleaching agents are used. Although in thin sections a fluorescent dye can be photoconverted to a dense label by illuminating the tissue in the presence of DAB (24), we have not had success with this procedure in 400-μm-thick slices of adult brain tissue.

Histological Procedures

Fixation and Visualization of Labeled Neurons

After the neurons have been filled and allowed to incubate for at least 1 hr, the slices are fixed in 1% paraformaldehyde and 2% glutaraldehyde in 0.1 M phosphate buffer (pH 7.3) for 10 to 16 hr at 4°C. They are then washed in 6 changes (10 min each) of phosphate buffer at room temperature. To increase

DAB diffusion into the middle of the thick slices (22), they are treated with 1% (w/v) Triton-X in buffer for 1 hr. The slices are then washed in 6 changes (5 min each) of buffer. All incubations are done on a shaker table.

To visualize the HRP, slices are reacted with DAB (Cappel, Durham, NC) and hydrogen peroxide (22). They are preincubated in 0.06% (w/v) DAB (filtered) in phosphate buffer for 1 hr, immersed in the same concentration of DAB containing 0.15% (v/v) H_2O_2 for 1 hr, and then washed in 6 changes (5 min each) of buffer. Because DAB is a suspected carcinogen, it should be handled in a glove box or confined to a limited area and inactivated with household bleach.

Clearing and Mounting

Slices are cleared in ascending concentrations of glycerol: 20, 40, 60, 80, 95% glycerol in distilled water for 10 min each, followed by 3 changes (30 min each) of 100% glycerol (22). They are mounted in 100% glycerol and the coverslip sealed with nail polish. After analysis of the filled neurons, the slices are stored in vials in 100% glycerol at 4°C. In most cases, the quality of the labeling does not deteriorate for at least 6 months.

Using the fixation and clearing procedures described above, the slices shrink less than 5% in overall size, and the dendrites are not distorted. (Measurement errors introduced by tissue shrinkage are discussed in Refs. 6 and 25.) We have not found a permanent mounting medium that does not cause severe shrinkage. When the slices are dehydrated with ethanol and cleared in xylene, the dendrites become wrinkled and distorted. The same phenomenon occurs with water-soluble mounting media such as glycol methacrylate; the dendrites wrinkle over a period of days as the medium hardens.

HRP Intensification

If the dendritic tree is completely filled, but the reaction product is light, the HRP labeling can be intensified with metals (L.-A. Coleman, 1991, personal communication; 26, 27). A slice is first cleared and examined as described above. If the labeling is light, the slice is rehydrated in descending concentrations of glycerol and washed in 3 changes (10 min each) of 50 mM Tris buffer (pH 7.4). It is then incubated in the following solution on a shaker table: 5 mg of DAB, 2.5 ml of 1% (w/v) cobalt chloride (filtered), 2.0 ml of 1% (w/v) nickel ammonium sulfate (filtered), and 10 ml of 50 mM Tris buffer (pH 7.4). The slice is incubated from 30 to 50 min, depending on its thickness and the age of the animal. After 30 min, it is examined every 5 min to ensure that the nonspecific tissue darkening does not obscure the filled neuron.

Next, 2 drops of 0.3% H_2O_2 are added to the above 14.5 ml of solution and the slice incubated for approximately 15 min. During this time, the HRP reaction product darkens. Again, tissue density is monitored closely. The slice is then washed in 6 changes (5 min each) of 50 mM Tris buffer and cleared as described above.

Digitizing Dendritic Trees

Camera Lucida Drawings

Before encoding, the neurons are drawn by hand using a camera lucida and a ×63 oil immersion objective with a 0.5 mm working distance (Zeiss Neofluar). These drawings serve as a guide during the digitizing process and as a means of examining the extent of the labeling. Strict criteria are used for the selection of complete trees. The staining must be uniformly dark throughout the entire tree. Furthermore, for each cell type, a maximum number of acceptable cut branches is determined. For example, a granule neuron is not analyzed if it has any dendrites severed in the proximal half of the dendritic layer or more than two cut branches in the distal half of the layer (15).

Computer–Microscope Systems

There are several computer–microscope systems now available for quantifying the dendritic tree structure of individual neurons directly from thick slices (4–7, 8–10). One is marketed by Eutectics Electronics, Inc. (Raleigh, NC), and is described by Capowski in a recent book (10). (This volume also provides an excellent summary of the use of computers in neuroanatomy.) An early model of the Eutectics system was used for portions of the work reviewed here (11, 22).

In our laboratory, we use a computer–microscope system designed by Dr. John P. Miller, with software written by Dr. R. H. W. Nevin (5–7).* The system designed by Miller incorporates two important features. First, optical encoders (with a resolution of 0.2 μm) are mounted on the microscope stage and are used to record stage position. In earlier systems, stage position was simply inferred from the commands sent to the stepper motors (see Ref. 6 for a complete discussion of this problem). Second, a video camera is mounted on

---

* Dr. Miller has kindly made the software and a list of the necessary hardware available to interested investigators. His address is 195 LSA, Department of Molecular and Cell Biology, University of California at Berkeley, Berkeley, CA 94720.

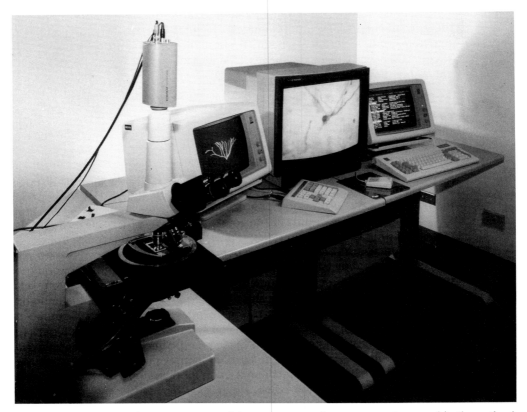

FIG. 1   Major components of the computer–microscope system used in the author's laboratory for analyzing dendritic trees in three dimensions (see text for description). The system was designed by Dr. John P. Miller, University of California at Berkeley.

the microscope, and the filled neuron is digitized directly from a monitor instead of through the microscope. This procedure reduces operator error and fatigue and makes it easier to train new operators.

The entire computer–microscope system is shown in Fig. 1. It consists of a microscope (Optiphot 66, Nikon, Garden City, NY), a motorized stage connected to an IBM AT computer through a series of indexers (not visible in Fig. 1), a CCD (charge-coupled device) camera (Cohu, San Diego, CA) mounted on the microscope, and three monitors. One monitor (PGA) displays the digitized neurons (monitor on far left in Fig. 1). A video monitor (Sony Trinitron) is used to visualize the filled neuron (middle monitor in Fig. 1), and a third monitor displays command menus and status conditions from the digitizing program. Also visible in Fig. 1 are a keypad (Touchstone, Rochester, NY), a mouse, and the computer keyboard.

The digitizing process can be understood more readily by first considering the interactions between the various components of the system. When a slice whole mount with a filled neuron is placed on the microscope stage, the ×63 objective described above is used to project an image of the neuron onto the CCD array of the camera. This image is subsequently sent as a composite black and white video signal to a video board (ITEX 100, Imaging Technology Inc., Woburn, MA) located in one of the card slots of the IBM PC-AT. This board allows graphics generated by the digitizing program to be overlaid on the video image of the neuron. The neuron image and the overlays are then conveyed by red-green-blue (RGB) signal lines to the video monitor.

The digitizing program (referred to as "DIG") is written in C, and when invoked initializes the video board and the three indexers (Series 2100, Compumotor, Rohnert Park, CA) that interface the computer with the stage and focus control motors via their drivers. DIG also runs a driver for the 3-button mouse. Before the operator digitizes a neuron, setup information (including magnification factors and display parameters) is requested by the program and provided by the operator.

DIG generates a colored cursor as an overlay on the video monitor. Using the mouse, the operator controls the position and diameter of the cursor. The operator moves the cursor along a dendrite and to encode a "data point," then presses a key on the keyboard or keypad, thereby recording the three-dimensional location and diameter of the dendrite. Each data point is encoded as one of several different types, such as soma, dendritic tree origin, dendrite, branch point, or termination point.

One limitation of all computer–microscope systems is that they lack the resolution necessary to measure dendritic diameters accurately. This results both from the limits of resolution of the light microscope, which makes it difficult to resolve dendrites that are less than 1.0 μm in diameter, and from the pixel size of the video monitor. The pixel size limits the available diameters of the cursor. In our laboratory, whenever the operator cannot exactly match the cursor diameter to a dendrite diameter, the next largest diameter of the cursor is always chosen. With relatively small dendritic trees such as the granule neurons, calibration tests have shown that this procedure does not increase surface area by more than 15% (15). However, for larger dendritic trees, the error introduced into surface area or volume calculations may be significant. The error could be reduced by using a ×100 objective on the microscope, or by measuring dendrite diameters at selected locations on the tree with a confocal or electron microscope and then using these measurements to correct the diameters recorded in the data file (6).

Digitizing Process

To digitize a neuron, the operator first centers a reference point, usually the cell body, on the monitor. Next he or she encodes the size and location of the cell body, records the origins of the primary dendrites, and selects one to digitize. The cursor is moved to the origin of this dendrite and the cursor diameter adjusted to match that of the dendrite. The operator then traces along the dendrite, continually adjusting the cursor diameter. A data point is entered approximately every 5 μm, or more frequently if the dendritic path changes abruptly. The locations of branch points and terminations are recorded, and, if desired, the locations of dendritic spines and varicosities can also be encoded.

Because only a portion of the dendritic tree is in the field of view at any one time, the stage and focus knobs on the microscope must be moved in order to visualize the entire tree. The operator controls the focus knob movements using the buttons on the mouse. The stage movements are also indirectly controlled by the operator. As the operator guides the cursor to a branch outside of the center portion of the field of view (i.e., outside of the optically flat portion of the field), DIG sends a command through the indexers to reposition the stage such that the branch is now centered on the monitor.

It is worth noting that much time was put into making the stage movements reliable. In our system Miller and Nevin found that the most reliable means of moving the stage was to move it in a large arc. Therefore, whenever a stage movement is initiated, the stage is first moved up and to the left of the point to be centered so that it approaches any point, regardless of its X, Y coordinates, from the same direction. This motion corrects for backlash in the stage and for the torques present in the motor drive shafts. The correction, together with the use of the linear encoders to measure the actual stage position, has resulted in the system having an error of less than 1% in absolute measurement over a scanning distance of 400 μm.

To ensure that all the branches of a tree are recorded and to reduce encoding time, the program assists the operator in returning to branch points. After one limb of a bifurcated branch is digitized and the operator signals its termination, DIG automatically sends a command to the stage to recenter the branch point on the monitor. A similar routine ensures that all of the primary dendrites and their branches are encoded: when a primary branch and its associated dendrites are completely digitized, one of the remaining dendritic origins is centered on the monitor. The operator proceeds to digitize that primary dendrite and its daughter branches.

Several other features of the Miller system improve operator accuracy and speed. For example, the operator can evoke graphic displays while digitizing. As each data point is recorded, a colored line indicating the region already digitized can be overlaid on the video image of the dendrite to help guide the operator. In addition, the completed portion of the dendritic tree can be displayed on the PGA monitor (see far left monitor in Fig. 1). The operator can then compare the display with the camera lucida drawing of the tree to ensure that none of the branches have been omitted.

Once the digitizing process is completed, the program prompts the operator to name the cell and enter any comments. The data file is then stored. Each recorded data point is numbered sequentially, is coded according to type (soma, dendrite origin, etc.) and has four values associated with it. The first two values, X and Y, represent the cartesian coordinates of the stage with respect to the reference point. The third, or Z, value represents the depth of the data point in the slice. The fourth value represents the diameter of the soma or dendrite.

In our laboratory, both granule neurons and pyramidal neurons from the hippocampal formation of the rat are routinely digitized. Approximately 80 neurons have been digitized. An experienced operator requires from 1 to 2 hr to digitize a granule cell (about 500 data points) and from 12 to 15 hr (3000–4000 data points) to digitize the more complex dendritic tree of a pyramidal neuron.

Morphological Data Analysis and Compartmental Models

After a dendritic tree has been digitized it is displayed as a three-dimensional structure on the PGA monitor (Fig. 1). The tree can be rotated in any plane and three-dimensional depth cues provided by shading routines. The data can also be presented as a two-dimensional projection showing the branching pattern of the neuron (Figs. 2, 3, and 4.) In addition, it is often useful to examine a ''schematic diagram'' (10) in which the branching pattern is reduced to a sticklike form (Fig. 2). Numerical values for the lengths of each branch can also be displayed on this diagram (Fig. 2).

The data are analyzed quantitatively using the IBM AT computer. Dendritic parameters, including branch length and order, total dendritic length, and the maximum extent of the tree in the X, Y, or Z plane, are obtained (5, 15). The program also calculates the surface area and volume of the dendritic tree; however, we exercise caution in using these parameters because they are based on diameter measurements that are not always accurate (see above).

With the availability of less expensive workstations, Miller and Nevin have developed a Neural Editor program (NED) that runs on a 32-bit workstation under X-Windows in a UNIX environment (5–7). Data files obtained under DIG are imported directly into NED. In addition to providing high resolution,

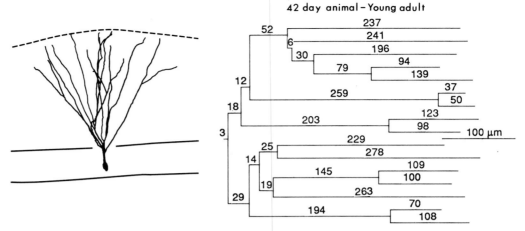

FIG. 2 Computer-generated representation of the dendritic tree of a granule cell from the dentate gyrus of a young adult rat. The tree was digitized in three dimensions using a Eutectics computer–microscope system (10). Schematic diagram of the same neuron, showing the lengths of the individual branches.

color displays of the neurons, and a number of morphological parameters (including those listed above), NED performs more sophisticated morphological analyses. These include neuronal scaling to correct for tissue shrinkage, calculations of volume overlaps of two or more neurons, and methods for analyzing portions of dendritic trees (5, 7).

It is also now possible to incorporate the data files generated under DIG into programs that generate electrotonic models of the digitized neurons. One such program was developed in Miller's laboratory by Dr. John Tromp (6). It takes data directly from NED, combines it with the electrophysiological properties of the cell, and generates a compartmental model of the neuron. In a collaborative project with Dr. Thomas Brown and colleagues at Yale University (New Haven, CT), we are using a program called NEURON, written by Michael Hines (28), to make compartmental models of hippocampal neurons. Data files from DIG are ported directly to NEURON on a Sun workstation, the appropriate biophysical characteristics of the neuron are incorporated, and a compartmental model of the dendritic tree is generated. Simulation results are displayed using a graphics program developed in Dr. Thomas Brown's laboratory (18–20).

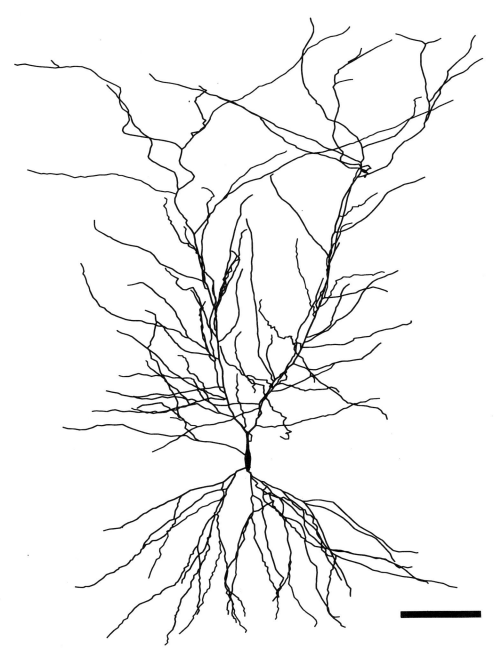

FIG. 3 Computer-generated representation of the dendritic tree of a CA1 pyramidal neuron from the hippocampus of a young adult rat. Spines were not encoded. The tree was digitized in three dimensions using the Miller computer–microscope system shown in Fig. 1. Bar: 100 μm.

A

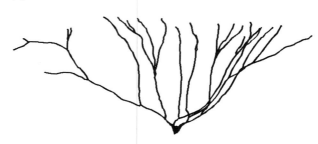

B

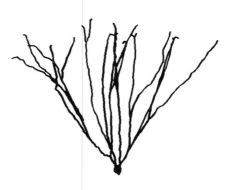

C

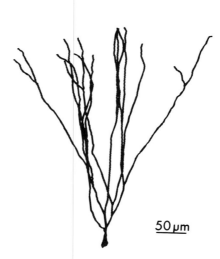

50 μm

Analysis of Hippocampal Neurons

The power of the techniques described above are perhaps best appreciated by briefly reviewing the results of our recent studies on neurons in the hippocampal formation of the rat. The three-dimensional data have revealed that distal dendrites of hippocampal neurons, those most often lost in reconstructions from serial sections, account for a large percentage of the dendritic tree. Approximately 40% of the total dendritic length of the granule cells in the rat dentate gyrus is located in the outer third of the molecular layer (11), and about 20% of the total dendritic length of CA1 and CA3 hippocampal pyramidal cells is found in stratum lacunosum-moleculare, that portion of the dendritic layer most distant from the cell body (29).

A comparison between granule neurons in the two blades of the adult dentate gyrus showed that neurons in the dorsal blade had greater total dendritic lengths, more dendritic branches, and larger transverse spreads than did neurons in the ventral blade (11). Further studies of the dorsal blade neurons during late postnatal development (15) showed that individual dendritic branches increase in length as the animal matures (Fig. 4). Concurrently there is an overall decrease in the average number of dendritic branches per neuron between postnatal days 14 and 60 (Fig. 5). The combination of both dendritic growth and branch loss appears to result in a conservation of total dendritic length during this time period (Fig. 5). These results are significant because they show that, in contrast to dentate granule cell development in the monkey, granule cells in rodents do not exhibit a decrease in total dendritic length as they mature. They also suggest that the total dendritic length of granule cells may be regulated during late postnatal development.

Recently we have incorporated the three-dimensional morphology of hippocampal CA1 pyramidal neurons (Fig. 3) into realistic compartmental models (see above). Simulations using these models have demonstrated the existence of steep voltage gradients within the dendritic trees (18–20). These findings suggest that it is not possible to voltage clamp transient synaptic events on the distal dendrites from an electrode in the cell body. The results also have important implications for the computational capabilities of pyramidal neurons; they suggest that portions of the tree may act as independent processing units.

FIG. 4   Computed-generated representations of the dendritic trees of granule neurons from the dentate gyrus of developing rats: (A) 14-day-old rat; (B) 40-day-old rat; and (C) 53-day-old rat. Spines were not encoded. The trees were digitized and analyzed using the Miller computer–microscope system shown in Fig. 1. [Reproduced with permission from L. L. Rihn and B. J. Claiborne, *Dev. Brain Res.* **54**, 115 (1990).]

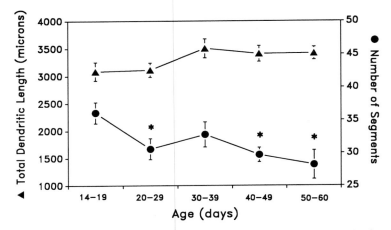

FIG. 5   Comparison of total dendritic length and number of segments (or branches) per granule neuron with increasing age of the rat. Note that total length is conserved, while segment number declines (see text). Asterisks indicate a significant difference from the youngest age group (two-tailed Student's t-test, $P < 0.05$). There were no significant differences in the number of segments among the four oldest age groups. The number of trees analyzed from the five age groups were 13, 8, 8, 15, and 9. Values are means ± S.E.M. The analysis was done using the Miller system. [Reproduced with permission from L. L. Rihn and B. J. Claiborne, *Dev. Brain Res.* **54,** 115 (1990).]

Limitations and Future Developments

There are several limitations to the approach described here. One limitation is the small number of neurons that are completely labeled after intracellular injections. We routinely reject at least 50% of the labeled dendritic trees because their distal tips are not filled. The use of smaller dyes (see above) and the use of patch pipettes for dye injection may reduce this percentage in the future.

A second limitation is that the dendritic trees of some cortical neurons may extend beyond the boundaries of a tissue slice. It is generally accepted that slices thicker than 500 μm do not survive well *in vitro*. If the spread of a dendritic tree is greater than the thickness of the slice, obviously dendrites will be severed when the slice is cut. Fortunately, hippocampal neurons rarely have cut dendrites on both surfaces of a slice. Although one could argue that neurons with cut dendrites are more difficult to fill and therefore are not included in a sample of filled neurons, this does not appear to be the case: filled neurons often have numerous cut dendrites on one side of a slice.

The one limitation of current computer–microscope systems is the lack of accurate diameter measurements (see above) (6, 10). This results from both the limits of resolution of the light microscope and the pixel size of the video monitor. One solution is to measure dendritic diameters at selected locations on a filled neuron using a confocal or electron microscope and then use these measurements to correct the diameters in the data file generated from the computer–microscope system (6). A more elegant solution would be to adapt the three-dimensional computer–microscope system described here to a confocal microscope. This would allow accurate diameter measurements and would also provide a method for digitizing fluorescent-labeled neurons; the light levels used in a confocal microscope generally do not cause severe quenching of the fluorescent signal.

Conclusions

Computer–microscope systems are now available for quantifying the dendritic structure of cortical neurons in three dimensions. In the approach described here, neurons are labeled by intracellular injection in slices of the cortex maintained *in vitro,* the slices are processed as whole mounts, and the entire dendritic tree is analyzed in three dimensions directly from the thick slice. This method avoids the laborious, and sometimes impossible, task of reconstructing dendritic trees from serial sections. It has been used successfully to classify dentate granule neurons on the basis of their morphological features (11), to quantify dendritic growth and regression during granule cell development (15), and to create realistic compartmental models of hippocampal neurons and their synapses (18–20).

Acknowledgments

I thank Drs. John P. Miller and R. H. W. Nevin for developing the computer–microscope system, and for help in setting it up in our laboratory. I also thank Michael P. O'Boyle for excellent technical assistance with the system. Rosie Reyes provided constructive comments on the manuscript, and Omid B. Rahimi assisted with the figures. This work was initiated in the laboratory of Dr. W. M. Cowan at the Salk Institute for Biological Studies, La Jolla, CA. Research in our laboratory is currently funded by the National Institute on Aging, the National Science Foundation, and the Office of Naval Research.

References

1. E. M. Glaser and H. van der Loos, *IEEE Trans. Biomed. Eng.* **12,** 22 (1965).
2. D. F. Wann, T. A. Woolsey, M. L. Dierker, and W. M. Cowan, *IEEE Trans. Biomed. Eng.* **20,** 233 (1973).

3. R. D. Lindsay, *in* "Computers in Biology and Medicine" (G. P. Moore, ed.), p. 71. Plenum, New York, 1977.
4. J. J. Capowski and M. J. Sedivec, *Comput. Biomed. Res.* **14,** 518 (1981).
5. R. H. W. Nevin, "Morphological Analysis of Neurons in the Cricket Cercal System." Doctoral Dissertation, University of California, Berkeley (1989).
6. J. P. Miller, J. W. Tromp, and R. H. W. Nevin, *in* "Neural Computation, 1990 Short Course 3 Syllabus" (A. I. Selverston, ed.), p. 10. Society for Neuroscience, Washington, D.C., 1990.
7. G. A. Jacobs and R. H. W. Nevin, *Anat. Rec.* **231,** 563 (1991).
8. D. P. Huijsmans, W. H. Lamer, J. A. Los, and J. Strackee, *Anat. Rec.* **216,** 449 (1986).
9. H. B. M. Uylings, A. Ruiz-Marcos, and J. van Pelt, *J. Neurosci. Methods* **18,** 127 (1986).
10. J. J. Capowski, "Computer Techniques in Neuroanatomy." Plenum, New York, 1989.
11. B. J. Claiborne, D. G. Amaral, and W. M. Cowan, *J. Comp. Neurol.* **302,** 206 (1990).
12. G. Leuba and L. J. Garey, *Dev. Brain Res.* **16,** 285 (1984).
13. N. T. McMullen, B. Goldberger, and E. M. Glaser, *J. Comp. Neurol.* **278,** 139 (1988).
14. J. E. Vaughn, R. P. Barber, and T. J. Sims, *Synapse* **2,** 69 (1988).
15. L. L. Rihn and B. J. Claiborne, *Dev. Brain Res.* **54,** 115 (1990).
16. M. J. Sedivec, J. J. Capowski, and L. M. Mendell, *J. Neurosci.* **6,** 661 (1986).
17. N. T. McMullen, B. Goldberger, C. M. Suter, and E. M. Glaser, *J. Comp. Neurol.* **267,** 92 (1988).
18. T. H. Brown, A. M. Zador, Z. F. Mainen, and B. J. Claiborne, *in* "Long-term Potentiation: A Debate of Current Issues" (J. Davis and M. Baudry, eds.), p. 357. MIT Press, Cambridge, Massachusetts, 1991.
19. B. J. Claiborne, A. M. Zador, Z. F. Mainen, and T. H. Brown, *in* "Single Neuron Computation" (T. McKenna, J. Davis, and S. Zornetzer, eds.), p. 61. Academic Press, Boston, 1992.
20. T. H. Brown, A. M. Zador, Z. F. Mainen, and B. J. Claiborne, *in* "Single Neuron Computation" (T. McKenna, J. Davis, and S. Zornetzer, eds.), p. 81. Academic Press, Boston, 1992.
21. N. L. Desmond and W. B. Levy, *J. Comp. Neurol.* **212,** 131 (1982).
22. B. J. Claiborne, D. G. Amaral, and W. M. Cowan, *J. Comp. Neurol.* **246,** 435 (1986).
23. R. Dingledine, "Brain Slices." Plenum, New York, 1984.
24. E. H. Buhl and J. Lubke, *Neuroscience* **28,** 3 (1989).
25. H. B. M. Uylings, C. G. van Eden, and M. A. Hoffman, *J. Neurosci. Methods* **18,** 19 (1986).
26. J. C. Adams, *Neuroscience* **2,** 141 (1977).
27. J. C. Adams, *J. Histochem. Cytochem.* **29,** 775 (1981).
28. M. Hines, *Int. J. Bio-Med. Comput.* **24,** 55 (1989).
29. D. G. Amaral, N. Ishizuka, and B. J. Claiborne, *in* "Progress in Brain Research" (J. Storm-Mathisen, J. Zimmer, and O. P. Ottersen, eds.), Vol. 83, p. 1. Elsevier, Amsterdam, 1990.

[20] Interactive Computer-Assisted Coding, Three-Dimensional Reconstruction, and Quantitative Analysis of Neuronal Perikarya, Processes, Spines, and Varicosities

M. Freire

Introduction

The quantitative study of neurons requires the use of a light microscope and added devices permitting the recording of the three-dimensional (3-D) disposition of neurons in nerve tissue. A practical and efficient way to do this work is to store in the computer memory the 3-D coordinates of points of the neuron, selected by the operator using ocular cross hairs or a computer-generated cursor. Tracing linear segments between these points, a simplified graphic representation of the neuron is obtained resembling a "stick" or "wire" structure. Topological identifiers or anatomical descriptors indicating the beginning, the branching points, and the end of the processes are necessary for plotting the neuron or making quantitative calculations (1).

The stick model allows us to make computer computations and to know the number and length of the neuronal processes. However, information dealing with the width of processes and with spines and varicosities is not included. A number of authors proposed to elaborate or implement in their computer-assisted microscopes an expanded stick model to include more information relevant to the quantitative study of the neuron and, at the same time, to obtain a more realistic graphic representation of the neuron (2–5).

We have developed a very simple computer-assisted light microscope system for studying all the morphological and structural characteristics of the neuron, including the perikaryon, axon, dendrites, spines, and varicosities, that produces a realistic graphic representation of the neuron (6–9). The hardware and the software developed to implement the system will be explained in this work. The purpose when writing this chapter was to give to any researcher sufficient background to reproduce this computer-assisted microscope or to take advantage of the ideas presented when making their own computerized system.

Interactive Approach for Codifying Neurons

Using an interactive computer-assisted microscope, the operator can see superimposed both the light microscope image of the neuron and the graphic representation of the neuron made by the computer. Any shift between the two images can be noted and corrected at that moment by the operator during the coding of the neuron, avoiding periodic checks and the use of computer-based error correction procedures (1, 10). Another advantage of the interactive approach is that the operator can choose the best fitting between the segment of the neuronal process that is being coded and the segment drawn by the computer. This segment is drawn between the last stored point and the computer-generated cursor each time the operator moves the cursor ("rubber technique"), before storing the 3-D coordinates of the cursor as the next selected point.

The mixing of the light microscope image and the computer-generated image can be made by electronic (video mixing) or optical (camera lucida) means (11). Video mixing is a more expensive procedure than optical mixing, and it usually has less resolution than that of the drawing tube; however, investigators are liberated from the slavery of the constant use of the eyepieces of the light microscope during the coding. In the design of the computer-assisted microscope described in this chapter, the drawing tube is used in the mixing. The procedure is as simple as directing the drawing tube to the screen of the TV monitor, where the computer draws the coded image. The operator looking thorough the oculars can see both the light microscope image and the computer drawing of the coded image.

Computer-Assisted Light Microscope System

The original system was made in 1986 using an Apple II microcomputer with 48K RAM, a standard light microscope with a drawing tube, and a wheel attached to the fine focus knob and to a digital planimeter (Tamaya) that acts as a curvimeter measuring the movement of the wheel, allowing us to obtain the z coordinate (6). The basic design is still working satisfactorily and remains unchanged. However, a QL-SINCLAIR computer with 896K RAM and two 3.5-inch drives is presently being used during the plotting and the quantitative study of the neuron. The coding is made employing the original system, and the data arrays are transferred to the QL computer using the serial ports of the two computers.

The design of this computer-assisted microscope system is inexpensive, and its implementation is simple because the system does not use special hardware to obtain the x and y coordinates. A flashing point cursor (FPC) plotted by the computer on the TV monitor determines these coordinates by its movement on the monitor screen, and they are stored automatically in the RAM memory of the computer. A rectangular box is also drawn by the computer on the TV monitor, and it corresponds to the graphic screen (CGS) of the computer (6). The CGS is drawn by the computer in the central part of the video monitor screen, far enough away to avoid the distorting edges of the monitor screen. The upper left-hand corner of the CGS is the coordinate origin ($x = 0$ and $y = 0$), and its bottom right-hand corner has the coordinates $x = 279$ and $y = 159$. The FPC moves to one of the 44,800 points of the CGS or computer graphic digits (CGDs) by whole incremental steps (minimum step $= 1$). The incremental steps in the x and y axes are made to move the FPC exactly the same distance by adjusting the horizontal and vertical alignments of the video monitor.

If the neuron to be codified is larger than the CGS, it is codified by parts, each one corresponding to one CGS. A "screen code" and a very simple algorithm allow the labeling of the distinct CGSs and reconstruction of the neuron from its parts (6). In short, the screen codes (SX, SY, SZ) indicate the translation in CGSs from the coordinate origin of an arbitrary CGS where the operator situates a very characteristic part of the neuron (cell body, beginning of a process, etc.), and these codes are stored together with the coordinates of each selected point (see below). The final 3-D coordinates of the selected neuronal points are obtained according to the BASIC statements: $X = X + (SX * 279 - 279)$; $Y = Y + (SY * 159 - 159)$; $Z = SZ * 10,000 + Z$.

There is no constraint on the size of the neuron that the system can codify. To economize the RAM memory, a packing procedure is used. The screen codes are stored together with the x, y, and z coordinates in the same dimensions of the data array, according to the BASIC statements: $A\%(Q, 0) = SX * C + X$; $A\%(Q,1) = SY * C + Y$; $A\%(Q,2) = SZ * T + Z$; $Q =$ array pointer; $C = 1000$; $T = 10,000$ (planimeter digits). Because the integer array overflow is 32,767 in the BASIC employed, the SX and SY codes range from 10 to 32 and the SZ code from 0 to 2. We have a group of $23 \times 23 = 529$ CGSs in the x–y plane, sufficient for most coding purposes. When we need more CGSs, a simple procedure is used. Each group of 529 CGSs is stored in a different file, indicating the displacement (DX, DY) with respect to the coordinate origin or an arbitrary CGS where a very characteristic part of the neuron was situated (soma, beginning of a process, etc.). This displacement is virtual if, for instance, we have codified the upper part of a neuron locating the soma in the screen $SX = 21$ and $SY = 32$ and we want to codify the bottom part of the neuron; then all we have to do is to draw the CGS with $SX = 21$ and $SY = 32$ and to call it, for instance, $SX = 21$ and $SY = 10$, and to initialize the pointer of the data array to 1. This makes a virtual displacement of the group of 23 × 23 CGSs of $Dx = 21 - 21 = 0$ and $Dy = 32 - 10 = 22$ CGSs, which corresponds to the translation of the neuron on the y axis a negative distance equal to $-22 * 159 = -3498$ CGDs.

Similar reasoning can be applied, if necessary, to the z axis. However, until the present, this type of virtual displacement was never needed. Using oil objectives, one complete turn of the device measuring the z coordinate equals 183 μm, and three turns implemented in the packing procedure mean 549 μm depth, sufficient for the neuronal types coded until now.

If the neuron is bigger than a CGS, when the operator reaches the border of a CGS and types the appropriate code, the computer program automatically translates the FPC to the corresponding point of the opposite border of the new CGS where the operator has to follow the coding of the neuron, because the computer program automatically increases or decreases the screen codes adequately. The operator only has to use the microscope stage controls for moving the light microscope image the necessary distance to make its alignment with the CGS border where the cursor is equal to the one that makes the computer-generated image with the opposite CGS border. This procedure is straightforward. At the same time the operator can see through the microscope oculars the alignments that both the coded and the light microscope images make with the respective border of the CGS, while the operator is moving the light microscope image in order to get the best fit with the border. If the new CGS has a part of the neuron coded previously, the operator can utilize the facility of the computer program for drawing CGSs, having the computer-generated image of the part of the neuron coded previously as an additional reference during the translation of the light microscope image.

During the coding, the system has to return to the branching points to follow the coding of the other neuronal branch, or to the beginning of another neuronal process. A list of these return points are stored in an array. When the operator codifies the end of a process, the computer program checks whether there are return points in the list. If the checking is positive, the FPC is moved to the last stored return point. If this point is in a CGS distinct from the present one, the computer program automatically draws the new CGS before moving the FPC to the return point. The operator has only to superimpose the light microscope image and the computer-generated image to continue the coding.

Calibration

If the computer-assisted light microscope is to be used to obtain quantitative measurements, it must first be calibrated. The standard procedure for calibration of the light microscope has been extended to the other components of the system: the video monitor and the measurement device for the z coordinate (6, 8, 9).

Calibration in Plane of Objective (x and y Axes)

As is well known, the calibration of the light microscope in the x–y plane of the objective is made using an objective micrometer and an ocular micrometer. The objective micrometer is a glass slide with a real microscopic ruler, usually 1 mm in length divided 100 times. Thus, one objective micrometer unit equals 0.01 mm or 10 μm. The ocular micrometer is a round glass with a microscopic ruler that can be inserted into the eyepiece. Because the objective micrometer has to be removed from the microscope stage in order to see the specimen, the only way to take real measurements is to compare the ruler of the ocular micrometer with that of the objective micrometer using the different objectives. It is useful to make a table indicating the values (in μm) of the ocular micrometer unit at the possible ocular–objective combinations. Thus, the ocular micrometer can be used to take real measurements in the x–y plane of the image.

The system uses the movement of the FPC to obtain the x and y coordinates; this movement is compared with the ruler of the ocular micrometer, superimposing both the FPC and the ruler, in order to obtain the value (in μm) of 1 CGD. Ten measurements of 20 units of the ocular micrometer, located at random in the CGS, are taken using the FPC. The number of CGDs obtained is compared with its equivalent in micrometers at all ocular–objective combinations. The value (in μm) of 1 CGD at the distinct ocular–objective combinations is incorporated in the computer programs. The operator has only to input the type of ocular and objective used during the coding, and the results are directly given in micrometers.

Calibration in Depth (z Axis)

There is a very easy and inexpensive way to calibrate any device attached to the microscope stage or to the microscope fine-focus control for measuring depth (6). A cover glass is fixed perpendicular to a glass slide using a drop of paraffin. The thickness of the top border of the cover glass is measured (in μm) using the previously calibrated (see above) ocular micrometer. At the point of the border of the cover glass where the measurement was taken, a blue spot is made on one side of the cover glass using a pen. A similar red spot is made on the other side of the cover glass. Then the cover glass is put on the glass slide. Because the distance (in μm) between the blue and red spots is known, we can now calibrate the movement of the fine focus control or any device attached to it or to the microscope stage, focusing first one of the two spots and then the other.

A simple and inexpensive way to increase the resolution of the fine-focus control is to attach to it a big wheel marked off as a ruler facing the operator. After the calibration, numbers (in μm) can be added to the ruler. Thus, the operator can read and introduce the depth (in μm) into the computer. To make this reading easy and more accurate, we use a digital planimeter that, acting as a curvimeter, measures the movement of the wheel attached to the fine-focus control of the microscope.

Accuracy of Measurements

In the x–y plane, the accuracy of the system depends on the distance covered by a single-step movement of the FPC that is directly proportional to the size of the video monitor and inversely proportional to the distance from the drawing tube to the screen of the video monitor (9). In other words, the accuracy of the system can be increased by using a video monitor of smaller size or by moving the light microscope and camera lucida away from the monitor screen. In both cases, there is a proportional decrease in the area of nerve tissue covered by one CGS and in the visibility of the FPC. If the amount of nerve tissue covered by one CGS is too small, the coding of the whole neuron becomes a huge task; hundreds of CGSs would be used and the operator would have to spend several days coding a neuron (9).

An equilibrium has to be established between the factors influencing the accuracy of the system and the scope of the coding. At present, a 9-inch video monitor is used, and the distance from the video monitor screen to the drawing tube is 78 cm. An 100× oil immersion objective (numerical aperture, N.A. 1.25), 15× oculars, and a drawing tube with an 1.25 intermediate lens, producing a total magnification of × 1875, are usually employed in the coding. The drawing tube has a zoom lens that is used at the maximum during the coding of the soma and dendritic and axonal trees, not including spines and varicosities. The accuracy of the system is 0.23 μm, sufficient for obtaining quantitative results on the length and area of the soma and the length, width, and area of the neuronal processes. The nerve tissue surface covered by one CGS is 2275.20 μm^2, allowing a rapid coding of the whole neuron. A graphic representation of the whole neuron (soma, dendrites, axon) is also obtained from the coding. Then, representative dendritic and axonal branches of the neuron are coded including spines and varicosities; now, the zoom lens of the camera lucida is at the minimum, resulting in a resolution in the x–y plane of 0.11 μm and in a surface covered by one CGS of 561.60 μm^2. This coding is used to obtain the graphic representation of the neuronal branches with their spines and varicosities, and also quantitative results on the size of spines and varicosities and their distribution along the neuronal processes.

The accuracy along the z axis was studied by a 10-fold measurement of the depth between the red and blue spots marked on the sides of a cover glass border of a known thickness (see above) using the different objectives of the light microscope (6). The accuracy along the z axis depends on the type of objective utilized (air versus oil immersion objectives) and not on the power of the objectives (6). Air objectives produced a 37% foreshortening of the light microscope image (6). The theoretical exactness of the device measuring the z coordinate was 0.0293 μm using air objectives and 0.0184 μm with oil immersion objectives. However, the resolution of the system along the z axis is limited by the precision of the optical system: 0.4 μm employing a $100\times$ oil immersion objective and a $15\times$ ocular.

Testing System Accuracy

Before using the computer-assisted light microscope system to obtain quantitative results, the precision of its measurements has to be treated with real observations of the smaller neuronal structures that the operator wants to code. The idea in mind for making the system was to be able to code all the structural characteristics of the neuron, including spines and varicosities. Three dendritic spines, visually different in size, were coded 10 times each using the greatest resolution of the system (9). Statistical analysis of the length and width of the pedicle and head of the three coded dendritic spines demonstrated that the system has enough accuracy and sufficient precision for making repetitive measurements for the quantitative study of neuronal structures as small as dendritic spines (9).

The Expanded Data Model

At present, a five-dimensional integer array is used to store the codes of the data model. The first three dimensions of the data array store the $x, y,$ and z coordinates. During the coding of the neuron a packing procedure, described above, is used to store the screen code and the coordinate in the same element of the array. After the coding, the screen codes are no longer needed, and an unpacking procedure is used (see above) to obtain the final coordinates of each selected point that are stored in the first three dimensions of the final data array.

The same packing procedure described above is utilized, storing together the topological identifiers and the width of the processes, spines, and varicosities in the fourth dimension of the data array, according to the BASIC statement $A\%(Q,3) = B * C + W$, where B is the topological codes (integers from 0 to 9), $C = 1000$, and W is the width in integer numbers (CGDs from 1 to 150). However, the unpacking procedure is totally different from the procedures described above. The rationale of the procedure is to convert the integer array element in a string $[B\$ = A\%(Q,3)]$ and to use the BASIC commands for slicing strings to obtain the first character of the string, which is the topological identifier, and the last three characters that are the width.

The "nature" and "shape" codes are stored in the fifth dimension of the data array. The shape code is a special part of the nature code developed for coding spines and varicosities, permitting both the quantitative study and the graphic representation of these neuronal structures (9). No packing procedure is used; only one of these codes is stored in a data array element.

The storing procedures described above save half of the RAM memory needed for the data array: ten codes are stored in an integer array of five dimensions. This is very relevant for using small RAM memory computers or for coding very complicated neurons.

Topological identifiers are used to add to the coded neuronal point the information of its situation in the neuronal process. A neuronal process has one beginning point, one end point, and can have several branching points; all the other coded points are sample points. The following integers are used as topological identifiers for neuronal processes: 1 (beginning), 2 (sample), 3 (branch), and 6 (end). Additional topological identifiers are employed: for the end points at the CGS border during the coding (4); for end points of perikaryon and dendritic spines (5); and for end points of multispines, spines with several heads but only one neck or pedicle (8). If the neuronal process width is greater than 1 CGD and we are coding a branching point or a dendritic spine contralateral to the neuronal process border form where we are moving the FPC in the coding, two more topological identifiers are needed: 7 (contralateral spine end) and 9 (contralateral branching point).

The nature code is used for identifying the regions of the neuron to which the coded points belong: nucleus, perikaryon, dendrite, axon. It is also employed to label coded points relevant to the location of the neuron in the brain tissue sections: border of brain areas, injection site of a tracer, blood vessels, etc. Integer numbers less than 32,767 (integer array overflow) are used for this code. For instance, the following nature code is used in the coding of a pyramidal neuron: 9 (perikaryon), 11 (basal dendrite), $11n$ (basal collaterals, n = branching order), 14 (apical dendrite), $14n$ (apical collaterals, n = branching order), 20 (main axon), and $2n$ (axon collaterals, n = branching order). Information about the process ends (natural, truncate, or lost end) is encoded in the last cipher of the nature code: 4 (truncate end), 5 (natural end), 6 (lost end). Thus, the nature code 11205 means that the neuronal process is a dendrite (1----) of basal type (-1---), a second order collateral branch (--2--), and the end is natural (----5). The last cipher of the nature code is also used for labeling the end of the segments of a neuronal branch with symmetrical branching.

The shape code is used to add to the coded point information on its location in a spine or a varicosity. Only the two points limiting the length of the varicosity are employed for its coding; the shape codes for these two points are 51 and 50. The width of the varicosity is automatically calculated by the system: the operator only has to put the FPC on the two points determining the varicosity width. This width is stored in the third dimension of the data array according to the packing procedure described above. Two points are also used for coding a pedunculated spine: the base and apex of its head; the codes employed are 53 and 52, respectively. The spine pedicle is determined by the code 53. If the pedicle is not a straight line, two or more points with shape code 53 are necessary in its coding. The width of each segment of the pedicle is automatically calculated by the computer and stored in the third dimension of the data array (see above). The spine head is determined by the shape codes 53 and 52 in two consecutive points. The width of the spine head is stored in the third dimension of the element with shape code 52. Only one point is used for coding sessile spines, the apex of the spine. The shape code used is 55.

Depending on the type of neuron, other neuronal structures (branched or nonbranched multispines) have been observed. These structures have more than one spine head, but only one neck, and they are stored in special data arrays; their beginning has a shape code 56, and their end has a topological identifier 8.

Appendage-ended spines have also been observed. These are like pedunculated or sessile spines, but end in an appendage that goes out from the spine head. The coded points belonging to these appendages are characterized by shape code 53, which follows the shape codes 52 or 55, indicating pedunculated or sessile spine heads.

Graphic Representation of Coded Neurons

The processes of the neuron are graphically represented by filled rectangles joining the coded points. The rectangle length is the distance between two consecutive coded points, and the rectangle width is the width of the neuronal process. If the neuronal process is curved, a gap between consecutive rectangles can occur when using the scale for obtaining a large increment in the size of the drawing. The computer program knows the existence of this gap, checking whether the coordinates of the four points defining the two sides in front of two rectangles are different. If this check is positive, the gap is also filled in with the ink of the drawing.

In recent years, we have developed two approaches for making a graphic representation of spines and varicosities (7, 9). In the first, three predefined shapes were developed (7) for representing pedunculated spine heads, sessile spines, and varicosities. The operator has to put the FPC on the two points that define the length of the spine head or the length of the varicosity, in order to store the 3-D coordinates and introduce the appropriate shape codes. Then, the computer draws the predefined shape according to the shape code stored and adapted to the length of the structure. This approach permits the scaling, rotation, and translation of these neuronal structures.

The second approach developed for graphically representing spines and varicosities (9) utilizes a closed plane figure, the ellipse. The major axis of the ellipse represents the length of varicosities and spines, and the minor axis the width of these neuronal structures. The practical use of this approach has demonstrated that the ellipse can accurately represent the spine head and varicosities at the resolution and magnification limits imposed by conventional light microscopy. The main reason for this is the flexible shape of the ellipse that can represent a continuous range of shapes from a circle to a very flattened ovoidal shape, depending on the relative size of the major to the minor axes (9).

Quantitative Calculations

An estimation of the perimeter and area of the neuronal perikaryon is obtained by coding the largest perikaryal profile that has been found focusing through the entire thickness of the perikaryon. All the coded points are in the same optical section of the perikaryon and are considered to have the same z coordinate and to belong to only one x–y plane. The calculation of the perimeter and area is done using the BASIC computer routines described by Pullen (12). A quantitative estimation of the shape of the perikaryon is obtained by the calculation of a form factor according to the BASIC statement: 4 * PI * AREA/PERIMETER * PERIMETER. If the form factor is 1, the shape is a perfect circle. When the form factor decreases toward 0, it indicates a decrement of the area with an increment in the perimeter of the shape.

The calculation of the length and width of neuronal processes and structures (spines and varicosities) is made according to the known formula of the Cartesian three-dimensional distance between two points, that is, using a BASIC statement: $D = SQR [(X2 - X1) * (X2 - X1) + (Y2 - Y1) * (Y2 - Y1) + (Z2 - Z1) * (Z2 - Z1)]$. D is the distance between the points with coordinates: $X1$, $Y1$, $Z1$, and $X2$, $Y2$, $Z2$; and SQR is the square root. In the calculation of the mean width of the processes the width of the varicosities is taken in account.

The area of the neuronal processes and spine pedicles is computed considering that the segments joining consecutive points are rectangles (area = length × width). The area of varicosities and spine heads is calculated considering the structures as ellipses, according to the BASIC statement AREA = PI * (LENGTH/2) * (WIDTH/2).

Computer Programs

Three main programs, written in BASIC and optimized for speed using a compiler, are used by the computer-assisted microscope system: coding, display, and quantitative programs (6–8). The coding program allows the interactive digitalization of Golgi-impregnated or tracer-filled neurons. The operator, looking through the microscope oculars, superimposes the FPC on the neuronal point to be codified using a mouse and, by "clicking" it, thereby stores the coordinates of the selected point. The packing procedures developed (see above) and the scanty amount of RAM memory needed for running this program allow the use of a solely dedicated, but inexpensive, small computer during the coding. However, the display and quantitative programs are needed for efficiently running a fast computer, with a large amount of RAM memory. Thus, the independence of the computer programs, imposed by the scanty RAM memory of the computer used in the original system (an Apple II, 48K), has demonstrated it to also be very useful, allowing one operator to code neurons while another can be displaying or quantitatively analyzing other neurons using a stronger computer independent of the computer-assisted microscope system. Also, the independent computer programs make less traumatic the incorporation of a new computer to the system, with, for instance, better graphic capabilities, because only the relevant computer programs have to be adapted.

The display program allows one to see the graphic representation of the coded neuron by using the known three-dimensional transformations: scaling, rotation, and translation. Because the graphic image is computer-generated, no focus problems arise when increasing the scale. The operator can play with the size of the image, the only constraint being the size of the video monitor screen. The scaling is also useful for obtaining a complete view of the neuron when it is bigger than the video monitor screen. The rotation allows the operator to have an accurate idea of the morphology of the neuron, viewing it from distinct vantage points. A useful and indispensable implementation using these transformations is the interactive choice of a new transformation center. The first point of the data array is the transformation center, and it usually corresponds to the center of the cell body of the coded neuron. If we want to scale and/or rotate a part of the neuron, we have to situate the transformation center in it; otherwise, we will be disoriented by the unexpected results. When the operator selects the ⟨NEW CENTER⟩ option, a flashing cursor runs over the graphic representation of the neuron on the video monitor screen, following the three-dimensional coordinates of the data array. When the operator hits a key, the three-dimensional coordinates of the point where the cursor is become the new coordinates of the transformation center.

The quantitative program is, by far, the largest of the three computer programs driving the computer-assisted microscope system. It has two main parts. The first one computes distances, numbers and size of spines, and varicosities (in μm), and orders the resulting information, storing it in arrays. The "nature code-Length-Width-Area" array (NC) has nine dimensions, and each row stores information related to one neuronal branch. The first element of each row stores the nature code that characterizes the type of neuronal branch and also carries information about the end of the process (see above). The other elements of each array row store the length, the width, and the area of the neuronal branch, and the number of varicosities, pedunculated spines, sessile spines, and branched and nonbranched multispines, successively.

Information about varicosities, pedunculated spines, sessile spines, and multispines is stored in independent arrays: VA (varicosity array), PE (pedunculate spine array), SS (sessile spine array), BS (branched multispine array), and NB (nonbranched multispine array). Each array row stores information pertaining to only one varicosity or spine. The first element in each row of these arrays stores the distance between each varicosity or spine and the beginning of the neuronal branch. The VA array has three dimensions and stores the distance, length, and width of varicosities, successively. The PE array has six dimensions, storing consecutively the distance, length of the spine pedicle, width of the spine pedicle, length of the spine head, width of the spine head, and area of the spine pedicle. The SS array has three dimensions and stores the distance, length, and width of sessile spines, successively, in each array row.

The BS and NB arrays have six dimensions, storing the distance, the length, mean width and mean area of the pedicle, and the number of heads and spine appendages. Because the number of heads is variable from one multispine to another, in order to economize array elements, the length and width of the multispine heads are stored in an independent array: VL for branched multispines and NVL for nonbranched multispines. The number of heads in the BS and NB arrays is the pointer to the VL and NVL arrays, respectively, that allows one to know the length and width of heads that belong to each multispine. The number of spine appendages is also utilized as a pointer to another array of three dimensions permitting one to know the length, the mean width, and the area of spine-appendages that belong to each multispine.

The same procedure is also used to determine what varicosities and spines belong to a neuronal branch. The number of varicosities and spines in the NC array is utilized as a pointer to the VA, PE, SE, BS, and NB arrays delimiting the varicosities and/or spines belonging to a given neuronal process.

The second main part of the quantitative program utilizes the arrays described above and, making the appropriate calculations, outputs the following data: (1) the length and area of the perikaryon and its form factor; (2) the number of neuronal branches, their codes, and their total length and area; (3) the total number of varicosities, pedunculated and sessile spines, and multispines; (4) the mean and standard deviation (SD) of the length, width, and area of the neuronal branches; (5) the mean and SD of the length, width, and area of sessile spines, varicosities, and the pedicle and head of pedunculated spines and multispines; (6) the length, width, and area of each neuronal process and the density of varicosities, pedunculated and sessile spines, and multispines per segment determined by the operator (the size of these neuronal structures is also computed per neuronal branch and per segment); and (7) the distribution of the calculated size (length, width, and area) of varicosities and spines per neuron.

Practical Use of Computer-Assisted Light Microscope System

A neuron from the nucleus hyperstriatum ventrale of a 3-day-old chick has been codified, displayed, and quantitatively analyzed for illustrating the practical use of the computer-assisted light microscope described in this chapter (Figs. 1–5). The neuron (Fig. 1A) was intracellularly injected by iontophoretic injection of a solution of Lucifer Yellow (LY) in distilled water, using fixed brain slices under a fluorescence microscope. The photooxidation of the LY-filled neuron by the excitation light for LY in the presence of a 3,3'-diaminobenzidine (DAB) solution produced a dark reddish impregnation of the soma, processes, spines, and varicosities of the neuron (13). Figure 1B is a photographic view of the screen of a high-resolution color video monitor, where the computer-generated graphic representation of the coding of the neuron shown in Fig. 1A was made. The graphic resolution of this display is 512 × 256 pixels and with four colors. The coding was made at a final magnification of ×1875, employing a 100× oil immersion objective, 15×

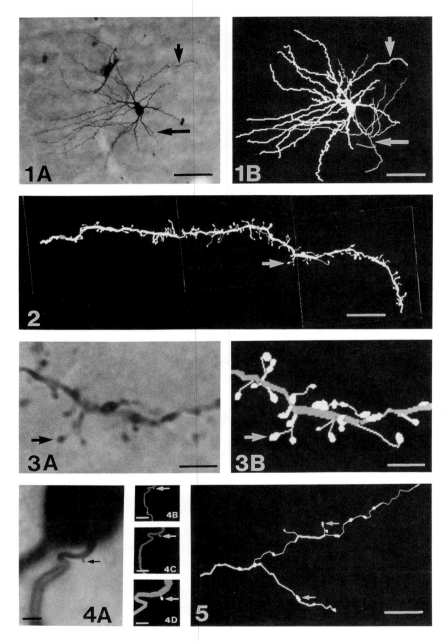

eyepieces, a 1.25× intermediate lens, and the zoom of the camera lucida at the maximum. In these conditions, the resolution of the system is 0.23 μm (see above). Spines and varicosities were not codified. Twenty-four CGS were employed in the coding of this display with a total of 1722 points. The neuron was scaled, diminishing the size until reaching the size that it has in Fig. 1A.

The dendrites of this neuron have a symmetrical branching. The length, width, area, and number of the different order dendritic segments are shown in Table I. The axonal tree (long arrow, Fig. 1) has a nonsymmetrical branching, and the length, area, and number of the distinct order branches are also shown in Table I.

The dendrite labeled with the short arrow (Fig. 1) was codified using the same conditions as above, but with the drawing tube at the minimum, getting an accuracy of 0.11 μm. In this case, spines and varicosities were codified for studying their types, size, and density distribution along the dendrite. Seven CGS were needed in the coding with a total of 499 points. The dendrite (Fig. 2) was scaled and rotated for obtaining its complete view. The arrow

---

FIG. 1   Photographic view of one focal plane showing the soma and part of the dendritic and axonal trees (A). (B) Three-dimensional reconstruction of the neuron, made using the computer-assisted microscope system and scaled to the size of the neuron shown in (A). The long arrow points to the axonal tree and the short one to the second-order dendrite shown magnified in Fig. 2. Bar: 58 μm.

FIG. 2   Three-dimensional computer reconstruction of a second-order dendrite showing dendritic spines and varicosities (photomontage of three video monitor screens). The arrow points to a branched spine shown magnified in Fig. 3. Bar: 10 μm.

FIG. 3   Part of the dendrite shown in Fig. 2. The coding for studying spines and varicosities was made at the resolution shown. (A) Photomicrograph of the focal plane passing through one of the heads of the branched spine labeled by the arrow. (B) Three-dimensional computer reconstruction of the dendritic segment shown in (A); the arrow points to the branched multispine, also labeled in (A). Bar: 3.25 μm.

FIG. 4   A tiny pedunculated spine going out from the proximal part of the axon is shown. (A) Photomicrograph of the focal plane through the beginning of the axon and its tiny spine (arrow). The neuronal soma is out of focus in the upper part of the photograph. Bar: 2.7 μm. (B, C, and D illustrate the scaling procedure of the computer-assisted microscope system. The arrow points to the axonal spine, also labeled in (A). In (B), the scale bar is 5.1 μm; (C), 2.6 μm; and (D), 1.3 μm.

FIG. 5   A part of a second-order axonal branch (the thin process going to the upper right-hand side) and a part of a first-order axonal branch (the thick process going to the bottom center) are shown. The arrows point to axonal varicosities. Bar: 7.6 μm.

TABLE I Quantitative Characteristics of the Neuron Shown in Figure 1[a]

Neuron	Dendritic Tree				Axonal Tree			
	Length	Width	Area	Number	Length	Width	Area	Number
Segment order (dendrites) or branch order (axon)								
1	6.74(2.75)	2.08(0.22)	11.93(7.59)	5	132.15(0.00)	0.48(0.00)	57.50(0.00)	1
2	12.27(3.71)	1.04(0.17)	12.75(4.14)	8	98.05(45.21)	0.23(0.00)	22.73(10.30)	6
3	96.16(49.21)	0.35(0.07)	33.78(16.56)	13	45.66(3.17)	0.23(0.00)	10.51(0.73)	2
4	59.77(59.35)	0.34(0.09)	22.59(22.86)	8	3.87(2.06)	0.23(0.00)	0.89(0.47)	2
5	30.64(3.55)	0.29(0.02)	9.11(0.95)	2				
Total	2277.51		982.89		819.48		216.64	
Spines[b]								
Pedunculated								
Head	0.61(0.21)	0.42(0.16)	0.21(0.13)	184	0.51(0.06)	0.28(0.08)	0.11(0.02)	2
Pedicle	1.15(0.83)	0.14(0.06)	0.17(0.16)	184	0.16(0.00)	0.11(0.00)	0.02(0.00)	2
Appendage	0.72(0.27)	0.15(0.06)	0.11(0.07)	6				
Multispines								
Branched								
Head	0.52(0.19)	0.40(0.11)	0.17(0.09)	27				
Pedicle	3.53(0.76)	0.13(0.02)	0.47(0.13)	9				
Nonbranched								
Head	0.60(0.21)	0.45(0.23)	0.24(0.21)	32				
Pedicle	1.84(1.36)	0.13(0.04)	0.25(0.23)	13				
Appendage	1.04(0.00)	0.11(0.00)	0.11(0.00)	1				
Sessile								
Head	0.65(0.25)	0.48(0.18)	0.26(0.16)	24				
Appendage	0.53(0.19)	0.17(0.08)	0.07(0.02)	2				
Varicosities	0.99(0.39)	0.63(0.23)	0.53(0.37)	19	0.77(0.22)	0.46(0.17)	0.29(0.16)	38

[a] All measurements are in microns. Standard deviations are given in parentheses. Perikaryon length, 58.47; perikaryon area, 192.77. Form factor (circle 1), 0.71.

[b] The figures in this section were obtained from a part of the neuron (388.88 μm of dendritic tree and 348.05 μm of axonal tree) codified at the highest resolution of the computer-assisted microscope system.

DISTRIBUTION OF DENDRITIC SPINES

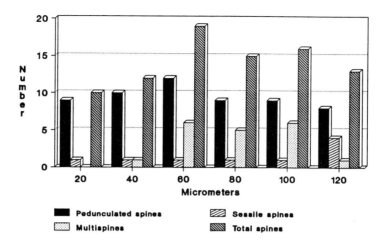

SIZE OF VARICOSITIES

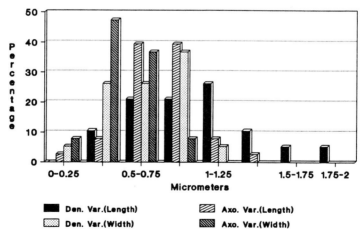

FIG. 6   Two bar graphs showing the density distribution of dendritic spines per 20-μm segment in a second-order dendrite (top) and the distribution of the size of the varicosities (bottom). Axo., Axon; Den., dendrite; Var., varicosity.

SIZE OF DENDRITIC SPINES

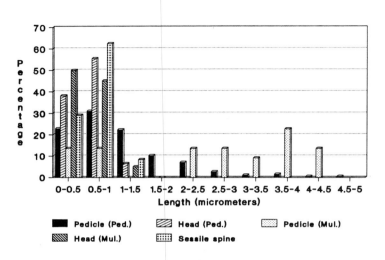

SIZE OF DENDRITIC SPINES

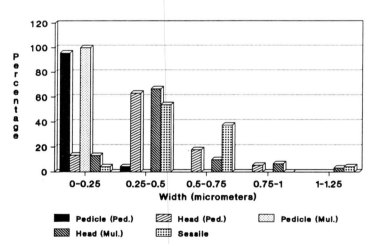

FIG. 7   Bar graphs showing the size distribution of dendritic spines, length in the top graph and width in the bottom graph. Ped., Pedunculated; Mul., multispine.

points to a branched multispine that is also shown in Fig. 3, which illustrates the resolution utilized during the coding. The spine density distribution of this dendrite is shown in Fig. 6, top graph.

Two tiny pedunculated spines were seen at the beginning of the axon of the neuron shown in Fig. 1. One of these spines is shown in Fig. 4. Three scalings of this spine are shown in Fig. 4B–D. A part of the axon codified at the highest resolution of the system is displayed in Fig. 5. Different size and shape varicosities are shown (arrows). The type of spines and varicosities found in this neuron, and their main quantitative characteristics obtained using the computer-assisted light microscope system, are summarized in Table I. The size distributions of the spines and varicosities of the coded neuron are shown in the bar graphs of Figs. 6 and 7.

Acknowledgments

I wish to thank J. Picazo for skillful technical assistance, C. Diaz for able photographic assistance, and S. Jones for reviewing the manuscript.

References

1. D. F. Wann, T. A. Woolsey, M. L. Dierker, and W. M. Cowan, *IEEE Trans. Biomed. Eng.* **20,** 233 (1973).
2. R. D. Lindsay, *in* "Computer Analysis of Neuronal Structures" (R. D. Lindsay, ed.), p. 71. Plenum, New York, 1977.
3. J. Yelnik, G. Percheron, J. Perbos, and C. François, *J. Neurosci. Methods* **4,** 347 (1981).
4. J. J. Capowski, *J. Neurosci. Methods* **8,** 353 (1983).
5. J. J. Capowski, *in* "The Microcomputer in Cell and Neurobiology Research" (R. R. Mize, ed.), p. 86. Elsevier, Amsterdam, 1985.
6. M. Freire, *J. Neurosci. Methods* **16,** 103 (1986).
7. M. Freire, *J. Neurosci. Methods* **20,** 229 (1987).
8. M. Freire, *J. Neurosci. Methods* **25,** 143 (1988).
9. M. Freire, *J. Neurosci. Methods* **37,** 71 (1991).
10. M. Lowndes, D. Stanford, and M. G. Stewart, *J. Neurosci. Methods* **31,** 235 (1990).
11. E. M. Glaser, M. Tagamets, N. T. Mcmullen, and H. Van der Loos, *J. Neurosci. Methods* **8,** 17 (1983).
12. A. H. Pullen, *J. Neurosci. Methods* **12,** 155 (1984).
13. E. H. Buhl and J. Lübke, *Neuroscience* **28,** 16 (1989).

[21] Computerized Three-Dimensional Analysis of Shape of Synaptic Active Zones in Rat Spinal Cord

W. Terrell Stamps and Claire E. Hulsebosch

Introduction

The computer-aided analysis of neuroanatomical data, both two- and three-dimensional, has been growing steadily over the past several years, owing primarily to the increased availability of affordable computer systems and software. Three-dimensional (3-D) analysis in particular has become an important tool in visualizing and understanding neuroanatomy. Much of the 3-D analysis has emphasized the size, shape, and distribution of neurons and neuronal assemblies and the distribution of synapses and synaptic connectivity patterns among groups of neurons (1). One area that has received less attention is the ultrastructural aspects of neuroanatomy, namely, visualizing structures much smaller than whole neurons or neuronal assemblies. This may be due in part to the difficulty in obtaining a series of sections thin enough to reconstruct ultrastructural objects in the nervous system. This chapter explains tissue preparation and sectioning as well as the reconstruction and analysis which can be applied to a variety of ultrastructural objects. In our investigations, we chose to study the synaptic active zones since this structure is important in neuronal cell-to-cell conduction and communication. It should be noted that the techniques used are applicable to other structures in the nervous system as well as structures in other anatomical areas.

Background

The dorsal horn of the mammalian spinal cord is an important area for somatosensory integration. Consequently the synaptic architecture of this region is a subject of importance. Synapses in the dorsal horn, as elsewhere, consist of presynaptic terminals, membrane and cytoplasmic specializations which are referred to as active synaptic zones (2), and postsynaptic structures which are usually a dendrite or a soma (2, 3). The shape of the active zone and in particular the amount of curvature are aspects of dorsal horn synaptic architecture that have not been extensively examined. Because the cyto-

Methods in Neurosciences, Volume 10

plasmic specializations that form the active zone are the sites where synaptic transmission presumably occurs, the size, shape, and extent of curvature of the active zone should be determined and related to fiber type if possible. To examine this issue, the present study has devised an index of curvature for the active zones in laminae I–IV of the rat dorsal horn and then used serial reconstructions and computer modeling to facilitate the gathering of the data. Our goal is to describe quantitatively the shapes and sizes, particularly the exact extent of curvature of the active zones in laminae I–IV of normal animals, and to show that a particular class within this grouping is affected following dorsal rhizotomy. This is a step toward correlating the geometry of spinal synapses with meaningful functional data.

Methods

Animal Perfusion, Tissue Preparation, and Sectioning

Four adult male Sprague-Dawley rats are anesthetized with sodium pentobarbital (Nembutal, 35 mg/kg ip). When anesthesia is deep, a laminectomy is done, and dorsal roots L3-Ca2 are cut unilaterally midway between the ganglion and spinal cord. Seventy-two hours later, the animals are reanesthetized as above and perfused intraventricularly with 0.9% NaCl containing 200 IU heparin and 0.02% sodium nitrite. When the vascular system is free of blood, the perfusion mixture is changed to 3% glutaraldehyde, 3% paraformaldehyde, and 0.1% picric acid in 0.1 M cacodylate buffer, pH 7.4. Following fixation, the S2 spinal segment is removed and divided transversely into three approximately equal slices with a razor blade.

The slices are postfixed in 1% osmium tetroxide and 1.5% potassium ferricyanide. After embedding in plastic, the block is trimmed in plastic to a single dorsal horn. Because the dorsal horn is a laminar structure when viewed in cross section and the functionally different neurons project to specific laminae (4–6), it is important to retain the orientation which allows ease in laminar identification. Thus, semithin (0.25 μm) sections are prepared and stained with 0.5% toluidine blue in a 1% borax aqueous solution, and the laminae are identified by criteria previously described (5). After laminar identification the block is trimmed further to include laminae I–IV, and 14–24 serial thin sections of laminae I–IV of each dorsal horn are cut, placed on Formvar-coated grids, and stained with 0.1% lead citrate. The use of the C-hole or slot grid allows visualization of the entire section surface without obstruction by a grid bar. The grids are then examined on a Jeol 100CX set at 60 kV.

Section Photography and Preparation for Reconstruction

The dorsal horn is identified on a toluidine blue-stained semithin (0.25 μm) section under light microscopy or by printing a low-power electron microscopy (EM) photomontage (approximately $\times$ 12,000) of the section. Areas are randomly chosen using a grid thrown on the low-power photomontage to establish points which serve as centers for high-power pictures. The areas identified are found in the first section on the electron microscope, noting vessels, nuclei, or other structures to be used as landmarks in finding the same areas throughout the series. The same areas are identified in every section before photographs are taken to ensure the lack of folds or dirt over the desired structures. A fold in the tissue and a calibrated grid are photographed at the same magnification as that used to photograph the random areas from the serial sections. This is the magnification on the electron microscope. Photographs are taken of the same area in each section, developed, and printed (8 $\times$ 10 inches). The prints are numbered sequentially.

The 3-D program requires section width to recreate accurately digitized objects in three dimensions, and a magnification factor is necessary to accurately determine lengths and areas of digitized objects. The section width is determined by photographing a section of the tissue that contains a fold. One-half the width of the fold is the section thickness. It is necessary to choose a fold carefully, as some folds are the result of creases in the Formvar and not folds in the tissue. A fold of uniform width that does not extend beyond the edge of the tissue into the Formvar is ideal.

The magnification is determined by photographing a calibrated grid (1260 lines/mm) and printing it at the same magnification as the synapse prints. By measuring the millimeter distance between lines and multiplying it by the number of lines per millimeter, the magnification may be obtained. In this case, the magnification is approximately 40,000. The section thickness and magnification need to be entered only once initially with this particular program, though the values may be modified at a later date, if necessary.

Before preparing the prints for input, several definitions are necessary to understand 3-D reconstruction as it applies to the Boulder HVEM 3-D Reconstruction software (other programs use the same or similar terms and procedures). Planes refer to serial sections. A series of stacked planes creates a 3-D reconstruction. Objects (also called contours or polygons) are the items traced. A synaptic active zone is an object, as well as the presynaptic terminal and postsynaptic terminal. Each object or class of objects has a numerical type associated with it. The type of an object will later determine its color and fill style. Fiducial marks are points in every plane that are used to align the planes properly when reconstructed three-dimensionally. The centers of mitochondria are used in our reconstructions, though any points of reference that occur at the same position throughout a set of serial sections can be used. Artificial fiducial marks can be created in larger sections by making needle holes or cuts in the tissue before sectioning. Such artificial marks, while very precise, are difficult, if not impossible, to create for ultrastructural object reconstruction.

In our study, the prints are examined, and desired structures are identified and numbered with the same number throughout the series. Two to ten structures to use for fiducial points are identified throughout the series and numbered.

Shape Input

The Boulder HVEM Three-Dimensional Reconstruction Program, Version 1.2 (distributed on a nonprofit basis by the Laboratory for High Voltage Electron Microscopy, University of Colorado, Boulder, 1987), is used to reconstruct synaptic structures. Though this discussion of methods involves this particular program, most of the terms and procedures apply to other manual input 3-D reconstruction systems [see Huijsmans *et al.* (7) for a review of 3-D reconstruction software].

The Boulder HVEM 3-D Reconstruction software requires a minimum of computer hardware and can be added to an existing computer system without disrupting other software programs, making it a practical alternative to higher priced, dedicated systems. For greater resolution and such features as complex shading to indicate three dimensions, more powerful systems are a necessity. The software requires a personal computer (PC), two monitors (a monochrome and an EGA), a math coprocessor, and a digitizing tablet. A hard disk is recommended but not required. Several digitizing tablets are supported; we use a Summagraphics 18 × 12 inch tablet. Because of the complexity of 3-D reconstruction, a 80286 or higher computer is recommended. Most 3-D software will require a computer with at least a graphics monitor and a digitizing tablet. The software program Sidekick (Borland International) is necessary to create procedure files, which determine the color and fill options of the 3-D reconstruction. Other memory resident text editors might work as well. Any text editor that can produce an ASCII file should work, though this would entail exiting and reentering the program to change parameters of the reconstruction.

Once the structures are identified and numbered on the micrographs, the micrographs are secured to the tracing tablet, fiducial marks are entered, and the objects appearing on the planes are traced. Each serial section is given a sequential plane number. Synaptic active zones, presynaptic structures, and postsynaptic structures are given different type numbers to differentiate them when the synapses are reconstructed.

The 3-D reconstruction of serial sections compensates for biases due to sampling in a single plane with respect to the random orientation of synapses within the dorsal horn, and thus is in accord with stereological principles. A total of 270 active zones are examined in normal rats ($n = 4$) and 210 in rhizotomized rats ($n = 4$). Reconstructed active zones are divided into populations using an index of curvature (I_c) which is determined by dividing the longest boundary trace length (L) of the active zone cross section by the distance between the end points (D) of the zone (Fig. 1). The 3-D program returns lengths of lines and perimeters of closed objects as these items are digitized. The length is the actual length of the line as drawn, and the perimeter is the length plus a distance generated by the program necessary to make the item drawn a closed object. For our purposes, we use the length measurement. The length (L) of the synaptic active zone is measured by tracing it on the digitizing tablet. The distance between the end points (D) of the synaptic active zone is obtained by entering only the end points themselves on the digitizing tablet and allowing the program to draw a straight line between them. This method takes advantage of a convenience feature of the 3-D program: any gaps in digitized lines resulting from the lifting of the digitizing pen will be connected by a straight line automatically.

The index of curvature works well for lines that curve uniformly in only one direction, as is the case with synaptic active zones, and yields the smallest standard deviations for our calculations. A perfectly straight line will have an I_c of 1 ($L = D$), a half-circle would have an I_c of 1.57 [$L = (\pi D)/2$], and as the object approaches a perfect circle, D approaches 0 and the I_c approaches infinity. Another method of obtaining a curvature index is to divide the difference between length and distance by the sum of the length and distance [$(L - D)/(L + D)$]. This formula yields numbers between 0 (straight line) and 1 (postsynaptic structure completely surrounded by presynaptic membrane) (8, 9). We choose an I_c of 1.6 to divide slightly curved from very curved active zones for two reasons: (1) 1.6 is as close an approximation of 1.57, the I_c of a perfect half-circle, as the accuracy of the tracing will allow, and (2) after plotting a histogram of the I_c values, the populations of active zones appear bimodal with an I_c of 1.6 falling as a natural division between them.

In the dorsal horn, distinctions are made between simple axodendritic synapses, where one presynaptic element contacts one postsynaptic element, and glomerular terminals, where one presynaptic element contacts two or more postsynaptic elements. For this reason, we separate the active zone measurements into simple terminals and glomerular terminals to determine if a zonal type is associated preferentially with either of these synaptic terminal types.

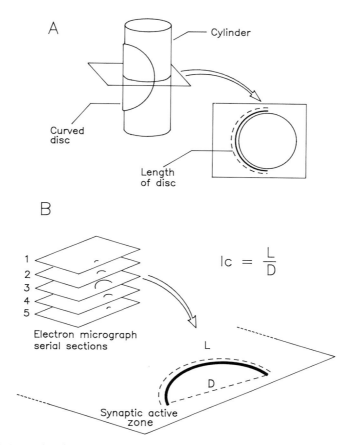

FIG. 1   Schematic of the determination of the index of curvature (I_c). (A) A curved synaptic active zone can be visualized as a flexible disc wrapped around a cylinder. Slicing through the center of the curved disc and cylinder gives a circle with an arc curving partially around it. The length of the arc is the diameter of the disc. (B) A synaptic active zone is identified in a series of electron micrograph serial sections. In this example, the active zone appears in five sections. The actual length (L) of the active zone in the third or middle section is traced, and the distance between the end points (D) of the active zone is determined. The diameter or length (L) is divided by the distance between the end points (D) to give an index of curvature (I_c).

Reconstruction

For the 3-D reconstruction program, the parameters of the reconstruction are determined by a procedure file. This file is created with the program Sidekick, supplied by the user. The advantage of this method is the ability to create a single procedure file for several reconstruction files and vice versa. Procedure files, in the form of a series of numbers, contain information for line thickness, fill color, whether objects are open or closed, and the type of fill for each object type, as well as information about which planes and objects are displayed. Fill type refers to the opaqueness of an object. Some objects must be transparent to allow viewing of interior solid objects. In the reconstruction of the synaptic active zones, the presynaptic structure is transparent, allowing the opaque synaptic active zone and postsynaptic structure to be seen. Color plays an important role in visualizing reconstructions in three dimensions. With this software, 136 colors are possible, and the outline color of an object can be different from its fill color.

The procedure file can display the objects in two ways: (1) all of the selected objects are displayed as each plane is drawn on the screen, or (2) objects are displayed in multiple cycles, in which one or more types of objects are displayed in all of the planes before other types of objects are displayed. Combined with various fill types, multiple cycles allow objects that normally would be obscured by other objects being drawn on the plane to appear on the final reconstruction. This flexibility of the procedure file allows the user great freedom in displaying and visualizing 3-D reconstructions. Other reconstruction programs have similar options in terms of hidden line removal, color selection, and transparency.

For the reconstruction of the synaptic active zone and associated structures, the postsynaptic structure and the active zone are drawn as opaque objects in one cycle and the presynaptic structure is drawn as a transparent object in a second cycle (i.e., the presynaptic structure is drawn after all of the planes of the other structures are drawn). The best view of the structures is obtained by tilting the reconstruction down 30° and rotating it to the left 60°. The program can also create stereo pairs to better visualize reconstructions in 3-D space.

Helpful Hints and Pitfalls to Avoid in Reconstruction

1. Preparation is the key to smooth input and reconstruction. It is essential to prepare the prints thoroughly by carefully choosing fiducial marks and identifying structures before any computer work is done.

2. Before entering data, trace a series of circles of decreasing diameter to form a three-dimensional cone when reconstructed. Then change the different variables of the reconstruction (e.g., colors, fills, rotation values) to see the effect on the cone. This helps visualize these effects on actual data. Also, read through and follow the examples in the user's manual, if available. The Boulder 3-D program has a concise and well-written manual with examples.

3. If numerical values are needed from the data, trace several objects of differing sizes but of known length and area to determine tracing error. A series of accurately drawn circles or boxes works well. If two or more persons will be tracing, have each person trace objects of known area or length and compare the values to assure accuracy. One rule of thumb is that the smaller the object is, the greater the error in tracing it, so try to print the objects to be reconstructed as large as possible.

4. If the program has a "movie" feature that allows the image to rotate around an axis in any of the three dimensions in real time, use it to determine the best viewing angle of the reconstruction, instead of entering various degrees of rotation manually.

5. Choose lighter colors for fills of transparent objects and darker colors for opaque objects. Choose bright colors for small structures within other structures.

6. Moderate amounts of rotation (30–60° off axis) work the best for viewing with this program. Extreme amounts of rotation or too little rotation tend to "flatten out" the reconstruction.

7. Missing sections can be replaced by retracing a section on either side of the missing one. The missing section can also be approximated by aligning and tracing (including the fiducial marks) the sections on either side of the missing one onto a single piece of paper. The objects can then be added to the reconstruction by "eyeballing" and tracing between the pairs of objects from either side of the missing ones. If dramatic changes in the shapes of objects occur between planes, then this method is not valid and the serial sections must be discarded, but in most cases gradual size differences occur and missing sections can be approximated.

Results

Active zones appear as straight or curved profiles in electron micrographs (Fig. 2). Although the degree of curvature varies, the zones always curve such that the surface of the presynaptic terminal is concave. Three-dimensional reconstructions reveal subpopulations of the disk-shaped active zones (Fig. 3). In particular the large majority of active zones are relatively flat or slightly curved (I_c values of 1.00–1.60). There are also a significant number that are

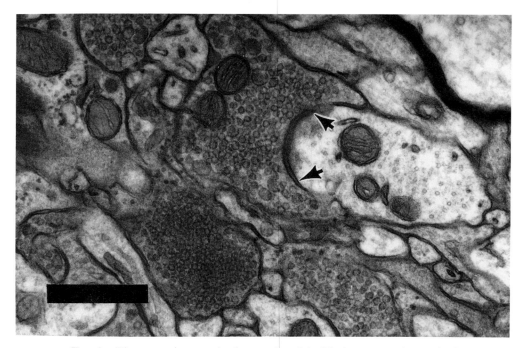

FIG. 2   Electron micrograph of several vesicle-filled presynaptic terminals. A curved active synaptic zone is present near the center of the picture, between the two arrows. Bar: 0.5 μm.

very curved, with I_c values of 1.61–6.50 (10). Very curved active zones make up 4% of the total disc population; 60% of the very curved active zones occur in laminae IIi and III. In normal rats, 28% of the total active synaptic zones were on glomeruli, but only two of these active zones were highly curved. In the highly curved active zone population, 30% synapsed on dendritic spines.

Following dorsal rhizotomy, the major changes are a significant reduction in the numbers of highly curved synaptic zones ($p < 0.05$, Fisher's Exact Test) and a significant reduction in the number of glomeruli ($p < 0.05$, Fisher's Exact Test). Very curved synaptic zones make up only 1% of the total active zone population on the rhizotomized side, and no very curved synapses are found in laminae IIi and III. Of the total synaptic zone population in the rhizotomized animals, only 9% are associated with glomeruli and none of these are highly curved.

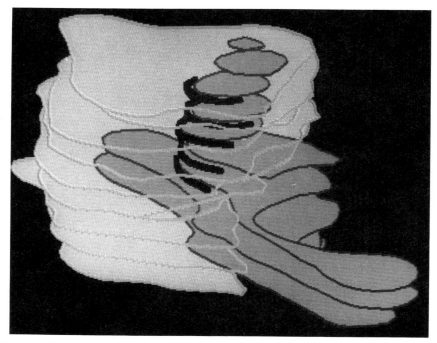

FIG. 3   Three-dimensional reconstruction of a presynaptic terminal region (light gray), a postsynaptic dendrite with a small dendritic spine extending toward the top of the picture (medium gray), and the very curved active synaptic zone which is associated with these two structures (dark lines).

Discussion

The typical "chemical" synapse consists of a presynaptic terminal, a postsynaptic element, and one or more active synaptic zones which are the membrane and cytoplasmic specializations that indicate where actual synaptic transmission occurs. Most studies on the morphology of the synapse focus primarily on the presynaptic terminal (11, 12). Studies that consider the active zones deal with important questions as to whether any of the active zones are perforated or with advanced techniques which reveal previously unknown facets of the organization of the cytoplasmic thickening that make up the active zones (13–16). Our data indicate that active zones are disk-shaped and not perforated. The extent of curvature of the zones has received less attention, presumably because such data depend on reconstructions and also because no obvious correlations of active zone curvature with other structural or functional variables were apparent.

The present study deals with the extent of curvature of the active zones in the rat dorsal horn and correlates the pattern in normal rats with that seen after dorsal rhizotomy. An index of curvature was devised to facilitate the comparisons, and computer-aided reconstructions of the active zones make the actual shapes of the active zones less laborious to attain. In normal dorsal horns, the synaptic zones fall into two categories on the basis of curvature: the large majority are flat or slightly curved, and a smaller number are very curved. It is the second population, the very curved zones, that is basically lost following dorsal rhizotomy. If the reasonable assumption is made that dorsal rhizotomy removes the primary afferent fibers ipsilaterally, the conclusion is that the very curved active zones are associated primarily with primary afferent synapses.

Although the glomeruli that characterize laminae I–III of the mammalian dorsal horn are a major focus of previous studies on primary afferent innervation of the dorsal horn (6, 17–20), the very curved synapses are not specifically associated with these structures. Instead they are associated primarily with dendritic spines and simple axodendritic synapses.

An obvious question that follows from this finding is whether the very curved active zones are involved in transmission of a different type of information than the more numerous flat or slightly curved active zones. Because the very curved active zones seldom occur on central elements of glomeruli, which are often labeled for substances that characterize small dorsal root ganglion cells (21–24), it might be suggested that they are associated with presynaptic terminals which arise from large cells that give rise to rapidly conducting myelinated primary afferent fibers. If this is the case, an experiment of interest will be to determine the numbers and proportions of very curved active zones in rats that received capsaicin at birth since these animals preferentially lose fine primary afferents. It is also of interest that the very curved terminals are associated with small dendritic spines or dendritic "bumps," a finding which presumably has implications for the integration of information passed to the postsynaptic cells.

There are several suggestions in the literature which may account for the degree curvature of the active zone. Some authors suggest that the degree of curvature is a reflection of the functional state of the synapse. For example, in a preparation in which it is possible to stimulate identified neurons, the Muller and Mauthner axons of the lamprey spinal cord, unstimulated axons have flat synapses, whereas electrically stimulated axons have highly invaginated curved synapses (8). This suggestion would be in agreement with the concept of synaptic vesicle release or exocytosis which would result in an increase of the area of the presynaptic membrane owing to the fusion of synaptic vesicle membrane to presynaptic membrane (8, 25). Some suggest that there is a correlation of synaptic structure with different species; however, this remains controversial (26).

We demonstrate that the degree of curvature may be correlated to function; that is, in the dorsal horn the highly curved active zones are from primary sensory fibers. Two spatial considerations result from highly curved active zones. First, it is geometrically possible to cluster more spheres around a convex surface than a flat or concave surface. Because synaptic vesicles are spheres and the active zones are convex in shape on the cytoplasmic surface of presynaptic membrane, this geometric construction will permit more vesicles to fuse simultaneously. Thus, a highly curved synapse may permit a greater rate of release of synaptic vesicles than a flat synapse. Second, highly curved synapses in our study are associated with dendritic spines or bumps, which results in a relatively smaller postsynaptic cytoplasmic volume in juxtaposition to the synaptic disc than that of a flat synapse. Assuming that the amounts of transmitter release, transmitter receptor distribution, G or G-like proteins, ionic species, etc., involved are equal, the result of synaptic release from a highly curved synapse, when compared to a flat synapse, will be a greater unit density of ionic flow in the postsynaptic cytoplasm. In addition, this will result in an increased unit density of molecular interactions of membrane-associated G or G-like proteins in the cytoplasmic microenvironment of the postsynaptic element near the active site. This would support the hypothesis that a sudden increase in concentration of a specific molecule could trigger a cascade of reactions that may be involved in trophic signaling which is dependent on the structure of the synapse (8).

References

1. E. R. Macagno, C. Levinthal, and I. Sobel, *Annu. Rev. Biophys. Bioeng.* **8,** 323 (1979).
2. T. M. Mayhew, *J. Neurocytol.* **8,** 121 (1979).
3. H. J. Ralston III, *J. Comp. Neurol.* **184,** 619 (1979).
4. K. Chung, D. L. McNeill, C. E. Hulsebosch, and R. E. Coggeshall, *J. Comp. Neurol.* **283,** 568 (1989).
5. D. L. McNeill, K. Chung, C. E. Hulsebosch, R. P. Bolender, and R. E. Coggeshall, *J. Comp. Neurol.* **278,** 453 (1988).
6. W. D. Willis and R. E. Coggeshall, ''Sensory Mechanisms of the Spinal Cord.'' Plenum, New York and London, 1978.
7. D. P. Huijsmans, W. H. Lamars, J. A. Los, and J. Strackee, *Anat. Rec.* **216,** 449 (1986).
8. W. O. Wickelgren, J. P. Leonard, M. J. Grimes, and R. D. Clark, *J. Neurosci.* **5** (5), 1188 (1985).

9. P. Kershaw and B. N. Christensen, *J. Neurocytol.* **9,** 119 (1980).
10. W. T. Stamps, R. E. Coggeshall, and C. E. Hulsebosch, *Exp. Neurol.* **108,** 151 (1990).
11. K. Akert, K. Pfenninger, C. Sandri, and H. Moor, *in* "Structure and Function of Synapses" (G. D. Pappas and D. P. Pupura, eds.), pp. 67–86. Raven, New York, 1972.
12. J. E. Henson, T. S. Reese, M. J. Dennis, Y. Jan, L. Jan, and L. Evans, *J. Cell Biol.* **81,** 275 (1979).
13. Y. Geinisman, F. Morrell, and L. de Toledo-Morrell, *Brain Res.* **423,** 179 (1987).
14. M. Nieto-Sampedro, S. F. Hoff, and C. W. Cotman, *Proc. Natl. Acad. Sci. U.S.A.* **79,** 5718 (1982).
15. A. Peters and I. R. Kaiserman-Abramof, *Z. Zellforsch. Mikrosk. Anat.* **100,** 487 (1969).
16. L. E. Westrum and E. G. Gray, *Proc. R. Soc. London B.* **229,** 29 (1986).
17. A. Coimbra, A. Eibeiro-da-Silva, and D. Pignatelli, *Anat. Embryol.* **170,** 279 (1984).
18. H. J. Ralston III and D. D. Ralston, *J. Comp. Neurol.* **184,** 643 (1979).
19. H. J. Ralston III and D. D. Ralston, *J. Comp. Neurol.* **212,** 435 (1979).
20. A. Ribeiro-da-Silva and A. Coimbra, *J. Comp. Neurol.* **209,** 176 (1982).
21. S. M. Carlton, D. L. McNeill, K. Chung, and R. E. Coggeshall, *Neurosci. Lett.* **82,** 145 (1987).
22. A. Coimbra, B. P. Sodre-Borges, and M. M. Magalhaes, *J. Neurocytol.* **3,** 199 (1974).
23. B. Csillik and E. Knyihar-Csillik, "The Protean Gate: Structure and Plasticity of the Primary Nociceptive Analyzer." Akademiai Kiado, Budapest, 1986.
24. E. Knyihar-Csillik and B. Csillik, "FRAP: Histochemistry of the Primary Nociceptive Neuron." Gustav Fischer Verlag, Stuttgard and New York, 1985.
25. J. E. Heuser and T. S. Reese, *J. Cell Biol.* **57,** 315 (1973).
26. J. C. Kinnamon, T. A. Sherman, and S. D. Roper, *J. Comp. Neurol.* **270,** 1 (1988).

[22] Statistical and Computational Methods for Quantal Analysis of Synaptic Transmission

Dimitri M. Kullmann

Introduction

The aims of this chapter are to describe some of the statistical problems which arise in quantal analysis of synaptic transmission, and to outline some numerical methods which can be used to circumvent them. When applied in the appropriate circumstances these can either improve the resolution of the quantal parameters or minimize the degree of nonuniqueness of the solutions.

Quantal analysis is a potentially powerful tool for the study of synaptic transmission. It can be defined as the measurement of trial-to-trial amplitude fluctuations in postsynaptic potentials or currents, with the goal of separating two processes underlying synaptic transmission: on the one hand, the presynaptic mechanisms which determine the release of neurotransmitter, and, on the other, the mechanisms which determine the amplitude of the "quantum" reflecting the postsynaptic action of a single packet of transmitter (1). Several recent reviews have dealt thoroughly with its application to both the peripheral and central nervous systems (2–4), so we shall restrict our attention in this chapter to some of the difficulties presented by the method. These arise principally from attempts to extend the use of quantal analysis to synapses other than the neuromuscular junction, where the signal-to-noise ratio may be poorer and multiple presynaptic fibers innervate the target cell.

Conventional symbols will be used to denote the quantal parameters: n, number of presynaptic release sites; p, probability of release from each site; m, quantal content, that is, the average number of quanta released, equal to the product of n and p; Q, quantal amplitude. According to the classic description developed at the neuromuscular junction, the amplitude fluctuation is described by the binomial model. The frequency f_j of different numbers of quanta released ($j = 0, \ldots, n$) is given by

$$f_j = \frac{n!}{(n - j)!j!} \, q^{n-j}p^j$$

where $q = 1 - p$. The Poisson model is a limiting case where $n \to \infty$ and $p \to 0$, so only m is required to describe the system. In this case,

$$f_j = e^{-m} \frac{m^j}{j!}$$

In ideal circumstances the quantal parameters can be estimated from the measured amplitude fluctuation as follows. The presynaptic fiber is stimulated a large number of times, and the amplitude of the synaptic signal is measured on each trial. If the quantal hypothesis holds, then the signal fluctuates among integral multiples of a minimal amplitude. This minimal increment is the quantal amplitude, Q, which can also be obtained from the amplitude of spontaneous synaptic signals presumably representing the presynaptic release of individual vesicles of transmitter. The value of m is given by the average number of quanta released, and if the binomial model is assumed, n and p can be obtained from the simultaneous equations, $m = np$ and the expression for the trial-to-trial variance in quantal content, $\text{Var}(m) = np(1 - p)$.

In practice, quantal analysis is generally considerably more involved, especially in the central nervous system: the signal-to-noise ratio is often poor, it cannot always be assumed that the quantal amplitude is constant between different synapses contributing to the signal, and there is often no evidence that an *a priori* probabilistic model of transmitter release (such as the binomial model) can be applied. The steps which need to be taken to deal with these problems are dictated by the precise circumstances which pertain, and an exhaustive list of scenarios is beyond the scope of this chapter. Instead, we consider three important problems which often arise in attempts to measure quantal parameters, and give some guidelines as to how they can be addressed by computational methods.

Noise Deconvolution

The first problem often encountered in quantal analysis is that the measured amplitude fluctuation is generally not exclusively due to probabilistic transmitter release, but also to other sources of variability of the postsynaptic membrane potential or current. These include electrode noise, membrane channel activity, spontaneous synaptic activity, nonstationarity in recording conditions, and noise introduced by the amplifier circuit. These will be collectively referred to as noise, and in most cases they can be measured separately. As long as they interact linearly with the underlying synaptic signal, it should still be possible to recover the quantal parameters. If Q is known, for instance, because representative spontaneous events can be measured, and the binomial model is assumed, then this is relatively straightforward: m is given by the ratio of mean synaptic signal to Q, and $\mathrm{Var}(m)$ is obtained by subtracting the noise variance from the measured variance in the synaptic signal, and then dividing the result by Q. In general, however, Q is not known or varies from synapse to synapse; nor can a simple binomial model always be assumed. In the most extreme case, it is not even known whether transmission is in fact quantal at all, that is, whether it fluctuates among amplitudes with a fixed increment. The effect of noise can then be addressed by a variety of computational techniques, usually referred to as noise deconvolution (4).

The deconvolution approach operates as follows. The observed synaptic signal amplitudes are a finite sample drawn from a probability density function $f(v;\theta)$, where v is the measured variable (potential, current, charge) and θ contains the unknown parameters describing the underlying quantal parameters. For convenience, and to simplify exposition of the method, the amplitudes can be binned to form a histogram $\mathbf{a}$ (a_i, $i = 1, \ldots, N$, where a_i is the number of trials resulting in a signal of amplitude $v_i \pm \delta/2$, δ is the binning interval, and N is the number of bins). Clearly, as the sample size increases $\mathbf{a}$ becomes an increasingly faithful representation of $f(v;\theta)$. At the same time, noise amplitudes can be measured in the same way to give a noise histogram $\mathbf{g}$ (g_i, $i = 1, \ldots, k, \ldots, N$, where k indicates the bin at 0), drawn from a noise probability density function $g(v)$. Since several noise measurements can generally be obtained for every evoked synaptic signal, $g(v)$ can be sampled more than $f(v;\theta)$ and therefore estimated with high precision. In practice, $\mathbf{g}$ can be fitted by a continuous unimodal function, for instance, a Gaussian integral, or more generally the sum of two Gaussians (5). If the synaptic signal and noise fluctuate independently and sum linearly, $f(v;\theta)$ is effectively a convolution of two functions: $s(v)$, describing the underlying synaptic signal fluctuation in the absence of noise, and the noise $g(v)$. The goal is to find $s(v)$ from $\mathbf{a}$ and $\mathbf{g}$. A further requirement is that $s(v)$ should be positive everywhere and its integral should be 1 (since it is a probability density function). This makes it impossible to find an analytical solution, so the problem is not strictly speaking an inverse convolution. Instead it becomes one of optimization: to find $s(v)$ such that when convolved with $g(v)$ it gives the best fit to $\mathbf{a}$. For convenience again, let us define $\mathbf{s}$ (s_j, $j = 1, \ldots, N$) as the discrete form of $s(v)$ defined over the same interval as $\mathbf{a}$ and $\mathbf{g}$. That is, s_j is the integrated probability density between $v_j - \delta/2$ and $v_j + \delta/2$.

Several approaches have been proposed for this task. The most powerful one is to maximize the likelihood function, given in its logarithmic form as

$$L = \sum_i a_i \log \sum_j s_j g_{i-j+k}$$

($\Sigma_j \, s_j g_{i-j+k}$ may be recognized as the discrete form of the convolution integral). This can be done in a number of ways, and a very simple approach makes use of the Expectation-Maximization algorithm (6). This is an iterative algorithm which works by gradually updating **s**. At each cycle of the recursion, each element of **s** is altered according to:

$$\mathbf{s}_j = \left(1 \Big/ \sum_i a_i \right) \sum_i a_i h_{i,j}$$

where $h_{i,j}$ is the posterior probability that trials of amplitude $v_i \pm \delta/2$ in fact arose from an underlying component of amplitude v_j:

$$h_{i,j} = s_j g_{i-j+k} \Big/ \sum_j s_j g_{i-j+k}.$$

Starting with an arbitrary distribution (with the only proviso that each $s_j > 0$) the recursion is allowed to cycle until **s** changes by less than a very small amount (the convergence threshold). The algorithm is generally insensitive to starting values and usually finds the global likelihood maximum. (If the binning interval δ is sufficiently small this approach is effectively that described in Ref. 5.)

Simply maximizing likelihood on its own, however, can often give misleading results because the problem as posed here is underconstrained: some of the features of the observed amplitude histogram reflect finite sampling of the data rather than the "true" underlying model, and the maximum likelihood method cannot distinguish between them. As a result, the maximum likelihood solution tends to overfit the data: when **s** is reconvolved with $g(v)$ and compared to **a**, the agreement is better than would be expected from random sampling. An analogy may be helpful: in a coin-tossing experiment, imagine that a fair coin actually came up heads 42 times and tails 58. We would be overfitting the data to infer that it had a 0.42 probability of landing heads without considering the possibility that if the experiment were repeated many times it might actually come up heads and tails with roughly equal probabilities. The agreement between the model (0.42 probability) and the data (42 heads) is too good. This problem underlies some of the recent controversy surrounding the interpretation of deconvolution results (7). An example of how incorrect results can be obtained with maximum likelihood optimization is shown in Fig. 1C.

How can this problem be overcome? One approach is to collect more data: as the size of the sample increases the potential for error becomes smaller. This is, however, rarely practical in an experimental situation. An alternative approach which has been applied with some success is to evaluate each deconvolution result by relating it to Monte Carlo simulations: a range of different underlying models are used to generate random samples of the same size, and with the same noise distribution as for the data, and the maximum likelihood optimization is carried out for each sample. If several different models give rise to maximum likelihood solutions similar to that obtained for the experimental data, then the latter is rejected as nonunique. In this way unresolvable data are rejected, and only if there was a unique underlying model which could have given rise to the data is the result accepted.

A third solution to the problem of nonuniqueness is to incorporate more constraints in the optimization to restrict its solution space. Two constraints have already been incorporated: s must be positive everywhere and must sum to 1. What other constraints may reasonably be added? Clearly, this is determined by what other information is available about the system under study. If it is known that the quantal model is correct, then that constitutes a possible constraint: the solution must be made up of components separated by a constant amplitude increment. This is in fact a very weak constraint since it says nothing about their relative probabilities; for instance, a solution made up of three components with equal probability at 0, 1, and 3 satisfies this constraint if an extra component with a negligible probability is included at 2. It is more effective to apply what is known of the probabilistic behavior of the presynaptic terminals. If transmitter release is known to conform to the binomial model, then this can be incorporated into the analysis. This forces the probabilities of 0, 1, . . . , n quanta released to have a unimodal distribution. (The binomial model outlined above is in fact a special case of a more general and arguably more plausible class of probabilistic models where individual release probabilities are allowed to vary, i.e., the compound binomial model; see below.) A third reasonable constraint is to assume a finite but unknown quantal variability. If the postsynaptic signal produced by a single release event at a given site has an intrinsic trial-to-trial variability, then s should be made up not of discrete components, but of a sum of distributions whose variance increases linearly with the number of quanta released.

Two approaches to incorporate the compound binomial model and quantal variability into maximum likelihood methods have been described (5, 9). These algorithms are computationally less elegant than for the unconstrained maximum likelihood estimator described above, and so far they have only been developed for the case where the quantal variability is assumed to be the same at every site and is described by a normal distribution. These methods nevertheless do represent an important advance in attempting to cope with a more general class of quantal models (Fig. 1D).

A caveat to this approach must, however, be stressed: one must in all

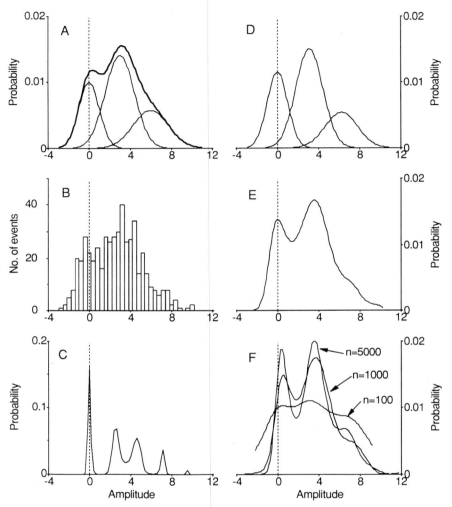

FIG. 1   Example of simulations illustrating different deconvolution approaches. A
quantal model was simulated, with two release sites with probability = 0.5, quantal
amplitude = 3, and quantal coefficient of variation = 33%. Gaussian noise with a
standard deviation of 1 was added. (A) The probability density function from which
the data were sampled is shown as a thick line, with the underlying components,
convolved with the noise, shown as thin lines. (B) Five hundred trials were randomly
sampled and binned to form a histogram. (C) Unconstrained deconvolution fails to
identify the underlying components correctly. (D) When the result was constrained
to correspond to a quantal, compound binomial model with quantal variability, the
solution was very similar to the underlying model. The free variables were the number

cases be aware of which assumptions can justifiably be made to restrict the range of possible solutions for the optimizations. If there is independent evidence that an assumption holds, then by incorporating it into the analysis procedure the resolution and robustness of the optimization are improved. If, on the other hand, assumptions are incorporated without justification, then the estimates of the quantal parameters can be severely biased away from their true values (see Ref. 5 for an illustration of this phenomenon).

Maximum Entropy Noise Deconvolution

What if there is no evidence in favor of a quantal model, that is, release events do not always give rise to approximately equal postsynaptic signals? This is in fact a plausible situation, especially when the active synapses are located at several dendritic sites which present different impedances to the propagation of the electrical signal to the recording site. How does one then apply noise deconvolution? The unconstrained solution may show peaks occurring at equal intervals, but how does one know if these are genuine? Clearly, the constraints mentioned above are unsuitable. One approach is to estimate confidence intervals about the maximum likelihood solution or to try to rule out the null hypothesis that the underlying distribution is in fact continuous. This is a cumbersome approach, and some guidelines for success are given below. An alternative approach is to apply maximum entropy noise deconvolution. This is a modification of the unconstrained approach outlined above, which attempts to find the smoothest, or most featureless, distribution which remains compatible with the data. That is, rather than finding the *best* fit to the data, the goal is to find an acceptable one, and the criterion used to choose between the different possible acceptable fits is to maximize the entropy of the solution. If the solution still contains equally spaced peaks, then this is strong evidence in favor of the quantal model.

Maximizing likelihood alone yields a solution made up of a number of discrete components, not all of which reflect the underlying model. Maximiz-

---

of components, the quantal amplitude and variability, and the release probabilities. The resolved components were reconvolved with the noise distribution to show the similarity to (A). (E) Result obtained when no assumptions were made but the entropy of the underlying solution was maximized. Peaks are clearly seen in the solution, indicating that the quantal hypothesis is correct in this case. (F) Maximum entropy solutions obtained from three other samples of sizes 100, 1000, and 5000. The resolution gradually improves as the sample size increases, with the peak at 6 more clearly resolved.

ing entropy alone, in contrast, simply makes the solution flat ($s_j = 1/N, j = 1, \ldots, N$). If entropy and likelihood are traded off appropriately, however, a unique result can be found, which is the smoothest solution still compatible with the data, given the size of the sample and noise distribution. In other words, if inflections are still present in the solution, then they must reflect genuine features of the underlying process [see Gull and Daniell (10) for a fuller exposition of how this can be applied to astronomical data]. To trade off entropy and likelihood, the recursion described above is modified to

$$s_j = \left[\left(1 \Big/ \sum_i a_i \right) \sum_i a_i h_{i,j} \right]^\lambda (1/N)^{1-\lambda}. \tag{5}$$

λ plays a role akin to the Lagrange multiplier, and serves to balance the degree to which likelihood or entropy is maximized. It can take a value between 0 (which gives the maximum entropy solution) and 1 (which results in the maximum likelihood solution). How does one choose the appropriate value? This is determined by the expected goodness-of-fit of the solution to the original data, given the size of the sample. If only a few synaptic signal amplitudes were measured, then λ must be small, to reflect the fact that one can say very little about the shape of $s(v)$. As the size of the data sample increases, there is more information about $s(v)$, and λ can then be made larger. A rational scheme for finding the correct value of λ was provided by Gull and Daniell (10). Starting with a low value of λ, **s** is obtained from the above recursion, convolved with **g**, and compared to **a** using the χ^2 test. Now, in an idealized random sampling situation the expected value of χ^2 is equal to the number of degrees of freedom (df, equal to one less than the number of classes used for the χ^2 test). If the measured χ^2 is greater than df, then λ is increased and the recursion is repeated. These steps continue until χ^2 becomes approximately equal to df.

In the above description the data were assumed to have been binned, but the maximum entropy method can also be applied to unbinned data. This is arguably a more rational approach, and the Kolmogorov–Smirnov test is substituted to choose the appropriate value of λ. (Because χ^2 = df corresponds to a probability of 0.5 of rejecting the hypothesis that **a** was drawn from the convolution of **s** and **g**, the expected value for the Kolmogorov–Smirnov statistic is chosen to satisfy the same condition.) An example of results obtained with this approach is illustrated in Fig. 1E,F.

Estimating Binomial Parameters When Quantal Fluctuation Pattern Is Known

Even when the relative frequency of 0, 1, $\ldots$, n quanta is unambiguously determined there is a problem in interpretation. At the neuromuscular junction, one of the earliest sources of uncertainty was whether the estimated binomial parameters, n and p, reflected a uniform population of available release sites, or whether they were simply a convenient approximation to a more complicated underlying distribution. This has remained a source of ambiguity: if the release sites are not homogeneous, then it may be misleading to identify n with the number of available release sites, and p with a single uniform release probability. In fact, it seems implausible that all the release probabilities should be the same, since the ultrastructural and biochemical parameters affecting exocytosis can be expected to vary considerably from release site to release site. Moreover, at both the neuromuscular junction and the spinal cord, it has been shown that a simple binomial model cannot account for some observed fluctuation patterns (reviewed in Refs. 4, 8, and 11).

It is not enough to redefine p as the average release probability: when there is appreciable nonuniformity of release probability, the estimated binomial parameters deviate systematically from their true value (12). Instead, the simple binomial distribution should be seen as only a limiting case in a more general probabilistic model of transmitter release. If each release probability p_k is allowed to be different ($k = 1, \ldots, n$, defining the vector $\mathbf{p}$), then the relative frequency of different numbers of quanta released ($f_j, j = 0, \ldots, n$, defining the vector $\mathbf{f}$) is governed by a compound binomial distribution. These frequencies are given by

$$
\begin{aligned}
f_0 &= q_1 q_2 \cdots q_n \\
f_1 &= p_1 q_2 \cdots q_n \\
    &\quad + q_1 p_2 \cdots q_n \\
    &\quad + \cdots \\
    &\quad + q_1 q_2 \cdots p_n \\
&\;\;\vdots \\
f_n &= p_1 p_2 \cdots p_n
\end{aligned}
$$

where $q_k = 1 - p_k$. They can be obtained more simply from the generating function:

$$
G(z) = \prod_k (q_k + p_k z)
$$

where z is a dummy variable: f_j is given by the $j + 1$th coefficient of the expanded polynomial. In an experiment the goal is the converse: to obtain $\mathbf{p}$ from $\mathbf{f}$. If $\mathbf{f}$ were known exactly, then $\mathbf{p}$ could simply be found from the roots of the polynomial. In a real experiment, however, $\mathbf{f}$ is finitely sampled, and root-finding algorithms generally fail: some rare events may never be observed during the recording period, and others may be spuriously over-represented. This has the effect of generating complex roots. An alternative approach is to maximize the likelihood function:

$$L = \sum_j f_j \log \phi_j$$

where ϕ_j is the $j + 1$th coefficient of the expansion of $G(z)$, that is, it is the predicted frequency of j quanta being released, as opposed to the observed frequency, f_j. The function reaches a maximum when $\boldsymbol{\phi}\ (\phi_j, j = 0, \ldots, n)$ is as similar as possible to $\mathbf{f}$. This can be done in a number of ways, but the approach given in Ref. 5 makes use of the E–M algorithm and is therefore well conditioned. This takes the form of a recursion. Starting with an arbitrary distribution, each element of $\mathbf{p}$ is updated by applying the recursion

$$p_k = \sum_j f_j h_{j,k} \qquad j = 1, \ldots, n$$

where $h_{j,k}$ is now the posterior probability that the jth quantum arose from the kth releasing site. This is given by

$$h_{j,k} = p_k \Psi_{j-1,k} / \phi_j$$

where $\Psi_{j-1,k}$ is the jth coefficient of the expansion of $G(z)$ obtained by setting $p_k = 0$. [The index j runs from 1 to n in this case because the information in f_0 is redundant, since it is equal to $1 - (f_1 + \ldots + f_n)$.]

The recursion is allowed to cycle until the estimates of $\mathbf{p}$ change by less than an arbitrary convergence threshold. This algorithm is insensitive to the starting values of $\mathbf{p}$, and it converges to the global likelihood maximum. In an experimental situation n is generally not known and can only be estimated from the largest number of quanta released on a single trial. Because there may have been some rare events which were not recorded where n was even larger, the recursion should be repeated with n gradually increased beyond the largest observed value, setting $f_j = 0$ for the events which were never observed. The solution will then asymptotically approach the global likelihood maximum, with any additional redundant p_k estimated to be 0. This algorithm can also be made to yield a solution which conforms to the simple binomial model by setting the starting estimates $p_1 = \ldots = p_n$. In this case, if the recursion is repeated with increasing n, different solutions will be obtained in each case, and the likelihood will pass through a maximum before decreasing again. This is in fact a more rigorous approach to estimating p than from the mean and variance, since it takes all the data into account, as well as accommodating the case where some events may not have been sampled.

To assess whether a given maximum likelihood solution is acceptable, the most natural goodnes-of-fit test in this case is the χ^2 test: the observed number of trials giving rise to $0, 1, \ldots, n$ quanta is compared to the expected number given by $N\phi_0, N\phi_1, \ldots, N\phi_n$, where N is the number of trials. Where necessary, adjacent classes can be grouped together to ensure that the expected number exceeds 5. Clearly, if the maximum likelihood solution obtained using the simple binomial model fails this test, one obtained using the compound binomial model may still pass because it has more free parameters. If both these binomial models fail, then it points to nonindependence of the release probabilities. That is, whether transmitter is released from one site is somehow dependent on whether it is released from another site, either because of some causal interactions between them or because they both depend on an upstream stochastic event such as failure of propagation of an action potential in the presynaptic fiber supplying them.

How Nonunique Is the Solution?

Finally, once estimates of quantal parameters have been obtained, the problem arises of how to determine confidence intervals, that is, to decide to what extent the optimization solutions are nonunique. This applies in two respects: first, there may be some leeway in the quantal parameter estimates, within which the data could still be accounted for, taking sampling error and noise into account; and, second, there may be several competing, possibly quite different, models of transmitter release, within which solutions can be found which are compatible with the data. These problems can be dealt with in several ways.

First, to test whether a given model is compatible with the data, goodness-of-fit tests can be applied, generally the χ^2 or Kolmogorov–Smirnov test. If, within the framework of a given model of transmitter release, the best fit to the data fails at a given level of significance, then the model can be rejected at that level. If, on the other hand, it passes, this does not indicate that the model is correct, but simply that it cannot be rejected.

Second, to obtain a "cloud" of acceptable values around the maximum likelihood results, akin to standard error bars in simpler models of statistical inference, analytical methods cannot be applied. Instead, a promising approach is to apply the bootstrap method (13): briefly, random samples are repeatedly drawn from the data, with replacement, and the optimization is applied to each sample. This generates a scatter of parameter estimates around the original one obtained from the data, and the breadth of this scatter gives an indication of the confidence interval. The danger of this approach is that some optimization methods have an implicit bias which may not be recognized: for instance, as mentioned above, unconstrained deconvolution approaches tend to resolve amplitude distributions into discrete components even when the underlying model is continuous. This possibility must always be borne in mind and can only be dealt with by extensive Monte Carlo simulations to assess how the optimization behaves in different conditions.

Finally, when several different models of transmitter release can fit the data, for instance, a nonquantal model as well as a quantal model, or with different estimates for m, Q, and quantal variance, comparing goodness-of-fit test scores such as the χ^2 value to choose the "best" model is both difficult and of limited value. It is difficult because the number of degrees of freedom is usually different in each model because varying numbers of parameters are constrained, and there is often some uncertainty as to how to make the appropriate correction. (This also applies to the likelihood ratio test, which might seem more natural in the context of competing maximum likelihood solutions.) This approach is also of limited value because even if it were possible to rank competing models in some order of probability on the basis of statistical tests, this ignores their relative plausibility from *a priori* biological considerations. For this to be taken into account, Bayes theorem needs to be applied, but detailed discussion of this is beyond the scope of this chapter.

Conclusion

The statistical problems encountered in quantal analysis may well present themselves in different forms from the examples listed in this chapter. Even so, it should still be possible to make use of the algorithms, either as described here or suitably modified to incorporate whatever other assumptions can reasonably be made.

References

1. J. del Castillo and B. Katz, *J. Physiol.* (*London*) **124**, 560 (1954).
2. H. Korn and D. S. Faber *in* "Synaptic Function" (G. Edelman, E. Gall, and M. Cowan, eds.), p. 57. Wiley, New York, 1987.
3. H. Korn and D. S. Faber, *Trends Neurosci.* **14**, 439 (1991).
4. S. Redman, *Physiol. Rev.* **70**, 165 (1990).
5. D. M. Kullmann, *J. Neurosci. Methods* **30**, 231 (1989).
6. A. P. Dempster, N. M. Laird, and D. B. Rubin, *J. R. Statist. Soc. B.* **39**, 1 (1977).
7. H. P. Clamann, M. S. Rioult-Pedotti, and H. R. Lüscher, *J. Neurophysiol.* **65**, 67 (1991).

8. B. Walmsley, *Prog. Neurobiol. (Oxford)* **36,** 391 (1991).
9. B. R. Smith, J. M. Wojtowicz, and H. L. Atwood, *J. Theor. Biol.* **150,** 457 (1991).
10. S. F. Gull and G. J. Daniell, *Nature (London)* **272,** 686 (1978).
11. R. Robitaille and J. P. Tremblay, *Brain Res. Rev.* **12,** 95 (1987).
12. T. H. Brown, D. H. Perkel, and M. W. Feldman, *Proc. Natl. Acad. Sci. U.S.A.* **73,** 2913 (1976).
13. B. Efron and R. Tibshirani, *Science* **253,** 390 (1991).

[23] Topography of Primary Muscle Afferent Neurons in Rat Dorsal Root Ganglia: A Three-Dimensional Computer-Aided Analysis

J. M. Peyronnard, J. P. Messier, and L. Charron

Introduction

Looking at impressively aesthetic three-dimensional (3-D) reconstructed images published in recent years, uninformed scientists may get the impression that computer technology has once and for all rendered obsolete laborious microreconstruction techniques using materials like plastic plates or wood sheets to assemble 3-D models from serial section images. The truth is that, despite considerable progress in hardware and software capacities, 3-D reconstruction remains a difficult task (1). The purpose of this chapter is to share the experience we gained in a project devoted to the topographical analysis of neuronal clusters in the rat dorsal root ganglion (DRG) with those wishing to embark on computer-assisted reconstruction with minimal knowledge of microcomputers and limited financial means.

Aim of Project and Selection of Computing System

We wanted to know whether cytochemically labeled primary afferent neurons supplying specific hindlimb muscles of the rat were randomly distributed in lumbar DRG, as it has been assumed so far, or were segregated in certain areas of the ganglion, and to what extent this intraganglionic arrangement remained constant from one animal to the next. We therefore required system analysis allowing (1) reconstruction of 3-D contours from serial two-dimensional (2-D) sections of a complex structure such as the DRG; (2) visualization of cells supplying specific muscle targets within the DRG; (3) rotation of the reconstructed figure in any direction in real time; (4) simultaneous display and animation of several reconstructed DRG for comparative purposes.

Complex modes of operation and prohibitive equipment costs precluded the use of mainframe systems equipped with huge mass memory resources of the kind recently employed to produce a 3-D digital atlas showing intricate

Methods in Neurosciences, Volume 10

details of the surface morphology of rat and monkey brains (2, 3). We also did not consider commercially available systems based on IBM microcomputers such as IBAS (Kontron, Munich, Germany) or the Quantimet 570 (Cambridge Instruments Ltd., Cambridge, England), for several reasons. Not only are these systems expensive but, being primarily designed for image analysis, despite some 3-D reconstruction capabilities, they afford limited image animation and cannot simultaneously display several reconstructed structures.

Having appreciated the user-friendliness and excellent graphic resolution of Macintosh microcomputers over the years, we searched for a 3-D software package which could make use of our Macintosh II microcomputer equipped with 5-megabyte random access memory (RAM), 40-megabyte hard disk, and 13-inch color monitor (Apple). Like others (4), we found that none of the existing packages, which for the most part have been reviewed previously (5), suited our needs either because of hardware incompatibility or defective programming language. For instance, some softwares did not permit fast image reconstruction and animation or failed to provide basic options such as hidden-line imaging, transparency of elements, perspective projection, and coordinate measurements of image elements.

At this point, our attention was drawn to a 3-D reconstruction software package (MacReco) developed by E. Otten and G. W. Dresden (Gröningen, The Netherlands) with three available versions: MacReco 3.2 which can run on Macintosh Plus or SE machines with 1-megabyte RAM; MacRecofast 3.2 which is faster because it uses the coprocessor of the Mac II; and MacReco 4.0 which is a color version of MacRecofast 3.2 but requires 2 megabytes of RAM and does not provide animation, a function only performed by black and white versions of the software. Since then, upgraded versions (MacReco 3.4 and 3.8) have been developed and include features of interest for those concerned with surface morphology, allowing one, for instance, to obtain a vivid 3-D representation of the external envelope of an object through interpolation of sections, gray scale shading, and contour smoothing. Later on, we became aware of another 3-D reconstruction package written by Y. Usson (Université Joseph Fourier, Grenoble, France), running on Macintosh Plus, SE, or II microcomputers with at least 1 megabyte RAM.

Both packages have two main functions, one concerned with data acquisition, and the other with image reconstruction. Input data are generated by drawing contours of image sections on graphic tablets. Realignment of serial sections is accomplished with the MacReco system by aligning three reference marks which have to be introduced on each section before digitization, whereas in Usson's system the last imported section image can be moved on the monitor screen and superimposed as best fit on the previously stored image, using the computer mouse. Three-dimensional images generated by MacReco or the Usson program

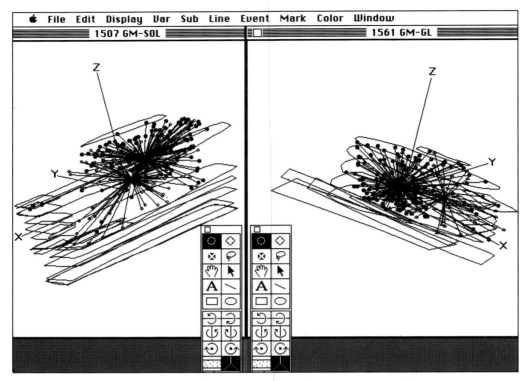

FIG. 1 Macintosh screen image showing two L5 DRG with GM-SOL and GM-GL neuronal aggregates reassembled with MacSpin software. The Apple MULTIFINDER program serves to display the ganglia in separate windows, each having its own tool window for image animation.

reconstructions are visualized on the computer monitor screen and can be printed along different view projections according to preselected X, Y, and Z rotation angles.

This type of image manipulation was too limited for our needs, which required simultaneous display and animation of several reconstructed ganglia in real time. Thus, through an option of the MacReco package, which gives easy access to the X, Y, and Z coordinates of digitized elements, all data pertinent to contours of the ganglia, roots, and DRG cell locations were transferred and reassembled by graphic data analysis software (MacSpin 2.0, D^2 Software Inc., Austin, TX, its latest 3.0 version being currently distributed by Abacus Concepts, Inc., Berkeley, CA). Of the many features offered by

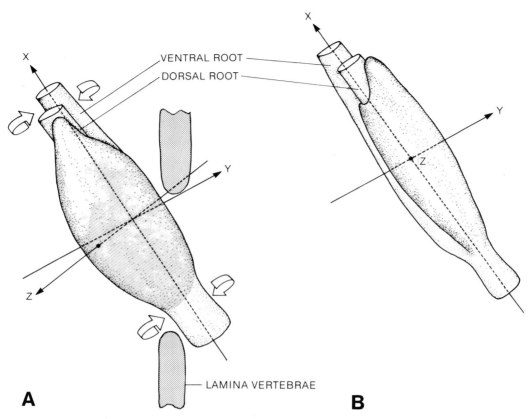

FIG. 2   (A) Oblique position of a L5 DRG *in vivo*. (B) Position given to all ganglia prior to processing and after a 60° lateral rotation along the ventral root (*X*) axis. [Reproduced with permission from Peyronnard *et al.* (7).]

MacSpin, the most interesting was the possibility of simultaneously opening up to nine plot windows, depending on the memory available (we routinely used no more than two in order to avoid crowding of data on the monitor screen). Each window has its own set of commands for rapid 360° rotation of any 3-D object in six different directions. The windows can be used to display either various configurations of the same object or different objects when the Apple MULTIFINDER program is activated (Fig. 1). The technical steps followed during the course of our study are presented in detail in the following sections.

Procedures

Cell Labeling

Nerves to the soleus (SOL), gastrocnemius lateralis (GL), and gastrocnemius medialis muscles (GM) are transected in 18 rats, and their proximal stumps are inserted for 2 hr in a sealed capillary tube filled either with a 20% solution of horseradish peroxidase (HRP type VI) (Sigma Chemical Co., St. Louis, MO) or a 3% solution of fluorogold (FG) (Fluorochromes, Engelwood, CO) in 0.9% saline. In some animals, HRP and FG are applied alternatively to the SOL, GL, and GM nerves; in other instances the two markers are used in the same animal to label afferent neurons supplying two muscles such as the SOL and GM or GM and GL. Seventy-two hours later, the animals are perfused with a solution of paraformaldehyde and glutaraldehyde. The L4, L5, and L6 DRG are dissected.

Tissue Processing and Analysis

Orientation of the DRG before freezing is standardized so that cutting could proceed as much as possible along the same angle in all instances. Thus, the anatomical position of the DRG is modified (Fig. 2) to bring the ventral roots to the bottom, with the dorsal roots surrounded by the bulk of the ganglia lying above. We elected to consider the ventral roots as the X axis, the Y axis being directed toward the external side of the ganglia and the Z axis extending from the ventral root to the top of the ganglia. Serial sections 20 μm thick are cut longitudinally with a microtome cryostat, starting from the top of the DRG and moving down to the ventral roots. At this stage, we discovered the need to take note of all manipulations affecting the spatial orientation of tissues such as the transfer of sections from the tissue block to the glass slide, a process during which right-sided structures become inverted and appear as left-sided, leading, if not corrected, to a mirror image during the reconstruction stage.

The sections are processed for HRP histochemistry according to standard techniques (6). In double-labeled ganglia prior to cytochemical treatment, fluorescent neurons are detected with an epifluorescence-reactive Zeiss photomicroscope. They are numbered and the diameter computed (Macintosh Plus) after projection on a digitizing tablet (Macintizer, GTCO Corp., Rockville, MD). The same procedure is then applied to HRP-labeled neurons, using bright-field illumination. All information pertinent to cell number and diameter is transferred to photographic prints of the sections. Using a light source to illuminate the prints, they are first superimposed manually and aligned by natural landmarks such as the contours of the ganglia and roots; subsequently, each section is labeled with three reference points. Owing to the small size of the DRG, we could not consider making a lesion or introducing objects (a small piece of nerve or vessel) into the embedding material as is often done to create external markers for matching the sections. During this process, cell profiles belonging to the same neuron but appearing on several adjacent sections are identified, and only the largest diameter is retained. The photographic prints are subsequently photocopied at such a size that the largest section would fit the active area of the digitizing tablet, and structures of interest, namely, the contours of roots and ganglia as well as the locations of HRP- and FG-labeled neurons, are outlined with a pencil.

Data Acquisition

The input device consists of a Summagraphic digitizing tablet with a Bit Pad Plus setup utility (Version 2.1, Summagraphics Corporation, Fairfield, CT) mapped to the monitor screen of a Mac II microcomputer. Each image section is positioned within the active area of the tablet and covered with a transparent plastic sheet allowing smooth displacement of the cursor used for tracing the outlines of the histological structures. Clicking on the icon ANY TABLET or MOUSE TABLET of the MacRecofast 3.2 program leads to an acquisition window with several fields. The REFERENCE field is first activated for digitization of the three reference points of each image section. Then, 26 ELEMENT fields named A to Z are accessed, each corresponding to a class object. For instance, we reserve the ELEMENT A field for entering data pertinent to DRG contours which are digitized at only four levels to avoid crowding of the reconstructed figures. In ELEMENT B and C fields, we put data relevant to contours of ventral and dorsal roots, respectively, these contours being drawn at every third to fifth section without trying to reproduce the complex intraganglionic pathway of the dorsal root fibers. Finally, data on HRP- and FG-labeled cells are entered in two other ELEMENT fields. Corrections are possible through deletion of the last entry or the whole section. Digitization of the approximately 50 sections cut from a L5 DRG takes approximately 4 hr.

Despite its conceptual simplicity, the acquisition process is frequently interrupted by sudden and inexplicable dysfunctions of the entry fields or occasionally leads to the acquisition of misaligned sections. As the program does not provide for a review option at this stage, aberrant sections could be detected only after 3-D reconstruction had been completed with the result that the digitization procedure had to be repeated. This deficiency has not been corrected in the newest versions of the software which, however,

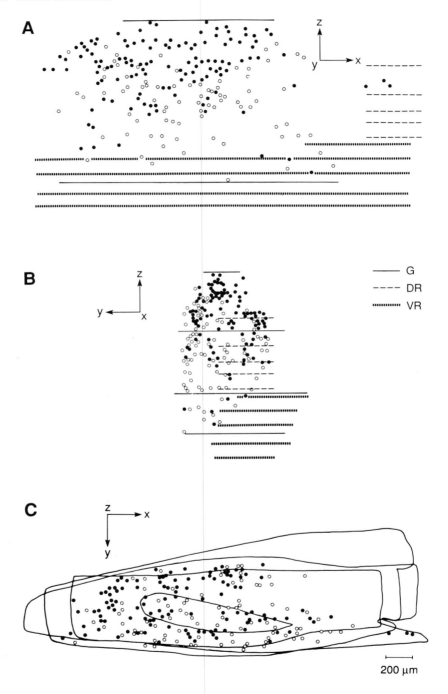

A

z

y   x

B

z

y   x

—————— G

- - - - - - DR

············· VR

C

z   x

y

200 μm

allow redrawing and repositioning of the reconstructed sections (see below). Acquisition is terminated after a CALIBRATION document had been made, using a section near the middle of the DRG. This document, consisting of the three reference points and of a vertical axis line running through the center of the section, is required by the software for aligning the digitized sections during the reconstruction process.

Three-Dimensional Reconstruction

The data are first translated in the MacRecofast 3.2 reconstruction package, and information relevant to magnification and section thickness is entered in the computer. After the various DRG elements have been labeled either with different tones of gray or a palette of 256 colors available only in the 4.0 version of the system, reconstruction is initiated. When it is completed, various options allow a review of the sections for corrections such as adding or deleting an image, and, as we mentioned earlier, MacReco 3.8 has the additional feature of making corrections with respect to contours or the position of image sections. We then use the MEASURE option to define the X, Y, and Z coordinates of multiple points along the ganglia, ventral and dorsal root contours, as well as the spatial coordinates of HRP- and FG-labeled neurons. These data are edited with a word processing program (FullWrite Professional, Version 1.0, Ashton-Tate, Torrance, CA). The X, Y, and Z numerical values are tabulated in a spreadsheet (Excel 2.0, Microsoft Corp., Seattle, WA) in sequences related to their anatomical origin and imported into MacSpin where they serve to recreate EVENTS appearing as dots. Some of these dots, aligned sequentially in rows along several X and Y planes, represented ganglia and root contours, while others with a scattered distribution corresponded to the neurons. The dots can be selected, labeled, united by lines to recreate outlines of ganglionic and root structures, or given different colors for better visualization of specific anatomical elements. A X, Y, and Z set of coordinates is inserted and serves to identify the direction of

FIG. 3   (A–C) Drawings made of computer prints showing the lateral, rostrocaudal, and dorsal views of a L5 DRG after labeling the GM (●) and SOL (○) neurons with FG and HRP, respectively. Contours of the ganglion (G), dorsal root (DR), and ventral root (VR) have been drawn at only a few levels. Despite significant overlap, the GM and SOL neurons tend to be segregated in different ganglionic areas, this being most apparent in the lateral and rostrocaudal projections. [Reproduced with permission from Peyronnard *et al.* (7).]

A

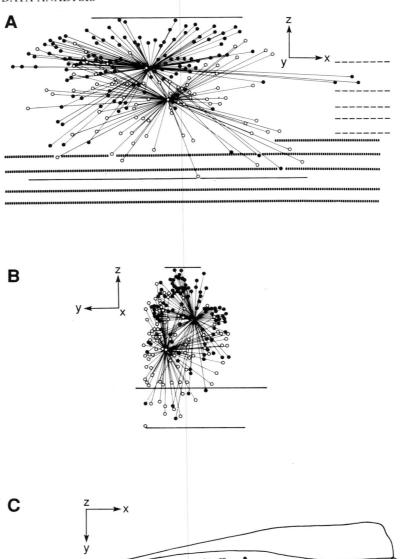

B

C

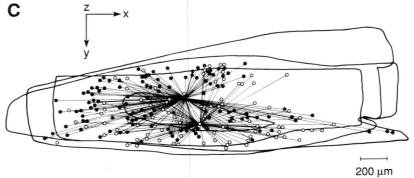

200 μm

displacement during animation as well as to examine different ganglia from the same view angle.

Treatment and Analysis of Reconstructed Ganglia

Being mostly descriptive, our work exploited only one of the many data analytical functions of the MacSpin program known as the centroid connection operation. It defines the center or centroid point of a neuronal cluster by averaging the X, Y, and Z coordinates of all cells belonging to the cluster and connects the neuron closest to the centroid point to every other related cell. We found this function most useful (1) to visually enhance the topographical location and spatial distribution of DRG cells innervating a specific muscle and (2) to quantitatively estimate the localization of muscle neurons, information otherwise difficult to obtain in DRG owing to the variability of anatomical landmarks such as ganglia or root outlines. This is accomplished by determining the centroid neuron of each SOL-GM or GM-GL neuronal pool which served as a reference point for positioning a set of X, Y, and Z axes within the ganglia. When the DRG are viewed along the root or X axis, the Y and Z coordinates delineate four quadrants: dorsal medial, dorsal lateral, ventral medial, and ventral lateral. Within the SOL-GM and GM-GL neuronal pools, the percentages of SOL, GM, or GL neurons located in each quadrant are determined and the values obtained for all animals averaged. The same procedure is followed after the DRG had been positioned laterally along the Y axis, in which case the Z coordinate serves to define a rostral and a caudal ganglionic sector. Selected views of the ganglia are printed on an Apple Laser Writer II NT and edited for illustration.

Results

Because data have been published in detail elsewhere (7), only those required to illustrate the possibilities of the methodology are presented. In our study, we took advantage of the fact that we were dealing with rather limited cell populations which, for the most part, were located in the L5 DRG. Indeed, the number of neurons found to innervate the SOL, GM, and GL muscles

---

FIG. 4   (A–C) Same as Fig. 3 but after tracing the centroid connections as a way of visually enhancing the separate locations of GM and SOL neuronal clusters in the ganglion. [Reproduced with permission from Peyronnard *et al.* (7).]

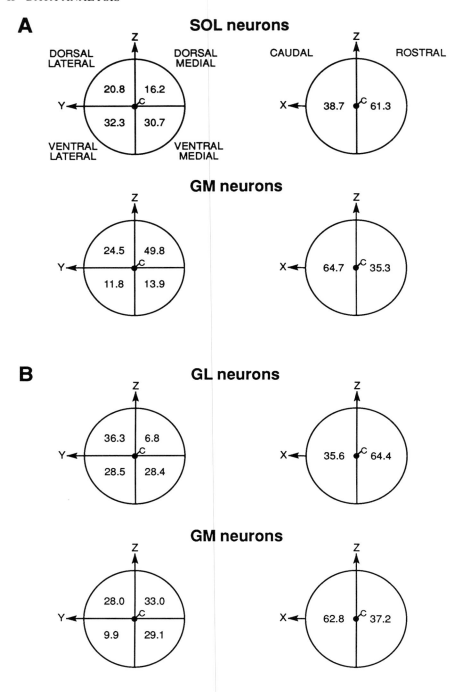

averaged 103, 136, and 101, respectively, after labeling with HRP. Although somewhat higher values (144, 162, and 146) were obtained with FG, the difference was not of sufficient magnitude to interfere with the validity of the descriptive work. The few neurons (3.9% on average) which were double labeled were excluded from the reconstructed figures. In the L5 DRG, 96% of the SOL neurons and 95% of the GM neurons were present, the remainder being scattered in the L4 and/or L6 DRG. The majority of GL neurons were also located in the L5 DRG, although a significant population (31%) was found in adjacent ganglia, usually at the L4 level. Thus, the L5 DRG was the structure of choice to demonstrate spatial relationships between cell clusters innervating specific rat muscles.

The three orthogonal views of a reconstructed L5 ganglion presented in Fig. 3 illustrate the fairly typical distribution of SOL and GM neurons in a dually labeled animal. Despite significant overlap, it can be especially appreciated that these two cell groups in the lateral (Fig. 3A) and rostrocaudal views (Fig. 3B) tend to be segregated in different regions of the ganglion. Drawing the centroid connections reinforced this impression (Fig. 4), showing more clearly that, contrary to SOL neurons, which are distributed in the ventral and external parts of the ganglion, most GM neurons are located dorsally and internally in addition to being closer to the spinal nerves than SOL cells. A visual comparison of ganglia belonging to different animals revealed that, despite some variations, the intraganglionic arrangement we just described was fairly constant. This was further substantiated by a quantitative analysis of cell location in three animals (Fig. 5A) showing that 83.8% of SOL neurons projected into the ventral and lateral sectors of the SOL-GM pools whereas 88.2% of GM neurons were concentrated in the dorsal and medial parts. Also demonstrable was a rostrocaudal segregation: 38.7% of SOL cells were situated in the caudal part of the SOL-GM pool as compared to 64.7% of GM neurons.

These techniques revealed that GL neuronal clusters (Fig. 6) had the same intraganglionic relationship with GM cells as SOL neurons, a fact perhaps linked to the common location of GL and SOL muscles in the posterior compartment of the rat leg, lateral to the GM. Quantitative data obtained on three animals (Fig. 5B) confirmed that most GL neurons occupied the

---

FIG. 5   Compartmental distribution of SOL, GM, and GL cells within SOL-GM (A) or GL-GM (B) pools, as seen in rostrocaudal (left) and lateral (right) views. The X, Y, and Z axes have been aligned on the centroid neuron (c) of the pools and serve to divide the neuronal population into defined compartments. The numerical values are averaged percentages of neurons located in a specific compartment in three different animals. [Reproduced with permission from Peyronnard et al. (7).]

A

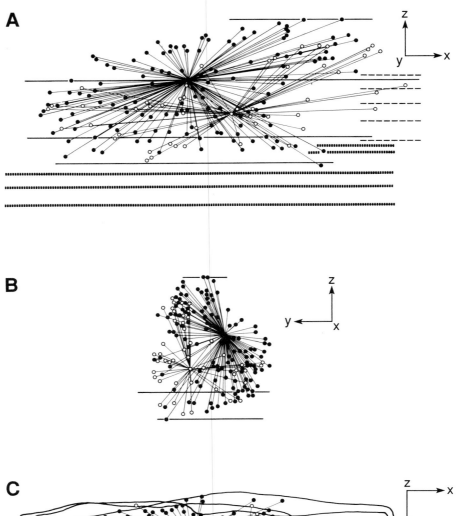

B

C

200 μm

ventrolateral and rostral sectors of the GL-GM neuronal pools in contrast to the dorsomedial and caudal location of the vast majority of GM cells.

Concluding Remarks

The image-processing system described above was instrumental in revealing that muscle afferent neurons of the rat are not erratically dispersed throughout the L5 DRG, but occupy distinct ganglionic sectors perhaps related to the anatomical location of muscles in the leg. Such a system could help those interested in 3-D analysis of biological particles. It was assembled around a medium-priced microcomputer and involved the use of commercially available software fairly accessible even to investigators unfamiliar with computing technology. The approach we followed could be applied to any reconstruction packages such as computer-assisted design (CAD), provided data files on numerical coordinates of the image elements are accessible. It can be expected that the development of power and memory capacity of microcomputers will simplify the cumbersome acquisition process we described, thereby allowing manipulation of digitized images taken directly from the microscope with a camera.

Acknowledgments

The authors are indebted to the Apple Company of Canada for technical support, and to Abacus Concepts, Inc., Berkeley, CA, for providing the 3.0 version of MacSpin Software. Gratitude is extended to C. Gauthier and G. Filosi for artwork, and to O. Da Silva for editing the manuscript.

References

1. W. F. Whimster, *Pathol. Res. Pract.* **185,** 594 (1989).
2. A. W. Toga, M. Samaie, and B. A. Payne, *Brain Res. Bull.* **22,** 323 (1989).
3. A. W. Toga (ed.), "Three Dimensional Neuroimaging." Raven, New York, 1990.

FIG. 6   (A–C) Three orthogonal views of the location of HRP-labeled GL neurons (○) and FG-labeled GM nerve cells (●) in the L5 DRG. In comparison with Fig. 3, it can be appreciated that GL neurons share a ventrolateral and rostral location with SOL neurons in the ganglion as opposed to the dorsomedial and caudal position of many GM neurons. [Reproduced with permission from Peyronnard *et al.* (7).]

4. R. K. S. Calverley and D. G. Jones, *Methods Neurosci.* **3,** 136 (1990).
5. D. P. Huijsmans, W. H. Lamers, J. A. Los, and J. Strackee, *Anat. Rec.* **216,** 449 (1986).
6. M. M. Mesulam, *J. Histochem. Cytochem.* **28,** 1255 (1978).
7. J. M. Peyronnard, J. P. Messier, M. Dubreuil, L. Charron, and F. Lebel, *Anat. Rec.* **227,** 405 (1990).

[24] Computerized Morphometric Analysis of Neurons in Culture

Erich Lieth

Introduction

The morphological development of neurons is under the control of a combination of genetic and environmental factors. Owing to the incredible complexity of neuritic branching and synaptic specificity in the brain, the mechanisms that give rise to neuronal circuitry are extraordinarily difficult to study in intact embryos. Consequently, investigators have turned to cell culture systems to study the events that govern the determination of mature neuronal morphology. Very specific parameters, such as the number of neurites, the number of neuritic branch points, and the size of neuritic arbors, can be defined and observed on individual neurons in the light microscope. Therefore, the effects of hormonal additives or coculture with other cell types (e.g., Refs. 1–3) on morphological correlates of neuronal differentiation (e.g., neuritic branching) can be measured.

Although cell culture reduces the complexity of neuronal morphology to two dimensions, it does not eliminate the inherent heterogeneity between individual neurons of a given phenotype. Under any given culture condition a large variety of neurite branching patterns and soma shapes can be observed. This variability tends to result in broad deviations of the data within any one class of samples. Therefore, the accurate measurement of possibly subtle effects on a heterogeneous population of cells poses a singular challenge. Invariably a large sample must be measured, and numerous morphometric parameters cataloged. A computerized data acquisition system is required to carry out such work expediently.

Several years ago I surveyed morphometric systems that might be used to approach the study of neuronal development *in vitro*. Of the several systems that have been described (4, 5) or are commercially available, none is dedicated to the rapid acquisition of two-dimensional data on a prescribed set of morphological parameters. Many are generic systems that are designed to measure lengths and areas on a digitizing tablet. These systems are necessarily cumbersome to use for neuronal morphometry because they provide no utilities to keep track of the complex data from numerous individual neurons that are represented by hundreds of straight line segments. Other systems are customized for three-dimensional reconstruction, making them expensive, since they require more hardware (such as stepping motors and expanded computer memory) than is warranted by *in vitro* morphometry. Only one system had been described for the acquisition of morphometric data from neurons in culture (6), making use of an antiquated computer and a very limited set of morphometric parameters. This suggested the need for an updated system that could accurately and efficiently gather data on a large number of neurons, keeping track of numerous morphometric parameters, and that was capable of transferring the data to software in which the data could be manipulated.

The paucity of a dedicated system to study changes in neuronal morphology in culture warranted the development of a new approach. I defined a comprehensive set of morphometric parameters pertinent to neurons and created new software to quantify these parameters and store the data. Importantly, the data are gathered during digitization, rather than afterward. This method is described in detail below. The software is for data acquisition only, as software for the manipulation and analysis of numerical data abound. The program is presented as a model for the design of other systems, although the source code can be obtained from the author and either used directly or customized.

System Description

Hardware and Software

The system was developed around the following equipment. Similar equipment can be substituted or adapted.

> IBM-XT with 512KB RAM and hard drive
> Color/graphics adaptor
> Monochrome and color monitor
> Two-button mouse and driver
> Light microscope with drawing tube, and $20\times$ and $40\times$ objectives

Although a floppy disk-based configuration can be used, I recommend a hard disk drive to facilitate program development, as well as data storage speed and capacity. Similarly color graphics are not necessary, but many monochrome monitor screens are too dim for comfortable viewing through the drawing tube. A high-resolution screen is preferred. I used a color/graphics adaptor in high resolution 640×200 mode which still serves beautifully. Today hardware with higher resolution is widely available, and the software described below can easily be adapted to it.

TABLE I   Definitions of Morphometric Parameters

Category	Parameter	Definition
Soma	Number	Sample count
	Area	Area within soma perimeter
	Form factor	Roundness
Neurite	Number	Number of neurites on the cell
	Average length	Mean total length of neuritic segments combined
	Field area	Area of smallest convex polygon
	Field form	Roundness
Initial segment	Length	Distance along neurite from soma to first branch point
	Bending	Length divided by shortest distance between soma and first branch point
Intermediate	Number	Number of intermediate segments on the neurite
	Length	Distance along neurite between branch points
	Bending	Length divided by shortest distance between branches
Terminal segment	Number	Number of terminal segments on the neurite
	Length	Distance along neurite from last branch point to terminus
	Bending	Length divided by shortest distance between last branch point and terminus

To draw an image, a cursor must be moved on the screen; a mouse was used to develop this system, but a variety of devices ranging from inexpensive joysticks to the more expensive digitizing tablets may be implemented. Any microscope with a focusing camera lucida attachment can be used. Quality objectives are essential to view the entire extent of fine neuritic processes.

Parameters Measured

Many morphological parameters can be measured on neurons growing in culture. In this system a defined set of characteristics of the soma (perikaryon) and neurites (axons and dendrites) is quantified. These parameters (defined in Table I) were chosen because of their relevance to the study of neuronal cell body and neurite development.

Each cell is numbered from 1 up to the number of cells in the sample. For each measured cell soma, the program calculates the area (an indicator of its overall size) and form factor (a measure of its roundness; Ref. 7) from the digitized data. The latter takes a value of 1 for a perfect circle and 0 for a straight line. Thus a very round cell body will have a form factor of close to 1, whereas a spindle-shaped cell body will have a form factor of 0.5–0.6.

For neurites, the system calculates the absolute field area over which they can influence their environment, together with the magnitude and complexity of their extension from the perikaryon. The absolute field area, the area of the smallest nonconcave polygon that can be fitted around the neuron cell body and its processes, gives an estimation of the sphere of physical influence of an entire cell. Because the borders of this area are selected geometrically, they cannot represent any absolute biological characteristic of the cell and should only be used in comparison between cells. The form factor, however, gives information about the distribution of processes around the soma. A high form factor (close to 1) generally means that the cell is multipolar, and a low factor (e.g., 0.2) indicates that the shape of the area is long and slender, as in a bipolar cell.

The segments that make up the neuritic branches are measured and categorized by their initial, intermediate, and terminal nature (see Table I). Variations in the length or bending of segments suggest that something in the environment of that process may have influenced its longitudinal or directional growth. The bending factor is a measure of the curvature of a segment. The greater the value, the sharper the curve. For example, a value of 1 indicates a straight line, and a semicircle has a bending factor of 1.6.

All individual measurements for each cell (i.e., soma data, number of neurites per cell body, and the data from individual neurites) are stored for later analysis.

Image Digitization Algorithms

The overall program design is shown in Fig. 1. First the program reserves RAM for a virtual screen coordinate set (an image plane transparent to the user) 12 times the size displayed on a monitor screen. After all variables have been initialized, a title page is displayed. The user than enters the file names for storage of output data and scale information for the calculation of appropriate units. As described below, the user digitizes individual cells by first drawing the soma, then the processes of the cell. The data are recorded both as Cartesian coordinates that define line segments rendering the microscope image as well as numeric measurements of morphometric parameters. Once the cell is completely digitized, the data are saved and the screen cleared in preparation for the next soma.

Scale Factors

Before any images can be digitized, the user must define horizontal and vertical scale factors so that the computer can correctly calculate lengths and areas in metric units. This is done by digitizing a horizontal and a vertical

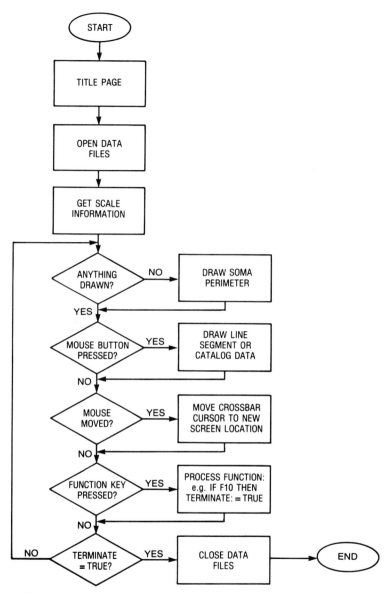

FIG. 1 Flow chart of morphometry program design. Once the program enters the input loop, it continuously awaits a mouse movement or key press. Every time the user acts (e.g., pressing a mouse button), the computer executes the appropriate set of instructions and returns to the input loop, until the user signals termination of the digitizing session.

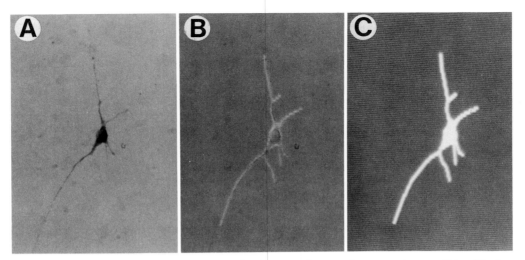

FIG. 2   Digitizing an immunostained neuron from a microscope image: (A) cell before digitizing, (B) digitized image superimposed on neuron, and (C) the digitized image.

line of known length from a stage micrometer on the screen in the manner which will later be used to take the actual measurements. Thus, separate horizontal and vertical scale conversion factors (e.g., pixels per micrometer) are calculated from actual lengths. All subsequent measurements are expressed in the specified units (e.g., micrometers). Scale information is stored and can be reused in future sessions if the investigator is certain of its validity (e.g., the monitor has not been moved with respect to the camera lucida).

Drawing Images

The user must digitize the portions of each cell in a predefined order. Thus the program can automatically keep track of all parameters, freeing the user to concentrate on drawing the parts of each cell accurately. This feature makes the program particularly easy to use. Figure 2 shows the view through the microscope (A) before, (B) during, and (C) after digitization of a neuron. First the outline of the perikaryon is drawn. This can be accomplished either at the same magnification (e.g., using a $20\times$ objective) or at two times ($40\times$ objective) the magnification in which the neurites of the cell will be drawn. In the latter case, the program shrinks the image and measurements mathematically by a factor of 2.

Next, the user draws entire neuritic processes one at a time, beginning each with the most proximal segment. The initial, intermediate, or terminal nature of the neurite segments (see Table I) is specified by pressing the mouse buttons in a coded sequence. The depression of individual mouse buttons elicits particular tones from the computer, so the user can be sure data are entered dependably without looking away from the microscope image. A similar method is used to specify the termination of input (e.g., neurite termination or neuron termination). Once all of the raw data for the current cell are saved to the data file, the screen clears and the user may begin drawing the next neuronal soma.

Drawing Cursors

The software separately accepts input of soma and neurite data. To cue the user on which type of input is expected, two different cursor shapes appear on the screen. In soma drawing the cursor is an arrow with the "hot spot" just off the tip. When drawing neurites, the user sees crossbars that extend the whole length and width of the screen.

Digitizing Coordinates Beyond Edge of Screen

Occasionally, the user will encounter a cell that does not fit completely onto a single monitor screen. When this situation arises, the user invokes a virtual screen in memory which increases the maximum digitizable cell size by a factor of 3 horizontally and a factor of 4 vertically. What is seen in the monitor is then a window, initially located over the exact center of this area. The size of the virtual screen is limited only by the amount of memory in the computer. I have found an array of 3 × 4 monitor screens sufficient, but requirements may differ depending on magnification and equipment configuration.

Measurement Algorithms

Lengths

The length of each neurite segment is calculated as the sum of all the smaller subsegments that the user defined while digitizing. Distances between scaled coordinates are calculated using the Pythagorean theorem [$c = (a^2 + b^2)^{1/2}$].

Bending

The length of each neurite segment is divided by the distance between the end points of the segment to obtain the bending factor.

Area

Two area calculation algorithms have been implemented, one to measure the soma areas and another to calculate absolute field areas. The soma area is calculated using the trapezoidal rule as given in calculus texts. The field area is calculated from the digitized array of coordinates in two steps: (1) determination of the coordinates of the smallest convex polygon which completely encompasses all coordinates and (2) calculation of the area of that polygon. To determine the perimeter of the field area, the program starts with a line (L) through the digitized point (P) at the extreme left of the image (smallest x value). Every line segment connecting P and one other point is checked to locate the coordinates of the segment that makes the smallest angle with line L. If more than one segment is located, the longest designates the next piece of the perimeter of the polygon. Its end point now becomes P (the pivot) while L is the line connecting the previous P with this new one. All subsequent points are selected in the same way until the polygon is closed at the starting point. Although the area of the polygon could be calculated using the trapezoidal rule, it is determined in this program by summing the areas of all the triangles formed by connecting one point on the polygon with all others. When applied to identical areas, the two algorithms produce identical results.

Form Factor

The form factor is defined as 4π multiplied by the area of the cell soma or absolute field, and divided by the square of the perimeter.

Output

The primary output from the program consists of two files. File 1 contains the raw coordinate and button click information of the points used to draw the picture on the virtual screen page. With the information in this file any cell can be reconstructed either for further analysis or for printing a hardcopy. File 2 contains a binary representation of the data resulting from the measurement of all of the cells in a given experiment. Also provided is a utility that produces a third file containing an ASCII character representation of the data in File 2 so that the results can be analyzed with existing commercial software such as electronic spreadsheets.

Caveats

A few idiosyncracies inherent in this kind of system should be mentioned. Since most CRT screens are curved, some distortion will be unavoidable as the edges of the screen are farther from the drawing tube than the center of the screen. In these areas the screen image, as viewed through the camera lucida, is not perfectly focused and the interpixel distances are reduced. This problem is compounded at the corners of the screen. Although the inaccuracy brought about by this factor could probably be alleviated by complex mathematical manipulations, the program does not incorporate these manipulations, and the user is advised to restrict measurements to the center of the screen. Newer flat screens should reduce this source of error.

Camera lucida attachments do not gather light efficiently. Therefore, the monitor image is not very bright superimposed on the image of the microscope slide. This inconvenience can be minimized by reducing the transmitted light from the microscope as much as posisble, and by using white, the brightest color on a color monitor. Also, if possible, it is best to work with the room lights off as this increases the foreground/background contrast on the screen.

All measurements have an inherent error due to the size of the screen pixels. The length of each segment is calculated from the distance between the Cartesian coordinates that define the beginning and end of that segment. These Cartesian coordinates lie at the center of each pixel. Thus, the line seen on the monitor is at least half a pixel longer on each end than the distance measured by the computer. This problem increases quadratically in area measurements, since the error is accumulated along the entire perimeter of the area. The user should always define lengths and areas using the exact center of each pixel rather than its edges.

To simplify programming, the total number of neurites which can be digitized is 20, and for each neurite 20 intermediate segments and 20 terminal segments can be drawn. Probably, neurites that exceed these limitations will never be encountered. If necessary, these restrictions can be relaxed by changing the program code.

Accuracy and Reliability

The accuracy of the system was evaluated by comparing computed parameter values with known values. A neuron was designed on paper, using arbitrary units. The image of the designed neuron was then photocopied onto paper and a transparency. The transparency was mounted on the screen of the computer monitor, so that the computerized morphometry program could be used to measure the neuron. For manual measures, standards were weighed on a balance. To measure length, a piece of 24-gauge copper wire was cut one arbitrary unit long and weighed. For area, copier paper was cut to one square arbitrary unit and weighed. Neurite lengths were measured by weighing lengths of copper wire and comparing them to the standard. Similarly, pieces of copier paper the size of the neuronal field area and soma area were weighed and compared to the standard.

TABLE II   Reliability of Morphometric Measurements

Parameter	Mean	Median	Range	Standard deviation	% Standard deviation
Soma area (μm^2)	134.02	132.92	27.64	5.567	4.15
Soma form factor	0.74	0.72	0.19	0.031	4.19
Field area (μm^2)	2454.84	2479.08	835.73	205.702	8.38
Field form factor	0.33	0.33	0.06	0.015	4.54
Length (μm)	22.52	21.80	5.90	1.058	4.70
Bending	1.00	1.00	0.01	0.002	0.20

Lengths measured on the computer erred from the weighed wire, on the average, by 1.9%. Soma area differed by 5.5% while field area deviated by only 0.6%. These statistics show that the program measures lengths and areas quite accurately when the traditional method of measuring weights is used as a standard.

To ascertain the reliability of measurement, a cultured neuron was measured 50 times. One neurite of this bipolar neuron had no branch points, while the other had one initial, one intermediate, and three terminal segments. Table II shows the means, medians, ranges, and standard deviations of the six classes of measurement. Areas and form factors for soma and field measurements were expressed separately since the program calculates these using different kinds of input. The error was estimated by calculating the relative magnitude of the standard deviation, expressed as a percentage of the mean. For soma area and both form factors, reliability was good (standard deviation less than 5% of the mean), but the field area was less reliable (over 8%). Since this area is calculated from the points specified while drawing neurite line segments, any minor errors in drawing line lengths are magnified to produce very large errors in area. Reliability can be maximized by determining the terminal points of neurites with as much care as possible. Segment lengths deviated by 1.058 μm. In the case of very short distances, this is a significant source of error. The longer the distance, the smaller the proportion of error. This is true for any kind of length measurement, indicating that using the highest possible magnification gives the most reliable result. Bending measurements deviated by 0.002, or 0.2% of the maximum value of 1.

Application

To illustrate the kind of information that can be obtained with this system, we measured the growth of immunostained monoaminergic neurons in culture between 1 and 4 days *in vitro* (DIV). Embryonic rhombencephalon (containing serotonin-immunoreactive neurons) is dissected and plated on polylysine-coated plastic coverslips as previously described (8). After the indicated days in culture samples are processed for immunocytochemistry using avidin–biotin peroxidase complex (9) and antisera raised against serotonin (10). Morphometric data are collected on each day *in vitro* by measuring the first 50 cells encountered. Cells are rejected from counting only if (1) one or more parts of the neuron are obviously missing or (2) cells are positioned such that a neurite of one cell cannot be distinguished from that of another. Growth cones and lamellipodia are measured only in length (cell body to farthest projecting point), disregarding their width. No attempt is made to distinguish axons from dendrites.

Twelve parameters of the development of these monoaminergic neurons are analyzed and compared: soma area, soma form factor, field area, field form factor, number of neurites, length of neurites, number of segments per cell, average segment length, and the number and length of initial, intermediate, and terminal segments (number of initial segments = number of neurites). Definitions of these parameters are given in Table I. For a more complete description of the growth of these cells in culture, see Lieth *et al.* (8). It is important to note that this is not an exhaustive analysis. Many other parameters, such as segment bending, segment density, percent terminal segments, and branch point density, can be extracted from the data. Even simply analyzing the maxima (e.g., of cell area or total neurite length) can give valuable information.

Figures 3 and 4 show the time course of development of the 12 measured morphometric parameters of serotonergic cell growth *in vitro*. The mean cell body area (Fig. 3A) is initially small and peaks at 2 DIV. The cell somas appear to shrink after 2 DIV. Soma form factors (Fig. 3B) decrease between 1 and 3 DIV. This reflects an initial rounded appearance, which eventually transforms into a more oblong and/or polygonal shape as cells develop polarity and acquire processes. After 3 DIV there is no further decline in form factor. A remarkably similar situation was seen for overall field form factor (not shown), where the perimeter of the field is marked by the terminal segments of the neurites, and long processes extending in one or two directions from the cell soma decrease the roundness of the field shape. The field area (Fig. 3C) continuously increases between 1 and 4 DIV, suggesting that even at 4 DIV processes on these neurons have not reached their maximum length. The number of neurites (Fig. 3D) increases minimally over the culture period, indicating that most of the neurites are initiated during the first day of culture. The slight dip in neurite number at 3 DIV suggests that there is a period during which neurites are pruned.

The overall length of neurites (Fig. 3E) increases linearly with time in culture. This again points to the conclusion that these processes have not

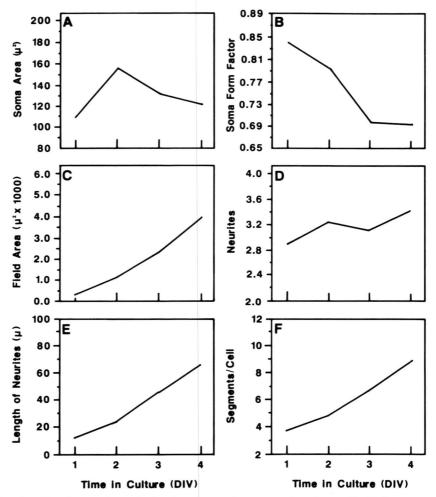

FIG. 3   Morphometric analysis of serotonergic neurons in culture. (A–F) Parameters relevant to overall cellular morphology.

ceased growing. The longitudinal growth appears to result from both the addition of new segments (Fig. 3F) and the elongation of existing segments (Figs. 4A,B). On neurites that have no branch points, the entire neurite is both the initial as well as the terminal segment. When nonbranching neurites grow, their initial segments grow, as reflected in Fig. 4B. The amount of growth of initial segments on branched neurites can be estimated with such

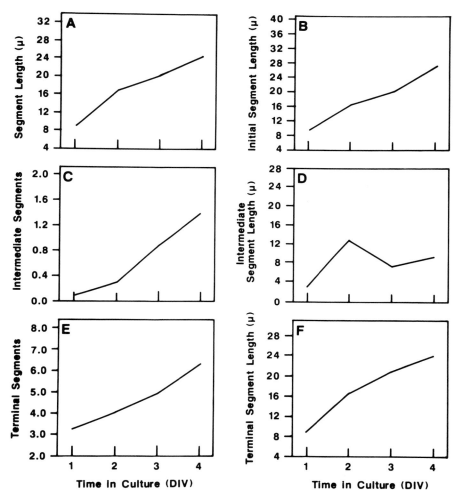

FIG. 4   Morphometric analysis of serotonergic neurons in culture. (A–F) Parameters relevant to individual neurites and segment.

data, but is not reflected in the current analysis. Intermediate segments (Fig. 4C) can only be found on processes with more than one branch point. Mean values less than one (e.g., 1–2 DIV) give an estimate of the number of cells that have an intermediate segment, so that 0.02 indicates that 1 in 50 cells has one. The low number of intermediate segments may have resulted in unreliable length data, as seen from the erratic nature of Fig. 4D. Generally the reduction in intermediate segments may reflect a removal of terminal segments (cf. Fig. 3D), but in this case this interpretation is unlikely since the number of terminal segments (Fig. 4E) rises consistently over the entire 4 days *in vitro*. The complexity of neurites is increased by branching, which causes a rise in the number of terminal segments. The mean length of terminal segments also continues to increase (Fig. 4F), suggesting that some of the processes stop branching and elongate instead.

Conclusion

The above sample analysis of the growth of serotonergic neurons presents the type of valuable information that can be obtained rapidly with a simple, dedicated morphometric analysis system. The important hallmarks of the system I have described here are its ease of use (manifested in the simplicity of data input) and its dediction to measuring neurons in culture, providing an output that is cataloged in terms of type of parameter. The system was inexpensive to create, yet achieves good accuracy and reliability. Also, it is flexible enough that the level of intricacy of the analysis can be adjusted by the investigator. The individual user can also extend the system to calculate other parameters. Documented source code (in PASCAL) or compiled versions of the program are available from the author.

Since its creation, this morphometry system has proven useful in the study of neuronal growth under normal and experimental conditions (8, 11–13). The method is applicable to any neuronal phenotype that can be grown in culture and visualized with the entire extent of its arborization. For instance, the equipment is easily adaptable to the tracing of neurons filled with diffusible cytoplasmic or lipid dyes. The method has also been applied to neurons visualized with radioisotopes in autoradiograms. Consequently, it is my hope that this approach to morphometric analysis will find use in a variety of scientific contexts.

Acknowledgments

I thank Dr. Jean Lauder, in whose laboratory most of this work was done. Dr. Norbert König provided valuable and stimulating discussions on morphometric measurement. Several algorithms and pieces of code were contributed by Dr. J. Heinrich Lieth.

References

1. S. Denis-Donini, J. Glowinski, and A. Prochiantz, *Nature* (*London*) **303,** 641 (1984).
2. N. König, V. Han, E. Lieth, and J. Lauder, *J. Neurosci. Res.* **17,** 349 (1987).

3. I. Reisert, V. Han, E. Lieth, D. Toran-Allerand, C. Pilgrim, and J. M. Lauder, *Int. J. Dev. Neurosci.* **5,** 91 (1987).
4. J. J. Capowski and M. J. Sedivec, *Comput. Biomed. Res.* **14,** 518 (1981).
5. C. A. Curcio and K. R. Sloan, *Anat. Rec.* **214,** 329 (1986).
6. M. Freire, *J. Neurosci. Methods* **16,** 103 (1986).
7. R. R. Mize, *in* "The Microcomputer in Cell and Neurobiology Research" (R. R. Mize, ed.), pp. 178. Elsevier, New York, 1985.
8. E. Lieth, D. R. McClay, and J. M. Lauder, *Glia* **3,** 169 (1990).
9. J. M. Lauder, J. A. Wallace, H. Krebs, P. Petrusz, and K. McCarthy, *Brain Res. Bull.* **9,** 605 (1982).
10. J. A. Wallace, J. M. Lauder, and P. Petrusz, *Brain Res. Bull.* **9,** 117 (1982).
11. E. Lieth, A. C. Towle, and J. M. Lauder, *Neurochem. Res.* **14,** 979 (1989).
12. J. Liu and J. M. Lauder, *Dev. Brain Res.* **62,** 297 (1991).
13. J. Liu and J. M. Lauder, *Glia* **5,** 306 (1992).

[25] Computerized Analysis of Polarized Neurite Growth

R. John Cork

Introduction

Neurons are some of the most highly asymmetrical cells found in nature. Intracellularly they have different regions with specialized structure and function. This intrinsic polarization is apparent from the beginnings of differentiation. Neurons also display extrinsic polarization during differentiation as they respond to their environment to grow toward specific targets. The growth cones can respond to a variety of environmental cues, including substrate molecules, diffusible chemotrophic factors, and electrical fields. Our laboratory has had a long-standing interest in the development of polarity in a variety of cells, both plant and animal. This chapter describes a computerized system for the analysis of polarized neurite growth. The system was specifically designed to meet the needs of investigators studying the effects of electrical fields on neurite growth (1). It is fortuitous that electrical fields have an inherent positive and negative direction, as this makes the analysis conceptually much simpler. Nevertheless, the general principles of the analysis and the equations can easily be adapted for other examples of polarized growth. Throughout this chapter I shall refer to the poles of the field as the anode (positive electrode) and cathode (negative electrode). In the case of other types of polarizing fields different terms would have to be substituted.

There are many areas of biological research where computers can be useful: controlling experiments, data collection and storage, data analysis, and theoretical modeling of results. Before implementing any computerization however, the necessity of the process must be carefully considered. The effort of programming, debugging, and maintaining computer programs can easily be as time-consuming as analyzing a few experiments by hand. If the experiments are not going to be repeated many times, or if only a small amount of data is going to be collected, then it is often not worthwhile embarking on a full-scale computerization.

The study of polarized neurite growth has traditionally been a very labor-intensive endeavor where comprehensive data about large numbers of cells need to be collected and analyzed (2–4). We considered this to be a prime example of a type of analysis that could be computerized. We are using the computer to both collect and store the data, and then to analyze it and provide

Methods in Neurosciences, Volume 10

a readable display of the results in tabular or graphic form. To do this requires very little computer effort but enables more data to be collected and analyzed than would otherwise be possible if it had to be done by hand.

The next few sections discuss the theoretical considerations used when planning the computerized system. That will be followed by a description of the practical programming required to implement the system.

Digital Morphometry

The initial step in any computer analysis is to enter the data into the computer. This involves converting it from whatever form it is collected in, to a form that the computer can manipulate. Our data are collected as photographic negatives of the cells at the end of an experiment. The negatives contain all the information about the cell morphologies, neurite lengths, and directions of growth. They also contain the information about how the cells are aligned in relation to the electrical field.

The process of "digitizing" can be done on many forms of raw data. For example, it can be done directly from the microscope by tracing camera lucida images of the cells on a digitizing pad. Alternatively, video images can be digitized using either a mouse or a light pen, either directly from the microscope or from tape. We consider that using photographic negatives projected onto a digitizer pad is the best compromise in terms of speed, expense, and resolution. It also provides a permanent record of the data which can be reused if necessary.

The first problem then is, What data need to be recorded for a digital "picture" of each cell? All the necessary data for the computer calculations should be stored so that the data analysis can be carried out later. Also, it is preferable to store enough data so that any parameters which might be needed for calculations that are added to the program at a later date are available. It would be possible to store an entire digital map of each cell in the computer. This could then be analyzed by the computer program, and whatever morphometric data were required could be extracted. This would, however, require inordinate amounts of disk space and computer memory. For example, data files for an average dish of cells currently occupy about 15 kilobytes of disk space. This would increase to somewhere around 1 megabyte if all coordinates from the digitizer were stored. Much of the data would be redundant, and data sorting and analysis would be severely slowed down. We therefore perform some preliminary calculations on the coordinate data as they are collected, and only store data that are considered useful. The only data that the computer can obtain directly from the digitizing tablet are (x, y) coordinates. Anything else, such as angles or lengths, has to be calculated

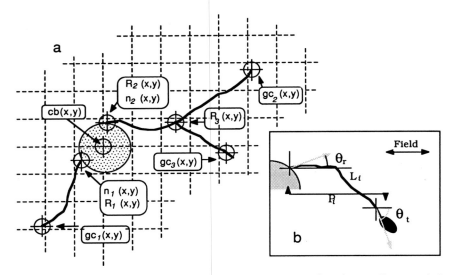

FIG. 1   (a) Coordinate data recorded for each cell. The $\{x, y\}$ coordinates of the center of the cell body ($cb\{x, y\}$) are entered first. Then the coordinates of all the neurite initiation sites on the cell body ($n_i\{x, y\}$) are recorded. Data are then entered for each neurite in turn. This set includes the coordinates of its growth cone ($gc_n\{x, y\}$) and origin ($R_n\{x, y\}$). (b) From the coordinate data the computer makes a few preliminary calculations about each cell. It then stores several parameters such as the angles defining the direction of growth for its tip (θ_t) and root (θ_r) and the traced length of the neurite in micrometers (L_i). Each neurite has a projection length (P_i); in our system this is simply the difference between the x coordinates of the growth cone and the neurite origin.

from the coordinate data. The basic data set that is stored for each cell during data entry is illustrated in Fig. 1.

From this basic data set most of the other morphometric cell parameters can be calculated. A summary of the ones that we use is given in Table I. Most of the parameters need no further explanation, but one concept that is especially useful when analyzing polarized growth is the field projection. Any polarizing field, whether it be electrical or chemical, has a principal axis. It is an important consideration when setting up experiments to investigate polarized growth that this axis should be well defined and preferably geometrically simple. The projection of the length of the neurite onto the field axis (Fig. 1b) is a measure of how much it grew in that particular direction parallel to the field axis. Thus, it is often preferable to use the projection rather than the total length.

TABLE I   Cell Morphometric Parameters
Calculated and Stored During
Digitization of Neurons[a]

Abbreviation	Definition
Nn	Number of neurons
Nt	Number of neurites (initiation sites)
Gc	Number of growth cones
Bp	Number of branch points (Gc − Nt)
BI	Branching index (Bp/Nt)

[a] Many of the other values that are calculated and stored are
defined in the text.

Theory of Asymmetry Analysis

Having stored the numerical data required to describe the cells it is then
necessary to devise methods to determine if they exhibit any polarized
growth. It is convenient to describe many of the polarization effects with
single values. We have made extensive use of asymmetry indices of the type
first applied to neurite growth by Patel and Poo (3). These have the general
form

$$\text{Asymmetry index} = \frac{M_a - M_c}{M_a + M_c} \tag{1}$$

where M is whichever morphometric parameter is being considered and the
subscripts a and c refer to the anodal and cathodal values of M, respectively.
There are many possible choices for M. The polarization of any parameter
that could be affected by the field can be assayed with such an index. Some
of the parameters that are commonly used are the number of neurites, the
number of branch points, the number of neurite initiation sites, the total
neurite length or total field projection length, and the mean neurite length or
projection length.

These indices are measures of the fractional difference in the observed
parameter on the two sides of the cell. As such, they reflect the strength or
magnitude of the polarizing response rather than just its direction. This is in
contrast to some other indices that are concerned primarily with the direction
of the response; for example, the neurite initiation site asymmetry index
(NISA) (3) is the average cosine of the initiation site angles. Our asymmetry
indices have a maximum value of 1 with a sign that reflects the direction of

the bias. Thus -1 would indicate that whatever parameter is being measured is completely biased toward the negative side of the polarizing field, in the case of an electrical field this would be the cathode. A value of 0 will usually indicate a random distribution with no bias in any direction. In certain cases, depending on how the classification of the parameter is defined, a value of 0 may indicate a strong effect perpendicular to the field.

It is important to remember that these indices are sample or population parameters and have very little meaning for individual cells. When N is small the asymmetry index can only have a few discrete values. In the extreme case when there is only one example of whatever parameter is being counted the asymmetry index can have only one of three possible values: $+1$, -1, or 0.

Indices of Neurite Initiation Loci

PMI (Plus/Minus Index)

The cell body is divided into two halves with a line running perpendicular to the field. The half facing the anode is designated the plus half, the other half being the negative side. As the data about a cell are entered into the files the neurite initiation sites are assigned to one half or the other.

AI (Angle Index)

AI is a variant of PMI where the cell body is divided into four segments instead of two. The cathode-facing and anode-facing segments both cover 120° at 90° ± 60° and 270° ± 60°, respectively. Neurites sprouting from the cell body in either of the two 60° segments at the top or bottom of the soma are designated "perpendicular." This index is considered preferable to the PMI as the definitions of anode- or cathode-facing initiation sites are more stringent. Furthermore, by increasing the size of the perpendicular category we have a clearer definition of "no effect." As this has been our preferred index of initiation site asymmetry its acronym has been redefined to stand for asymmetry of initiation.

Indices of Polarized Growth

NGA (Neurite Growth Asymmetry)

NGA was originally defined by Patel and Poo (3) to assay the asymmetry in neurite growth. They calculated the mean projection of the processes growing toward the anode as

$$\bar{P}_{a} = \frac{\sum\limits_{i=1}^{N_a} P_{a_i}}{N_a} \tag{2}$$

and toward the cathode as

$$\bar{P}_{c} = \frac{\sum\limits_{i=1}^{N_c} P_{c_i}}{N_c} \tag{3}$$

where P_{c_i} and P_{a_i} were the individual neurite projections toward the cathode and anode, respectively, and N_c and N_a were the numbers of projections facing each electrode. The projections of any subsidiary branches were added to the projection of the main neurite, and the category that the total projection was assigned to was based on the projection of the main neurite. Their asymmetry index (NGA) was then calculated by plugging the mean projection values into Eq. (4).

$$NGA = \frac{\bar{P}_{c} - \bar{P}_{a}}{\bar{P}_{c} + \bar{P}_{a}} \tag{4}$$

When we came to incorporate NGA into our analysis (1) we made two changes to its definition. First, to remove some of the confusion associated with data presentation, we reversed the order of the two terms in the numerator of Eq. (4) so that positive values of NGA represented polarization toward the anode and negative ones an effect to the cathode. This is intuitively how we would expect the values to be, as the anode is the positive pole of the field. In the cases of other types of fields, we would set up the equation for NGA in such a way that growth toward the signal, or up a chemical gradient, gave positive values of NGA, while growth away from the signal gave negative values. The other change that we made was in the definition of a neurite. We felt that each growth cone should be considered a separate entity that responds to the field independently from the others. Hence, we use the projections of all the branches, from their branch point (or initiation point) to their growth cone.

DNP (Difference in Number of Projections)

Although NGA is a good indicator of effects on the mean projection length, there are other possible effects that it does not reflect. For example, all the neurite projections may be approximately the same length, but there may be many more growing in one direction. To assess whether such effects were occurring I defined a different asymmetry index based simply on the number of projections facing each pole. The numbers used for this index are calculated as the data for NGA are collected. DNP is given by

$$\text{DNP} = \frac{N_a - N_c}{N_a + N_c} \tag{5}$$

DGI (Differential Growth Index)

If both NGA and DNP are calculated, together they can provide a reliable indication of whether there is any polarized growth. The two separate effects can be combined by using an overall asymmetry index which I defined as the differential growth index given by

$$\text{DGI} = \frac{TP_a - TP_c}{TP_a + TP_c} \tag{6}$$

where TP_a and TP_c are the total growth projections toward the anode and cathode, respectively:

$$TP_a = \sum_{i=1}^{N_a} P_{a_i} \tag{7}$$

$$TP_c = \sum_{i=1}^{N_c} P_{c_i} \tag{8}$$

Other Asymmetry Indices

In different versions of this analysis we have also developed and used two other asymmetry indices. Earlier studies of the effects of electrical fields on neurite growth had looked at effects on branching and turning responses. I have modified the procedures for assessing these effects for the computerized analysis, and both a differential branching index (DBI) and the deflection index (DFX) are calculated by our computer programs. Neither has proved to be particularly reliable or informative in their current forms. They both need some refinement before they will be used regularly.

DBI (Differential Branching Index)

We have calculated DBI by using the number of branch points as M in Eq. (1). While this method does indicate if there are more branches on one side than on the other, it is also sensitive to differences in the numbers of neurites on the two sides. Thus nonzero values may just reflect bias in the neurite growth, not an effect on the amount of branching. To remedy this, in future versions of the analysis programs DBI will be calculated using the branching index (BI) as the parameter M [Eq. (1)]. BI is the sample mean for the number of branch points per initiation site (Table I).

DFX (Deflection Index)

The deflection, or turning response, of growth cones has been analyzed in a variety of ways. Patel and Poo (3) did not quantify a turning response but reported curving of the neurites toward the cathode. Hinkle et al. (2) also reported a turning response toward the cathode and tried to quantify it by determining the numbers of neurites deflecting to each electrode. They also measured the mean angle of deflection in each category. McCaig (5) has used the approach of looking at neurites initially growing perpendicular to the field and then determining their deflection when the field is switched on. In our analysis we tried to adapt the method of Hinkle et al. (2) for the computer. The deflection angle of each neurite was calculated by determining its initial (θ_r) and final (θ_t) growth directions. The difference between these then gives the direction of the deflection and its magnitude. By categorizing all nonzero deflections as anodal or cathodal a deflection asymmetry index (DFX) can be calculated as

$$ \text{DFX} = \frac{F_a - F_c}{F_a + F_c} \tag{9} $$

where F_a and F_c are the numbers of anodal and cathodal deflections, respectively. The mean deflection in each direction is also determined from the actual deflection angles.

In the analysis of our experiments we have not found either of these approaches to be reliable indicators of turning responses. It is not known whether this is a result of differences in cultures, where we are not getting any turning response, or if it is due to problems in our concept of turning. The mean deflection angles are not usually significantly different for anodal and cathodal deflections, and there is much variability in the number of neurites curving in any given direction. By confining the analysis to neurites initially growing perpendicularly across the field, in the way that McCaig (5) has done, we are able to get a more reliable picture of the relative numbers deflecting in response to the field. In our present analysis system, however, this is one polarizing response that we have not yet been able to quantify satisfactorily.

Statistical Analysis

Much of the effort in the data analysis portions of the software is put into calculating confidence limits and statistical significance for the various indices. As the asymmetry indices are not means, we took the approach of looking at the underlying distributions and testing to see if they were significantly different from those that would be expected if there was no biasing effect present. These are tested most often with a z score using the binomial distribution standard deviation given by

$$\sigma = (Npq)^{1/2} \qquad (10)$$

where N is the number of observations and p and q are the probabilities of "success" and "failure," respectively. Equation (10) is only valid when both Np and Nq are greater than 5. For most of the asymmetry indices p and q are both equal to 0.5 because if there was no polarizing effect present a neurite would have an equal chance of growing toward the anode or the cathode.

Of course, many other forms of analysis can be performed by the computer. For normally distributed parameters, means and standard errors can be calculated and used for t-tests. It is especially useful to be able to get a graphic representation of the distribution of a parameter. The preferred formats are polar graphs for angular distributions and histograms for other frequency distributions.

Polar graphs such as that in Fig. 2 can be plotted by customized software, or the collected data can be transferred to one of the many commercial graphics programs available. We use our own routines or Cricket Graph (Computer Associates, San Diego, CA). The angular loci of parameters such as the neurite initiation sites or the growth cones are tabulated by the computer as it is sorting through the data. The frequencies in each 10° bin are then plotted on a circle whose radius is set to the highest frequency. To gain some information about the significance of these angular distributions we have employed the χ^2 test. When a comparison with a uniform distribution is required, that is, comparing a field treatment with a control, we simply calculate the expected frequencies by dividing the sum of the observed frequencies by the number of bins. If a comparison with another field-treated distribution is required, then a contingency table is set up to calculate the expected frequencies (6). For χ^2 tests it is required that all the expected frequencies be larger than about 5. If this is not the case, then either the number of bins has to be reduced or the analysis has to rely on the asymmetry indices. Other distributions are often displayed as histograms (Fig. 3). It is usually the case that these distributions of actual neurite length or neurite

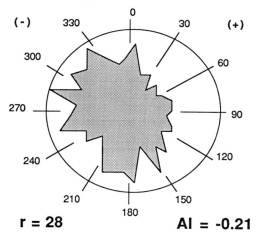

FIG. 2 Polar histogram of the neurite initiation site loci of field-treated cells. Such graphs provide a vivid representation of polarized growth. In this case the majority of the neurites sprouted on the cathode-facing ($-$) side of the cell bodies.

projection length are not normally distributed. χ^2 tests can be employed for these parameters, but we prefer to use one of the other nonparametric tests, the Mann–Whitney U test (6). The computer can easily rank the observed parameters and can calculate levels of significance using the same routines as used for the z score.

One final note that should be mentioned is that the whole of this discussion is based on the assumption that the neuronal growth and the polarizing field are two dimensional. This is quite acceptable for experiments dealing with tissue culture cells *in vitro*, but would probably not be the case in any *in vivo* investigations. In principle, most of the analysis described here could be extended to three-dimensional data; however, the art of analyzing neuronal structures in three dimensions is still in its infancy (7).

Practical Programming

In this section some of the practical considerations involved in implementing the computerization of our analysis system are outlined. Little of the actual programming is detailed as this would be different for each system designed. Some details of our programs can be found in an earlier paper (1).

The fundamental purpose of the computerization was 2-fold. First, it should enable the rapid entry of large amounts of data in a reproducible and error-

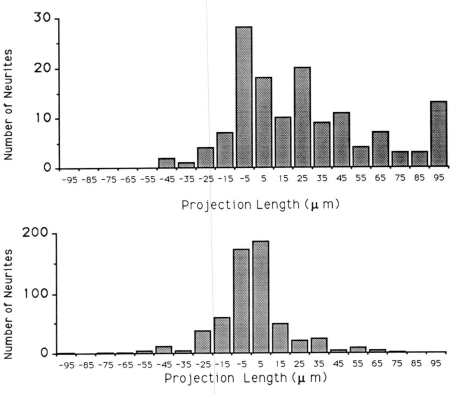

FIG. 3   Histograms of the distribution of neurite projection lengths. The graph at top shows data from field-treated cells, the bottom graph, from control cells. Projections toward the cathode are classed as negative. Differences between such distributions are tested using a Mann–Whitney U-test.

free manner. Second, it should be able to collate the data and present some analysis of its significance. In our system these two functions are largely separated into different programs. The overall design of the programs should be as modular as possible, to simplify modification and debugging. Over the course of several years the original ideas of what such a program should do change somewhat. We have several options that were designed into the original format that have never been fully implemented. These include analysis of time course data where sequential views of the same cell are digitized and the development of polarity, or growth rates, are calculated and displayed as graphs. We also have several debugging routines and utilities for reformatting data files or making backup copies of files.

The first thing to be done before entering any data is to set the experimental conditions and such things as scaling factors required to convert lengths from digitizer units to micrometers. These are all set in preliminary routines. The experiment is given an identifying label that will be used for all data files associated with the data set. All the relevant experimental conditions are entered and stored in a small file. As well as scaling factors for converting digitizer pad units to absolute lengths, we have also enabled the tracing resolution of the digitizing pad to be changed by the user. Thus, if neurites are all very short the distance between successive digitizer coordinates is set to be small to improve the accuracy of the tracings. If there are many very long neurites then the interval between coordinates is increased so that the volume of data does not get too large.

The data input program is an interactive one that is designed so that data entry is fairly rapid and uninterrupted. The entry of large data sets is improved if the operator can proceed at a natural rate without having to keep changing the type of operation being performed. For example, it slows things down considerably if the user has to keep switching between keyboard and digitizer pad for different forms of data entry. As our data entry has to be done in reduced light many of the prompts are accompanied by audible signals, and many of the control commands can be entered from the digitizer pad without having to move to the keyboard. We have also found it useful to have a continual printout of the parameters of each cell that can be used later for comparison with the computer-generated data. To enable interrupted data entry, the programs store intermediate totals in a small random access file. The input program should be as comfortable as possible to use because the largest source of error is operator fatigue. The operator is called on to make several subjective decisions regarding the center of the cell body and the initiation site locations. The quality of these decisions will become degraded if the program is too strenuous to use. If several people are going to be inputting data over a series of experiments, the criteria for making these decisions need to be agreed on beforehand.

Most of the calculations and display of results are done on combined data sets. The investigator can select a set of experimental conditions, or some neurite characteristic, and the computer will sort through all the data files collecting together all the appropriate data. Different levels of analysis and output format can be selected so that time is not wasted by getting comprehensive printouts about every preliminary search.

Acknowledgments

This work was carried out in the laboratory of Professor K. R. Robinson to whom I am indebted for encouragement and financial support. I am grateful to Ann Rajnicek, Steve Hall, and Lisa McKinney for entrusting me with their data. I thank Kevin

Hotary, Tom Keating, Kenneth Robinson, and Joe Vanable for constructive comments on the manuscript. This work was supported by National Institutes of Health Grant NS28514.

References

1. R. J. Cork and A. M. Rajnicek, *J. Neurosci. Methods* **32,** 45 (1990).
2. L. Hinkle, C. D. McCaig, and K. R. Robinson, *J. Physiol. (London)* **314,** 121 (1981).
3. N. Patel and M.-M. Poo, *J. Neurosci.* **2,** 483 (1982).
4. L. F. Jaffe and M.-M. Poo, *J. Exp. Zool.* **209,** 115 (1979).
5. C. D. McCaig, *J. Cell Sci.* **95,** 617 (1990).
6. N. M. Downie and R. W. Heath, "Basic Statistical Methods." Harper & Row, New York, 1974.
7. H. B. M. Uylings, J. Van Pelt, R. W. H. Verwer, and P. McConnell, *in* "Computer Techniques in Neuroanatomy" (J. J. Capowski, ed.), p. 241. Plenum, New York, 1989.

[26] Computer-Assisted Methods for Analyzing Images of Olfactory Bulb

Jean P. Royet

Introduction

Computer-assisted image analysis is useful in many scientific domains: materials sciences, robotics, geology, geography, and biological sciences. In neuroscience, two fields of research may require image analysis: autoradiography and morphometry.

Image Analysis in Autoradiography

The 2-deoxy[^{14}C]glucose (2-DG) neuroanatomical method represents a major technological advance in neurobiology. It permits one to obtain pictures of neural metabolic activity represented in autoradiograms of sequential sections of the nervous system. Neurobiologists are interested in the evaluation of the activity of anatomically defined cerebral structures, small [e.g., locus coeruleus or thalamic nuclei (1, 2)] or large [e.g., olfactory bulb or cerebral cortex (3, 4)]. In the first case, investigators have up until a few years ago preferentially used densitometric systems and the "quantitative" method developed by Sokoloff *et al.* (5). This method enables the determination of an absolute value of glucose utilization. In the second case, investigators have employed video systems and a so-called semiquantitative method in which one compares the relative quantity of labeling in different regions. Nowadays, video systems are unanimously used. They include essentially a television camera which allows the conversion of the televized image into a numerical matrix. They present many advantages. First, the data acquisition time in video systems is less than in densitometric systems. Second, because the image is entirely accessible by computer, it is possible to treat it with different algorithms for image processing (6).

In olfaction, neurobiologists are interested in the spatial distribution of neural activity in the olfactory bulb as a function of different odorous stimuli. With a quantitative evaluation of bulbar activation labeling, investigators (7) established a simple profile of optical densities (OD) along a line across the olfactory bulb (Fig. 1a). The analysis was local only and indicated variations of optical densities from one layer to the other. Researchers proposed per-

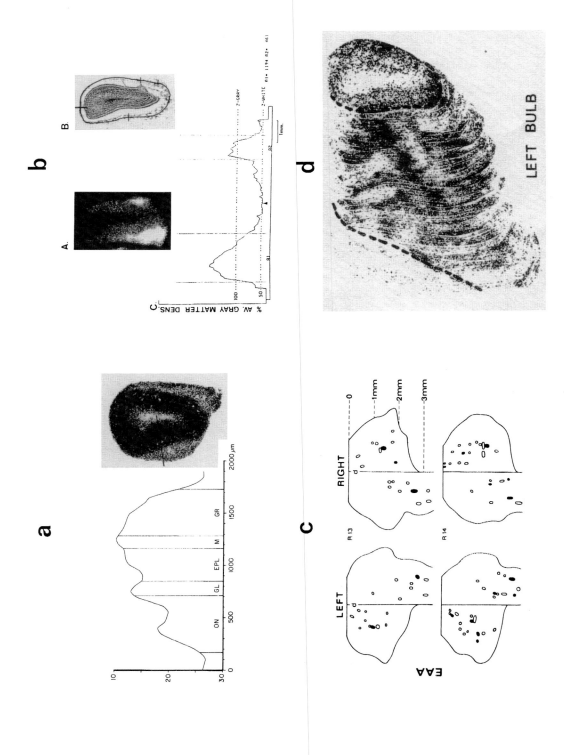

a

b

c

d

LEFT BULB

RIGHT

LEFT

R 13

R 14

EAA

0

1mm

2mm

3mm

B.

A.

2-GRAY

2-WHITE   G1= 1194 R2= 46.1

B2

B1

% AV. GRAY MATTER DENS.

50

100

C.

1mm.

ON   GL   EPL   M   GR

0   500   1000   1500   2000 μm

10

20

30

forming a more systematic sampling inside a given layer, for example, the glomerular layer. Thus, Kauer *et al.* (8) manually displaced a mobile probe all around this layer for each autoradiogram (Fig. 1b). All profiles were then represented by plane projection. By shifting profiles and modifying the orientation angle of the projection plane, a pseudo-tridimensional representation was obtained. Another commonly used method consisted of conserving only a part of the information by considering two or three levels of activity (Fig. 1c). The olfactory bulb was unfolded and only then were corresponding glomerular sites taken into account (9–12). Finally, Jourdan (13) showed the three-dimensional image of the glomerular labeling of a rat olfactory bulb. In these experiments, photographs of each autoradiogram were taken. The olfactory nerve layer was removed from the photographs, which were superimposed, slightly shifted, and glued to each other (Fig. 1d). The dark areas represent the spatial distribution of the most highly labeled glomeruli.

These previous studies either lacked a rigorous sampling procedure or did not allow subsequent efficient statistical analysis. The latter requires the adoption of a biometric approach including an efficient sampling strategy. The computer-assisted image analysis is well adapted for this purpose.

The first step is to choose adequate images. For example, a correct analysis of the spatial activity of the olfactory bulb from serial autoradiograms requires selection of autoradiograms regularly distributed along the rostrocaudal axis. The second step is to extract the pertinent information from the glomerular layer by using image analysis. To do this, data can be recorded either automatically or semiautomatically. Finally, the information extracted must be structured in order to perform a quantitative analysis.

Image-Analysis System

The analysis of autoradiograms is carried out with the aid of a Quantimet 570 image analyzer (Leica). It is equipped with a Sanyo CCD camera placed on a Zeiss Axioplan microscope using a Plan Neofluar objective, $\times 1.25$. As the

---

FIG. 1 (a) Laminar density profile (left) as measured across an autoradiogram (right). ON, Olfactory nerve layer; GL, glomerular layer; EPL, external plexiform layer; M, mitral cell layer, GR, granular layer. [After Sharp *et al.* (7).] (b) Density profile (bottom) as measured on the glomerular region of the autoradiogram (left) determined from the projected image of the corresponding histological section (right). [After Kauer *et al.* (8).] (c) Spread-out surface reconstructions of the glomerular labeling of the olfactory bulbs after exposure to ethyl acetoacetate (EAA) in two rats. [After Astic and Saucier (12).] (d) Three-dimensional reconstruction of the glomerular activity from photographs, superposed and glued to each other, of successive autoradiograms. [After Jourdan (13).]

entire histological section needs to be visualized, the optics of the microscope can be modified by suppressing the magnification changer (Optovar). The optical system then gives a magnification of $\times 47$ with each pixel corresponding to a 8.49-μm square. A pixel defines each picture point, or picture element, and possesses a numerical value which represents the average brightness of the corresponding image part. The system is calibrated with a micrometer scale which is superimposed onto the measure frame and allows the determination of length in pixels, thus the image resolution. The successive procedures of image analysis are controlled by a BASIC program running under the DOS operating system.

Owing to a lack of homogeneity of illumination, the center area of the video image was brighter than the peripheral region (14, 15). The final uniformity of light within the observation field (the shading correction) is directly controlled by the computer program. Obviously, any image analysis of transmittance is incorrect without perfect homogeneity of illumination.

Gray level data are calibrated in terms of optical densities (OD). Optical density expresses the reduction in light transmitted through different pixels of the image. Thus, transmittance (T) is related to OD values according to the following relations:

$$T = (\text{emergent intensity/incident intensity})$$

$$OD = -\log T$$

A simple linear equation allows one to relate the 255 gray level values (GL) to the transmitted light (Fig. 2):

$$GL = kT + C$$

To resolve the equation, two reference values are determined. The maximum of transmittance (T_{max}) is obtained with the background of the autoradiography film. T_{min} is the value at zero transmitted light, which is obtained by interrupting the light beam. Calibration is achieved by using a table which allows directly measured brightness to be converted to values proportional to optical density. It is performed before the analysis of autoradiograms for each animal.

Optical Density Measurements

To respect efficient sampling rules, measurements are performed on stained histological sections and the adjacent corresponding autoradiograms systematically distributed along the rostrocaudal axis of the olfactory bulb. The

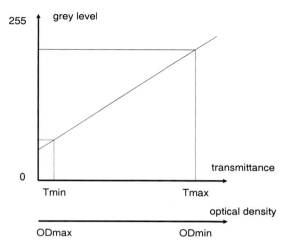

FIG. 2   Relation between gray levels and transmitted light T or optical density OD.

optical density measurements mainly included two steps: analysis of histological sections and analysis of the autoradiograms.

Analysis of Histological Section

The stained section is placed under the microscope (Fig. 3a). To position each section relative to one another and so to minimize geometric distortions during the reconstruction of the glomerular layer map, the median edge and the dorsal pole of the glomerular layer are chosen as geometrical references. To do this, the histological section is placed in such a way that the reference edges of the glomerular layer are adjacent to the left and dorsal sides of a first measure frame. The height and width of the measure frame are then adapted to the size of the glomerular layer. The measure frame defines the region of screen within which measurements are performed. The image frame delimits the part of the image over which image-processing operations take place. The maximal limits of these frames determine an array 512 pixels high and 512 pixels wide.

Outlines of sections are automatically selected as follows. A detect operation on the gray image extracts the section from the background (Fig. 3b). Thus, a brightness discrimination is used to select pixels belonging to features of interest. Lower and upper limits for the detection threshold are adequately preprogrammed, and all pixels whose brightness lies between the two thresholds are included. However, results vary according to the staining intensity of the section. Therefore, the threshold can be interactively adjusted to allow

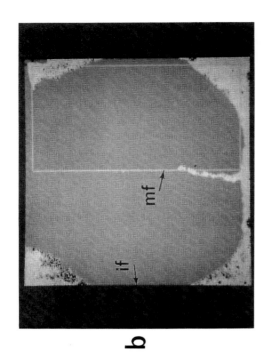

a

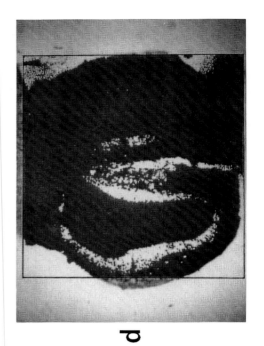

b

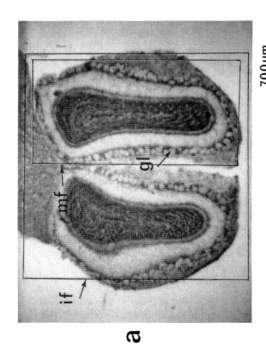

c

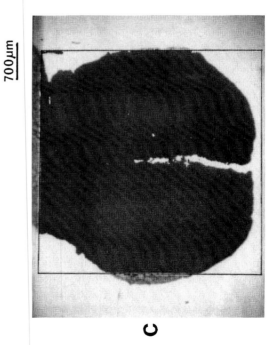

d

700μm

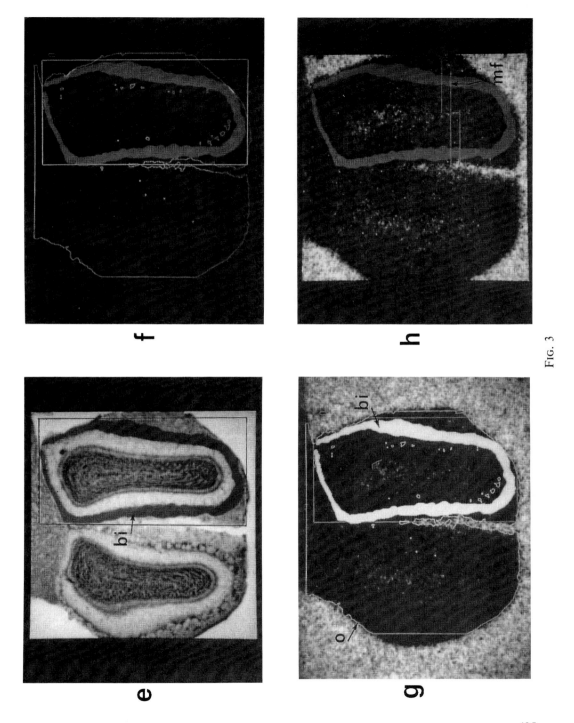

f

h

e

g

FIG. 3

the experimenter to pick out the zones of interest in the image provided by the camera. The "detect" function then ignores all pixels which lie outside the image frame (Fig. 3c). If the homogeneity of the light within the observation field is not controlled, good detection of the gray image of the histological section is impossible (Fig. 3d). Subsequent operations are then difficult, even unrealizable.

At the end of the detect operation, the gray scale image is converted to a binary one. Figure 3b shows that, in the image frame, many small features could also be automatically selected. These features represent noise and could be suppressed by performing an "opening" (Fig. 3c). This function, called a "morphological filter," consists conceptually of a succession of "erosions" and "dilations" (also named "edging" and "plating," respectively), operations which form the basis of mathematical morphology (15–19). The structuring element is the fundamental component by which morphological operations are performed (19). Owing to the absence of anisotropy of artifactual features (noise), the hexagonal structuring element is used. Finally, outlines of sections are automatically selected. This function is chosen by the operator and is conceptually obtained by performing an operation of "thinning" (19). The final binary image is stored in the bit plane of the computer.

To delineate the glomerular layer, the operator can interact with the image by using the "binary edit" function. It allows the creation of a new binary image on the monitor screen with the aid of a mouse (or light pen) when specific regions are extracted with difficulty. All pixels enclosed within a loop drawn defining the glomerular layer outlines are then transferred to a second bit plane. Finally, the two binary images (outlines of section and glomerular layer) are combined into a single image by performing the logical operation named "or" and transferred to a third bit plane (Fig. 3f). The Boolean function "or" is also termed "union" in mathematical language.

Analysis of Autoradiograms

The stained section is removed and replaced by the corresponding autoradiographic film, which is carefully aligned with the histological section outlines previously stored in the third bit plane (Fig. 3g). The gray levels of the autoradiogram are stored as OD values. That part of the gray image which

---

FIG. 3   Successive operations used in the autoradiographic study for analyzing images of the histological section and the corresponding autoradiogram. gl, Glomerular layer; if, image frame; mf, measure frame; bi, binary image of the glomerular layer; o, outlines of the binary image of the section. See text for details.

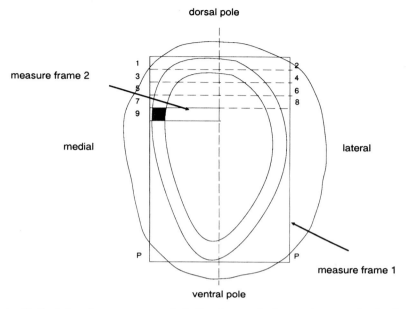

FIG. 4   Optical density measurements in the glomerular layer at one sectioning level. [After Royet *et al.* (3).]

corresponds to the glomerular layer is then masked by the binary image (Fig. 3g) which was stored in the third bit plane (Fig. 3f).

A second rectangular measure frame is automatically displayed on the image from the medial to the lateral side of the olfactory bulb and from its dorsal pole to its ventral pole (Figs. 3h and 4). The height of the frame is variable and is adapted to the maximal size of the glomerulus (say, approximately 150 μm in diameter for the mouse and 250 μm in diameter for the rat). For each position of the frame on the autoradiogram, the mean OD value is then calculated inside the outlines of the glomerular layer defined by the binary mask. A detailed diagram of the measure frame scanning is given in Fig. 4. For each autoradiogram analyzed, two ranges of P glomerular OD values (one on the medial side and one on the lateral side) are obtained and stored.

For a visual representation, the data are reanalyzed on a separate computer. The two ranges P of glomerular OD values are spread out on both sides of the dorsal pole of the olfactory bulb so as to obtain a linear profile of the glomerular activity (Fig. 5a). The juxtaposition of n profiles corresponding to n autoradiograms analyzed leads then to the reconstruction of the

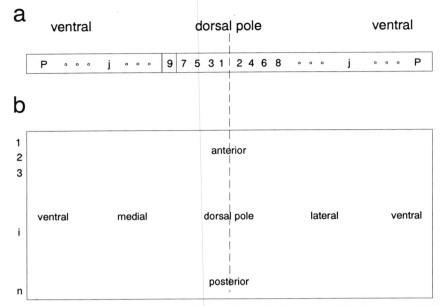

FIG. 5   Spread-out representation of the glomerular labeling. (a) Alignment of $2P$ OD values on both sides of the dorsal pole. (b) n sections are analyzed in each olfactory bulb. The corresponding OD values are arranged with reference to the dorsal poles, which are aligned vertically. [After Royet *et al.* (3).]

entire surface of the labeled glomerular layer (Fig. 5b). For each map, data X_{ij} are normalized into standardized Z scores according to the following transformation:

$$Z_{ij} = (X_{ij} - X)/\sigma$$

where X_{ij} is the jth OD value of the ith section, X the mean glomerular OD value, and σ the standard deviation. Each map of standardized OD values is converted to a graphic one (Fig. 6). Numerical data are grouped into 10 classes, from 0 to 9, each symbolized by a square whose side is proportional to the rank of the class. As a consequence, a two-dimensional unfolded representation of the glomerular layer is obtained which indicates the most highly labeled regions by the presence of the largest squares. It is also possible to perform an illustration by using color coding.

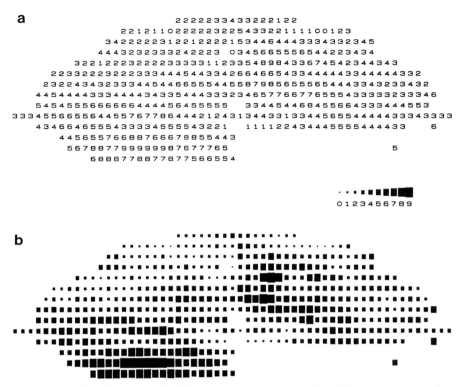

FIG. 6   (a) Visualization of 2-DG glomerular patterns. The OD values arranged as shown in Fig. 5 are grouped within 10 classes numbered from 0 to 9. (b) A graphic representation is achieved by replacing the numerical values by squares of relative sizes. [After Royet *et al.* (3).]

Quantitative Analysis

The next stage is to encode the information extracted from the image in order to perform a statistical analysis. The investigator can then compare the "maps" of neural activity resulting from animals submitted to different experimental conditions for statistical analysis (3).

Image Analysis in Morphometry

The estimation of the size or the number of various features in the olfactory bulb (such as mitral cells or glomeruli) is generally performed on histological sections, that is, two-dimensional images. To relate measurements of these

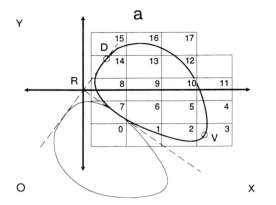
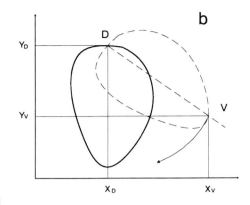

FIG. 7   (a) Scanning boustrophedon pattern on a histological section. (b) Illustration of moving a section around its dorsal pole (D) to vertically align the dorsoventral ($D–V$) axis. O, Origin of the coordinate system; R, reference origin of the coordinates of glomeruli or mitral cells for a section. [After Royet *et al.* (23).]

structures obtained on the section plane to their three-dimensional parameters, it is necessary to have recourse to stereology (15, 20–22). To perform measurements, it is necessary to use statistical procedures of sampling. This step in data acquisition can be performed manually (e.g., point-counting method) or by using the semiautomatic or automatic methods characteristic of image analysis systems.

Image Analysis System

Image analysis is carried out by means of a Quantimet 570 with a CCD camera placed on a Zeiss Universal microscope and using objectives ×6.3 for glomeruli and ×25 for mitral cells. This optical system gives a magnification for glomeruli of ×281 (each pixel corresponds to a 1.403-μm square) and for mitral cells of ×1075 (each pixel corresponds to a 0.349-μm square). A micrometer scale is used to calibrate the measurement. Measurements are made in a series of adjacent fields in the histological section by controlling the stage movement in both the X and Y directions. A program is developed in such a way that the series of fields is scanned in a rectilinear movement and describes a so-called "boustrophedon pattern" (Fig. 7a). This program allows a choice between analysis of the field visualized on the monitor screen or access to the next adjacent field on the same line or on the following line.

The control of X and Y directions requires the definition of a system of coordinates. The range of movement of the motorized stage is defined by its mechanical construction. However, with the aid of microswitches, the travel of the stage is restricted to the working area corresponding approximately to the size of slides (say, 75 × 25 mm). An origin (O) is then set up within this region (Fig. 7a), and all stage positions are related to this origin. In addition, because each section can be placed in very different areas of the slide, a reference coordinate system is determined which sets the same order of magnitude for the coordinates of all sections (say, within 4–5 mm only instead of 75 × 25 mm). To perform this, a reference origin (R) is chosen which is defined by the intersection of a line tangent to the medial edges of the olfactory bulbs with the line joining the two dorsal (D) poles (Fig. 7a, dashed lines). Finally, before analyzing each section, two other X and Y coordinates are also recorded, situated in the most ventral (V) and dorsal (D) parts of the glomerular layer. They subsequently allow each section to be vertically aligned with all other sections by correcting its orientation (Fig. 7b). To record all the reference coordinates, the measure frame size is reduced at a square of 100 pixels of surface (Fig. 8a). The image of the section on the monitor screen is then displaced by means of a joystick in order to place the measure frame in reference to these characteristic coordinates. The scanning boustrophedon pattern is started from the left extremity and the lowest part of the glomerular layer (field 0, Fig. 7a). To determine the coordinates of this field, the procedure of using a small measure frame (100 pixels) at the leftmost X coordinate and the lowest Y coordinate is again used.

When using image analysis for the measurement of particle size distributions, the resulting data must be protected from frame edge errors (15). That is, when a feature overlaps two scanning fields, it is visualized separately and partially in two adjacent frames. As a consequence, the number of features is overestimated (two features are counted instead of one) and the surface is underestimated. To avoid this error, it is necessary to define a "guard region" (Figs. 8b,d and 9) on the monitor screen: the limits of each field of visualization are given by the outlines of the image frame; the limits of the usual field of measurement are given by the outlines of the measure frame. Features are then taken into account by the analysis algorithm only if their lowest right-hand corner lies within the measure frame. For glomeruli, the guard region size is fixed at 70 pixels (i.e., a 98.21 μm width). For mitral cells, it is fixed at 50 pixels (i.e., 17.45 μm width). This size corresponds approximately to the diameter of the maximum size of the object under study. For the morphometric study of glomeruli, the size of the measure frame is a 620 μm width and for mitral cells, it is a 165 μm width. Finally, the program is arranged so that the X position of the next adjacent field is equal to the width, expressed in microns, of the measure frame (and not the image frame). The Y position of fields of the next line is a function of the measure frame

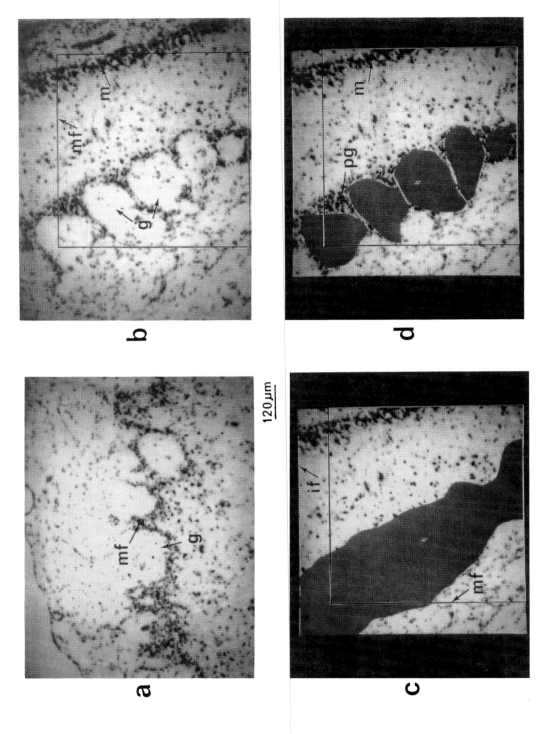

120μm

432

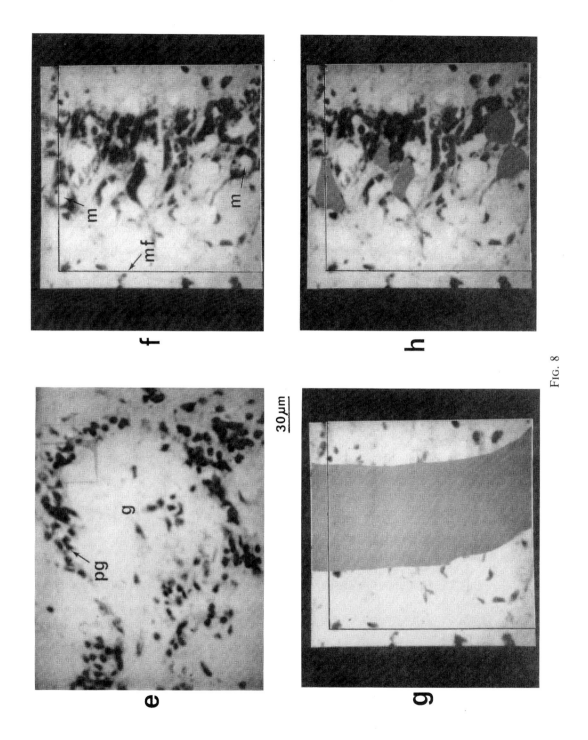

30µm

Fig. 8

433

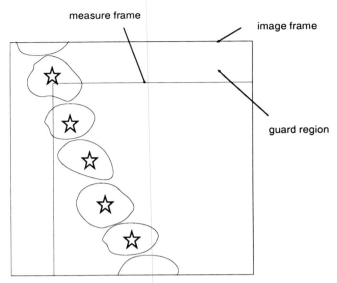

FIG. 9   The edge effect and the guard region. Parameters of features marked by a star are measured.

height. Inside each field, the glomerular layer or mitral cell layer outlines are manually delineated on the monitor screen by interacting with a mouse and the area of the layers is computed (Fig. 8c,g). All glomeruli or mitral cells are accurately delineated, and their areas, along with the X and Y coordinates, are computed (Fig. 8d,h). Figure 8e represents one glomerulus visualized with the $\times 25$ objective used for the analysis of mitral cells, therefore, with a high magnification.

The X and Y coordinates of each feature are defined as the point with the lowest Y value and largest X value. These parameters and the feature area are then used to visualize the spatial distribution of glomerular or mitral cell profiles on a separate microcomputer. To do this, the orientation of sections must be corrected in order to align vertically the dorsoventral ($D-V$) axis of the olfactory bulbs (Fig. 7b) as previously described (23). The corrected X_f and Y_f coordinates of each feature f are calculated as follows:

FIG. 8   Successive operations used in the morphometric study for analyzing images of glomeruli and mitral cells of the rabbit olfactory bulb. g, Glomerulus; pg, periglomerular cells; m, mitral cells; if, image frame; mf, measure frame. See text for details.

$$X_f = x_f \cos \alpha - y_f \sin \alpha$$

$$Y_f = y_f \cos \alpha + x_f \sin \alpha \qquad (5)$$

where x_f and y_f are the coordinates of the feature recorded on the image analysis system. The angle α is defined in radians as arctangent α, with tangent α:

$$\text{tangent } \alpha = (Y_D - Y_V)/(X_V - X_D) \qquad (6)$$

where X_V, Y_V, X_D, and Y_D are, respectively, the coordinates X and Y of the most ventral and the most dorsal glomeruli in each histological section.

Because of the small size of the field of analysis (say, $620 \times 620 \ \mu$m), the determination of the dorsal and ventral reference coordinates are only approximative. As a consequence, the orientation of sections can be over- or undercorrected. To resolve this problem, the two reference coordinates are again determined when all characteristics components of one section are visualized on the computer screen (Fig. 10). By managing the cursor function from the alphanumeric keyboard of the computer, a cross is displaced on the screen, positioned at the specific level (the most dorsal or ventral glomeruli), and the corresponding coordinates then recorded. The same procedure is used to correct the orientation of the mitral cell layer, also taking into account the previously established orientation of the glomerular layer. The spatial distributions of glomerular and mitral cell profiles can then be represented simultaneously (Fig. 11).

Comments

The image analysis method for mapping neural metabolic activity has been adapted for the analysis of autoradiograms of the olfactory bulb. Although it is described here for a given neuroanatomical method and a specific nervous structure, the method could be readily adapted to other structures and other neuroanatomical techniques. In the olfactory bulb, for example, other neuro-anatomical methods, also based on the principle of the measurement of transmittance, could be used, such as cytoarchitectonic investigations, immunocytochemistry, and quantitative autoradiography of transmitter binding sites. Such methods have already been developed for the analysis of other regions of the nervous system (24–26).

The method for mapping morphometric data as presented in this chapter can be used in any region of the nervous system. It allows the determination

FIG. 10   Accurate determination of coordinates (see arrows) of dorsal and ventral parts of one section on the microcomputer screen. Open circles mark bad positions of reference glomeruli, and crosses represent corrected positions.

of experimentally induced morphological changes by, for example, sensory deprivation, learning, or changes due to dynamic processes of neurogenesis. Owing to the systematic interative use of the mouse, this method, however, requires some time and effort during data acquisition. Therefore, it is not suitable for routine analyses which require automated analysis (27–29), indispensable, for example, for clinical studies. In general, the decision to develop a procedure for automated or semiautomated image analysis should be a function of the frequency of subsequent applications, the intrinsic "quality" of the biological material, and the power of available image analysis systems. For example, in fundamental research, the frequency of applications may be limited to one study only. The "quality" of the biological material depends on the contrast of cells relative to the background and the distinction of these cells from different neighboring cells.

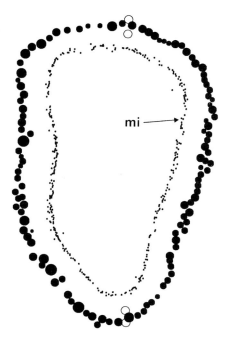

FIG. 11   Simultaneous spatial distributions of glomerular and mitral cell profiles of
the two bulbar sections (sixth sections) when they are correctly oriented. mi, Mitral
cells. Open circles mark coordinates of dorsal and ventral glomeruli in the two
sections.

In the present application, for example, glomeruli and mitral cells are not
very easily extractable. Finally, the power of an image-analysis system
depends on its capacity for image processing (transfer functions, filtering,
etc.). For this purpose, let us note the recent development of the gray-
level mathematical morphology by the Ecole des Mines de Fontainebleau
(France). Corresponding algorithms are sufficiently efficient to separate
the pertinent information in realistic time scales and are nowadays available
on the Quantimet 570. In brief, the question is, Must more time be spent
in analyzing the image or in collecting the data? The researcher must
therefore adapt a strategy which takes into account these experimental
constraints. In any event, stereological principles must also be taken into
account as they provide the means for a statistically nonbiased assessment
of the experimental data. For this purpose, a knowledge of recently
developed stereological methods is indispensable.

Acknowledgments

The work on the Quantimet 570 was carried out at the Centre de Quantimétrie de l'University Claude-Bernard Lyon 1, 8 Avenue Rockfeller, 69373 Lyon Cédex 8, France. The histological material used for the development on the Quantimet 570 of the image analysis method in morphometry were histological sections of rabbit olfactory bulbs furnished by Professor H. Distel of the Institut für Medizinische Psychologie, Universität München, Goethestrasse 31, 8000 München 2, Germany. I am very grateful to A. Mackay-Sim, A. M. Mouly, C. Souchier, and S. Weis for their very thorough remarks and assistance in writing this chapter.

References

1. T. Nowaczyk and M. H. Des Rosiers, *Eur. Neurol.* **20,** 169 (1981).
2. L. Astic and D. Saucier, *Dev. Brain Res.* **2,** 141 (1982).
3. J. P. Royet, C. Souchier, and F. Jourdan, *Brain Res.* **417,** 1 (1987).
4. A. Stein, S. Juliano, P. Karp, and P. Hand, *J. Neurosci. Methods* **10,** 189 (1984).
5. L. Sokoloff, M. Reivich, C. Kennedy, M. H. Des Rosiers, C. S. Patlak, K. D., Pettigrew, O. Sakurada, and M. Shinohara, *J. Neurochem.* **28,** 897 (1977).
6. C. Souchier, *Techniques de l'Ingénieur, Analyse Chimique et Caractérisation* **1,** 1 (1984).
7. F. R. Sharp, J. S. Kauer, and G. M. Shepherd, *J. Neurophysiol.* **40,** 800 (1977).
8. J. S. Kauer, D. Reddy, R. B. Duckrow, and G. M. Shepherd, *J. Neurosci. Methods* **11,** 143 (1984).
9. L. C. Skeen, *Brain Res.* **124,** 147 (1977).
10. W. B. Stewart, J. S. Kauer, and G. M. Shepherd, *J. Comp. Neurol.* **185,** 715 (1979).
11. L. Astic and M. Cattarelli, *Brain Res.* **245,** 17 (1982).
12. L. Astic and D. Saucier, *Dev. Brain Res.* **2,** 243 (1981).
13. F. Jourdan, *Brain Res.* **240,** 341 (1982).
14. M. T. Shipley, J. Luna, and J. H. McLean, *in* "Neuroanatomical-Tract Tracing Methods" (C. L. Heimer and L. Zaborsky, eds.), p. 331. Plenum, New York, 1989.
15. J. C. Russ, "Computer-Assisted Microscopy: The Measurement and Analysis of Images." Plenum, New York, 1990.
16. G. Matheron, "Random Sets and Integral Geometry." Wiley, New York, 1975.
17. J. Serra, "Image Analysis and Mathematical Morphology," Vol. 1. Academic Press, London, 1982.
18. J. Serra, "Image Analysis and Mathematical Morphology," Vol. 2. Academic Press, London, 1988.
19. M. Coster and J. L. Chermant, "Précis d'Analyse d'Images." Ed. C.N.R.S., Paris, 1985.
20. E. R. Weibel, "Stereological Methods: Practical Methods for Biological Morphometry," Vol. 1. Academic Press, New York, 1979.

21. E. R. Weibel, "Stereological Methods: Theoretical Foundations," Vol. 2. Academic Press, New York, 1980.
22. J.-P. Royet, *Prog. Neurobiol. (Oxford)* **37,** 433 (1991).
23. J.-P. Royet, F. Jourdan, H. Ploye, and C. Souchier, *J. Comp. Neurol.* **289,** 594 (1989).
24. A. Schleicher and K. Zilles, *J. Microsc.* **157,** 367 (1990).
25. A. Schleicher, K. Zilles, and A. Wree, *J. Neurosci. Methods* **18,** 221 (1986).
26. M. Zoli, I. Zini, L. F. Agnati, D. Guidolin, F. Ferraguti, and K. Fuxe, *Neurochem. Int.* **16,** 383 (1990).
27. P. Ahrens, A. Schleicher, K. Zilles, and L. Werner, *J. Microsc.* **157,** 349 (1990).
28. F. Rauch, A. Schleicher, and K. Zilles, *J. Neurosci. Methods* **30,** 255 (1989).
29. F. Rauch, A. Schleicher, and K. Zilles, *Anat. Embryol.* **181,** 373 (1990).

Section III

Data Modeling and Simulations

[27] Numerical Simulation of Cytosolic Calcium Oscillations Based on Inositol 1,4,5-Trisphosphate-Sensitive Calcium Channels

Stéphane Swillens

Introduction

Cuthbertson and Cobbold (1) have shown that the Ca^{2+} signal in stimulated oocytes consists of pulsatile oscillations. Numerous publications have generalized this concept to other electrically nonexcitable cells, for instance, macrophages, hepatocytes, fibroblasts, endothelial cells, chromaffin cells, and pancreatic and parotid acinar cells [see Jacob (2) for a review]. Hormone-induced Ca^{2+} mobilization from intracellular stores is mediated by inositol 1,4,5-trisphosphate (IP_3), which accumulates owing to the accelerated hydrolysis of phosphatidylinositol 4,5-bisphosphate by phosphoinositidase C (3). More recently, the cerebellar IP_3 receptor has been structurally and functionally characterized (4). It consists of four identical monomers which form an IP_3-sensitive Ca^{2+} channel embedded in the membrane of intracellular vesicles.

The oscillatory pattern of the Ca^{2+} signal is not unique. In hepatocytes, the frequency of the Ca^{2+} spikes, but not their amplitude, is modulated by the vasopressin concentration (5). In pancreatic acinar cells, the amplitude increases with cholecystokinin concentration (6). Moreover, a hormone stimulus produces various responses in different cells of the same cell population, but the same cell consistently exhibits the same response to a further stimulation (6, 7). The biochemical mechanism which produces the hormone-induced intracellular Ca^{2+} oscillations has not yet been demonstrated. Several explanations have been tentatively proposed and are based on the existence of feedback reactions involving cross-coupled components such as IP_3, free Ca^{2+}, and intravesicular Ca^{2+}. Because of the necessary complexity of such a system, it is quite difficult to predict intuitively that a proposed model may function as a Ca^{2+} oscillator. Even in the simplest cases, the ranges of parameter values for which an oscillatory pattern may be observed cannot be determined without the help of numerical simulations requiring computer facilities.

Theoretical simulations usually consist of four steps: (1) definition of the model deduced from current concepts and experimental observations, (2) translation of the biochemical reaction network in mathematical equations generally based on the laws of chemical kinetics, (3) elaboration of a numerical procedure for solving the equations, and (4) execution of this procedure by a computer using different sets of parameter values. At this stage, an apparent compatibility between the data generated by the theoretical simulation and the experimental results is in fact a demonstration that the proposed model is an acceptable description of reality. However, that does not prove that the proposed description is correct. Further experimental tests are required in order to validate the model. It must be emphasized that this theoretical approach is the only means to substantiate a claim that a proposed explanation can support the observations.

In the case of the pulsatile IP_3-induced Ca^{2+} mobilization, several possible explanations have been proposed in the literature, but only five have been tested by numerical simulation (8–12). Three models rely on a feedback action of free Ca^{2+} on IP_3 metabolism (8, 10, 11) leading to concerted oscillations of IP_3 and Ca^{2+}. The two other models require continuous and bidirectional Ca^{2+} exchanges between the cytosol and the extracellular medium (9, 12) but do not need any IP_3 oscillation. The aim of this chapter is to illustrate the theoretical simulation approach by studying a new model of the IP_3-sensitive Ca^{2+} channel accounting for the biphasic kinetics of IP_3-induced Ca^{2+} release.

Elaboration of Model

It has been demonstrated that the binding of IP_3 to the receptor induces the opening of the Ca^{2+} channel (4), allowing a passive transport of Ca^{2+} across the vesicle membrane. Thus, the rate of Ca^{2+} efflux depends at least on the IP_3 concentration and on the gradient of free Ca^{2+} concentration across the membrane. Experimental observations have shown that IP_3-induced Ca^{2+} release exhibits biphasic kinetics (13, 14): a short phase of fast Ca^{2+} release is followed by a prolonged phase of slow efflux which is mainly accounted for by an IP_3-independent leak across the membrane. Moreover, successive additions of IP_3 provoke repetitive bursts of Ca^{2+} release (15). Thus, the IP_3 receptor seems to desensitize as long as it is submitted to a constant IP_3 concentration, but is able to resensitize quite rapidly when the IP_3 concentration is abruptly increased.

To account for this kind of desensitization, a model has been proposed and simulated (16). It consists of a IP_3 receptor–Ca^{2+} channel molecule which can exists in two interconvertible forms termed A and B. The IP_3-bound forms are noted A_I and B_I, respectively. The A_I form represents the open Ca^{2+} channel, whereas the B_I form is inactive despite the presence of IP_3.

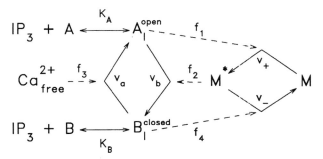

FIG. 1   Model of the IP$_3$-sensitive Ca^{2+} channel. IP$_3$ binds to both forms of the Ca^{2+} channel. The channel interconversion may occur only when the binding site is occupied by IP$_3$ (reactions v_a and v_b). The IP$_3$-induced Ca^{2+} release is dynamically controlled by the memory molecule M which exists in two interconvertible forms (reactions v_+ and v_-). The inactivation of the open channel form A$_I$ is accelerated by the active molecule M* (reaction f_2), the formation of which is promoted by A$_I$ (reaction f_1). Cytosolic Ca^{2+} favors the accumulation of A$_I$ (reaction f_3). A further control consists of the inactivation of M* by the inactive channel form B$_I$ (reaction f_4).

To account for the special kind of desensitization described above, the mechanism presented in Fig. 1 has been proposed (16): a memory molecule M is slowly converted by the open channel A$_I$ to an active form M* (reactions v_+ and f_1). The M* form accelerates the conversion of A$_I$ to B$_I$ (reactions v_b and f_2) leading to receptor desensitization. Thus, the model involves a dynamic control of the Ca^{2+} channel opening. For a submaximal IP$_3$ concentration, only a fraction of the receptor is in the A$_I$ and B$_I$ forms. The steady state corresponding to the desensitized situation is characterized by low concentrations of A$_I$ and M* and a high concentration of B$_I$. Further IP$_3$ addition displaces the binding equilibrium in favor of B$_I$. Because the B$_I$ concentration increases, the rate v_a of A$_I$ formation may overwhelm the rate v_b of B$_I$ formation; thus, the excess B$_I$ molecules are converted to the active A$_I$ form, allowing an increase of Ca^{2+} efflux. More active M* molecules are then slowly produced, leading to a further delayed inactivation of the Ca^{2+} channel.

It has been suggested that biphasic IP$_3$-induced Ca^{2+} release could explain sustained Ca^{2+} oscillation (15). As shown in the proposed model, the dynamic regulation of channel activity involves a negative feedback loop between the active forms of the channel (A$_I$) and the memory molecule (M*). Theoretical studies on the nature and multiplicity of stationary states came to the conclusion that sustained oscillations may be produced by an adequate combination of a negative feedback loop and an autocatalytic process (17). Therefore, the model defined above could contain a Ca^{2+} oscillator if an autocatalytic loop is introduced in the system. Recent experimental studies have shown that free cytosolic Ca^{2+} may act as a coagonist of IP_3-induced Ca^{2+} release in synaptosome-derived microsomal vesicles and in endoplasmic reticulum vesicles from cerebellum (18, 19). To account for this observation, the proposed model assumes that free cytosolic Ca^{2+} accelerates the conversion of B_I to A_I (reactions v_a and f_3). This positive control may be interpreted as an autocatalytic process since the cytosolic Ca^{2+} favors the formation of open channels and thus increases the Ca^{2+} efflux from the vesicle.

On the basis of the feedback loops defined so far, it can be expected that the model could generate sustained Ca^{2+} oscillations, at least for certain combinations of parameter values. Numerical simulation now has to confirm this prediction and to characterize the oscillation patterns.

Mathematical Description of Model

To describe the time course of the different components of the system, the laws of the chemical kinetics are used. It is assumed that the total intracellular concentrations of IP_3 receptor–Ca^{2+} channel molecules, memory molecules, and Ca^{2+} are constant.

Inositol 1,4,5-Trisphosphate Binding to Receptor

For the sake of simplicity, IP_3 binding is assumed to obey fast Michaelian kinetics. Thus, the binding of IP_3 to both forms A and B of the receptor is always at equilibrium. Defining A_{tot}, B_{tot}, and Ch_{tot} as the total concentrations of the A form, of the B form, and of both forms, respectively, one has

$$A_I = A_{tot}/r_a \tag{1}$$

$$B_I = B_{tot}/r_b \tag{2}$$

$$Ch_{tot} = A_{tot} + B_{tot} \tag{3}$$

with

$$r_a = 1 + K_A/[IP_3] \tag{4}$$

$$r_b = 1 + K_B/[IP_3] \tag{5}$$

where K_A and K_B are the equilibrium dissociation constants of the IP_3 binding to each channel form.

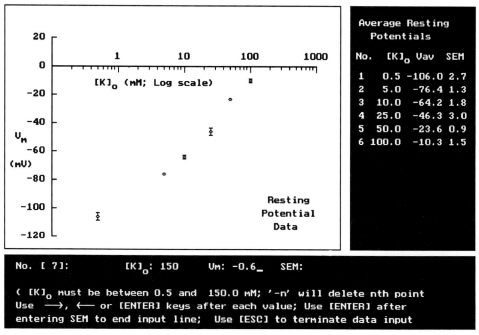

FIG. 5 Monochrome reversed-video graphics dump of the MEMPOT program screen showing the semilogarithmic axes for reentering the data points previously determined during the Resting Potential Measurements exercise. Details of the data points plotted are given in the right-hand panel. Points may be edited by again entering data with the same data point number or deleted by entering a negative data point number.

termination of the program, it is then suggested that users replot their data manually on the special paper provided with the program. A separate program was developed to produce this semilogarithmic graph paper on a Hewlett-Packard* LaserJet printer, in order to vary line thicknesses and styles, so that the resultant graph paper would be of a type that could be readily reproduced by a monochrome photocopier (e.g., Fig. 8).

* Registered trademarks, Hewlett-Packard (Hewlett-Packard Company); IBM (International Business Machines Corporation); Epson (Seiko Epson Corporation); Hercules (Hercules Computer Technology); Macintosh (Apple Computer Inc.); and Turbo C (Borland International).

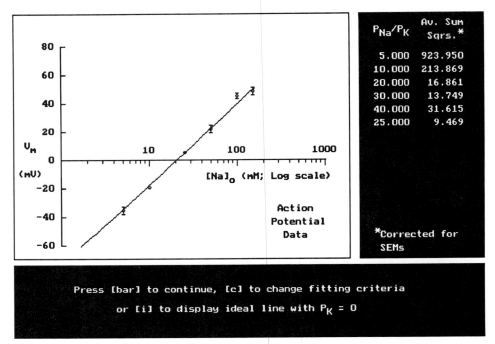

FIG. 6   Monochrome reversed-video graphics dump of the MEMPOT program screen showing the trial permeability fitting of data points previously determined during the Action Potential Measurements exercise. Relative permeability values (P_{Na}/P_K) are chosen and the predicted curve calculated using the Goldman–Hodgkin–Katz equation. The averaged sum of the squares of the deviations of the points from the predicted line are given in the right-hand panel, along with the accompanying P_{Na}/P_K value (generally this sum of squares value is corrected for the SEMs of the points, although this default option may be changed using the setup program or by pressing the [c] key option, shown here, at run time). Thus, in addition to visual judgment, goodness of fit may be gauged more objectively by choosing the relative permeability value that minimizes the sum of squares value. The data can also be compared with an ideal line, for which $P_K = 0$, by pressing the [i] key option.

Program Requirements

The program requires an IBM* PC/XT/AT, 80386SX, 80386, 80486, or compatible computer with a color or monochrome monitor. The graphics cards can be either CGA/EGA/VGA or Hercules* (the CGA graphics are displayed in their highest resolution black and white mode). Best results are obtained on an IBM AT-compatible (or 80386, 80386SX, or 80486) computer with an

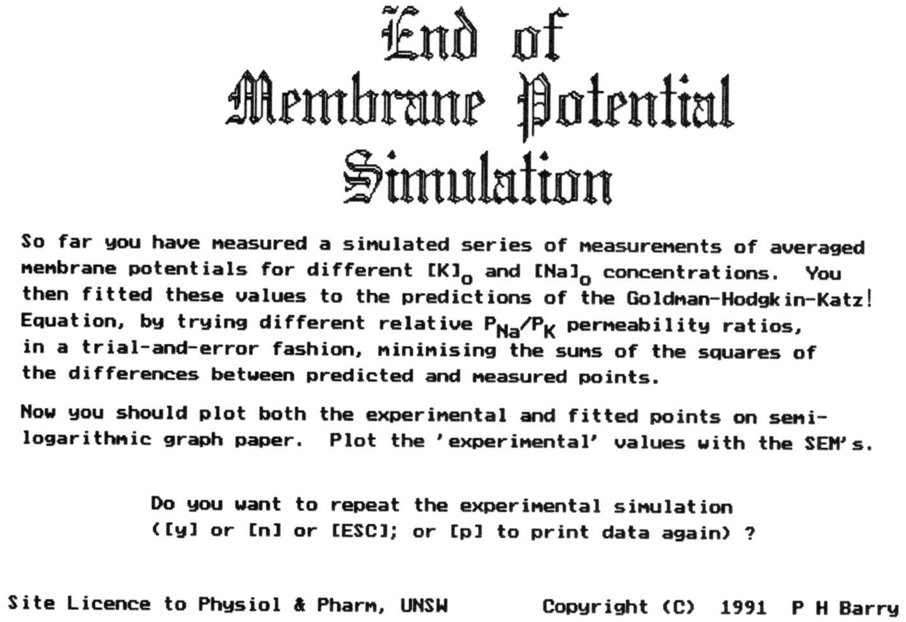

End of
Membrane Potential
Simulation

So far you have measured a simulated series of measurements of averaged
membrane potentials for different $[K]_0$ and $[Na]_0$ concentrations. You
then fitted these values to the predictions of the Goldman-Hodgkin-Katz!
Equation, by trying different relative P_{Na}/P_K permeability ratios,
in a trial-and-error fashion, minimising the sums of the squares of
the differences between predicted and measured points.

Now you should plot both the experimental and fitted points on semilogarithmic graph paper. Plot the 'experimental' values with the SEM's.

        Do you want to repeat the experimental simulation
        ([y] or [n] or [ESC]; or [p] to print data again) ?

Site Licence to Physiol & Pharm, UNSW   Copyright (C) 1991 P H Barry

FIG. 7   Monochrome reversed-video graphics dump of the final MEMPOT program
screen, at which point the user can repeat the whole program with the other exercise
option, print out both the data and best-fitted permeability predicted values, or
terminate the program.

EGA or VGA card and a color monitor. For printouts, an Epson-compatible*
printer or virtually any other printer handling MS DOS print commands can
be used. The program will also run on the Macintosh* II series of computers
operating under SoftPC emulation software. It will take advantage of a maths
coprocessor, but its presence is not mandatory. The program requires the
presence at run time of a department-copyable MEMPOT.KEY registration
file, which is supplied with the program.

   The program also checks for errors in printer communication. In the event
of a communication problem, an error message is displayed and the user is
given the opportunity of checking the printer and if necessary redirecting the
print data to a disk file. In particular, MEMPOT has been designed to also
work in a network situation. A ''waiting to print'' message is displayed once
the print option has been chosen followed by a ''printing'' message, once
successful printer communication has started, which remains until all the
data have been completely sent to the printer. In addition, the keyboard

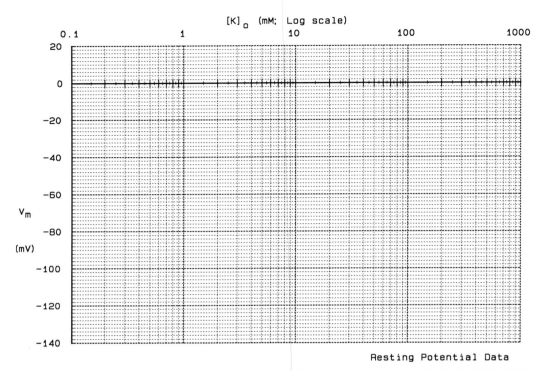

FIG. 8   Example of the special semilogarithmic graph paper provided with MEMPOT for manually plotting both data points and the predicted curve with the best-fitted permeability value. A customized C program was developed to produce such a semilogarithmic graph directly on an HP LaserJet printer, in order to vary line thicknesses and styles, so that the resultant graph would be of the type that could be readily copied by a monochrome photocopier.

buffer is kept cleared and will not process any more input during this time, to avoid the inadvertent printing of multiple copies of output by an impatient user. Furthermore, the MPSETUP program will allow the allocation of an identification number to be assigned to each computer, which will subsequently be printed at the top of each printout, to allow networked printouts to be identified by users of different computers.

Subsidiary Associated Programs

Apart from an automatic installation program, there are two other subsidiary programs written to accompany the MEMPOT program. MPSETUP was designed to modify a data file to reset default data values, to allocate the computer identification number, and to alter some default program options. The default data values include temperature of simulated experiment; values of internal concentrations $[K]_i$ and $[Na]_i$; initial values of $[K]_o$ and total external cation concentrations; and the minimum value of $[Na]_o$ used in the action potential measurements. In addition, it can also alter certain default decisions. In the resting potential measurements, either the total $[K]_o$ + $[Na]_o$ value can be kept constant or $[Na]_o$ can be set at a certain lower than normal value, the balance being made up by choline, which can be adjusted depending on the particular value of $[K]_O$ chosen during execution of the program. In addition, two other default options can be altered. One of them results in either applying a correction for the SEMs of data values when fitting the data or not, whereas the other results in either the display of equilibrium potentials after every cell impalement or just after each set of cell values has been averaged.

MPVIEW is a simple textfile reading program, designed to allow the data print files to be easily read in the event of a printer not being readily available.

Generation of Simulated Data

To simulate the basic membrane potential properties of an excitable cell, the program relies on the applicability of the Constant Field (Goldman–Hodgkin–Katz) Equation (2, 7). Equation (1) gives the membrane potential of a

$$V_m = \frac{RT}{F} \ln \left(\frac{[K]_o + P_{Na}/P_K [Na]_o}{[K]_i + P_{Na}/P_K [Na]_i} \right) \tag{1}$$

cell during zero net current flow. The equation, as originally derived, assumed that the electrical potential gradient across the membrane was constant and that the individual ions moved independently of one another. The above form of the equation assumes that no other permeant cations are present and that all anions are essentially impermeant, a reasonable approximation, particularly in the case of many nerve cells (or, as in the case of amphibian skeletal muscle fibers, that any permeant anions, like Cl^-, are in equilibrium). Under such conditions, a large number of a different permeation models still predict the same equation (e.g., Ref. 8).

In particular, Hodgkin and Horowicz (1) showed that the equation fitted the resting membrane potential dependence of amphibian skeletal muscle fibers very well. They also confirmed their fitted P_{Na}/P_K value of about 0.01 agreed with other different flux measurements. It would also be expected that the equation would be applicable to the (zero-current) peak of the action potential, and this has also been demonstrated by Hodgkin and Katz (2) for the squid axon and Nastuk and Hodgkin (3) for frog skeletal muscle. In particular, the data of Nastuk and Hodgkin for the peak of the action potential may be seen to fit the predictions of the Goldman–Hodgkin–Katz Equation very closely over the range of 17.5 to 116.5 mM [Na]$_o$ with a P_{Na}/P_K value of about 8.6. It could be argued that the presence of an electrogenic component from the Na/K pump contributing to the membrane potential might invalidate the use of this equation. However, it has been shown by Mullins and Noda (9) [see also Sjodin (10)] that such electrogenic contribution could be incorporated into the above equation without changing its form. The real passive P_{Na}/P_K value would simply be multiplied by a flux ratio factor ϕ (active K$^+$ flux/active Na$^+$ flux) to give the operational "P_{Na}/P_K" $= \phi P_{Na}/P_K$. Such a contribution would simply affect our interpretation of the relative permeability term and not the ability of the equation to fit the data.

The action potential was generated from an array of values chosen to depict a somewhat typical action potential waveform. For a particular cell in a specific solution, the values of the peak and after-hyperpolarization of the action potential are first calculated and then the array is scaled to fit those values. This produces a very fast display of the action potential waveform.

To allow for a range of different cell types and to present the students with a permeability ratio, whose precise value was unknown for these cells, the mean value of P_{Na}/P_K was randomly varied between 0.004 and 0.016 for the resting potential measurements and between 10 and 30 for the peak of the action potential. This randomization of the mean P_{Na}/P_K was achieved at run time by first randomizing the program and then by the application of a pseudo-random generator routine. In addition, to provide some apparent "physiological" cell-to-cell variation, randomness was built into the measurements in the following ways. First, the precise value of the internal composition of the cells was varied randomly within a small range ([K]$_i$ ± 5 mM and [Na]$_i$ ± 2.5 mM) to simulate possible variations in metabolic activity within each cell. In addition, the precise permeability ratio for each cell was obtained by taking the mean P_{Na}/P_K already determined and randomizing that value by ±40%. In addition, a random potential increment of up to about ±6 mV was added to each simulated potential measurement, the increment being scaled down as the predicted uncorrected potential approached zero. Furthermore, ±2 mV of electrical noise was added to each of the individual simulated oscilloscope traces. Averaging the traces during program operation could, of course, be seen to radically reduce this simulated electrical noise, as in the real laboratory situation. The whole randomization procedure means that each program user should get different values of permeabilities each time the program is run, a feature designed to make the exercise both more realistic and more interesting.

Typical internal and external ionic concentrations (e.g., Ref. 11) were used as default mammalian values, but these can be changed using the MPSETUP program.

Programming Principles

Program Overview

The program was written in Turbo* C (Version 2 and later compiled using the Borland C++ compiler, Ver 3.0; Borland International, Scotts Valley, CA) in eight separate program modules and compiled and linked using the Huge Memory Model. In every case throughout the program (apart from the final print screen and when optionally setting up printfile redirection), all screen display and text generation is done in the graphics mode. The size of the executable file is about 196 kilobytes.

The main sequence of program steps is indicated in the flow chart shown in Fig. 9. Various command line switch arguments can be used to force the program to work in CGA mode ('/c'), EGA mode ('/e'), VGA mode ('/v'), black and white ('/w'), to suppress sound ('/n') or only use beeps ('/b'), to suppress temporary file generation ('/s'), or to redirect print output to files ('/f') instead of to a printer. The program then reads a default setup parameters file (if available, otherwise the original default values are used). This file can be modified and regenerated separately with the MPSETUP program. A random seed value is then created for future random number generation, in order to make the program produce different sets of values each time it is executed. The program then checks whether or not the graphics card is EGA, VGA, or Hercules and whether the monitor is color or monochrome. (Note that CGA has to be specified in the command line switch, i.e., '/c'; similarly, VGA graphics will normally be set to EGA resolution and even in forced VGA mode will display main text screens in EGA resolution to obtain optimum text size and clarity). Dynamic memory is then allocated at this stage for many of the parameter arrays. The simulated action potential shape array is then set up and the underlying relative P_{Na}/P_K permeabilities for the resting potential, the action potential, and the after-hypolarization are each set up randomly (using the Turbo C pseudo-random number generator) within a reasonable range of values appropriate for each parameter. This is followed by a display of the title and (optional) instruction screens.

The program user then chooses which of the two exercises should be executed (thus setting either the resting potential or action potential flags). The graphics display of cells and electrodes are then set up (with the reference electrode only being displayed at this stage), together with the oscilloscope (CRO) and solution panels (with details about the ionic composition of the

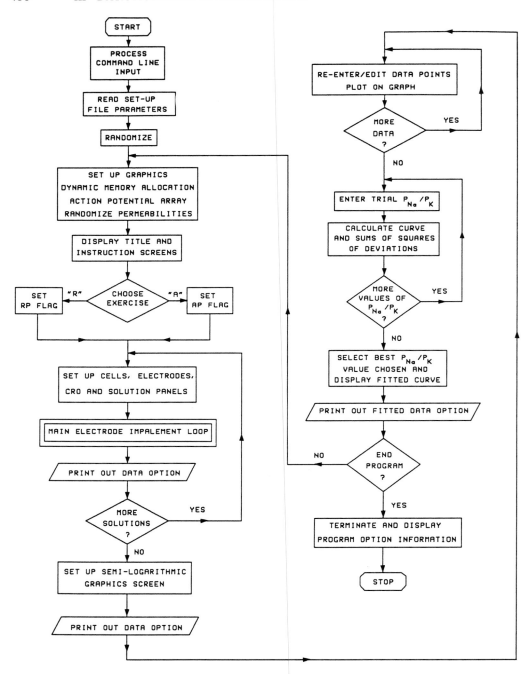

cells). The computer then waits for the keyboard entry of a solution value (Fig. 2), after which the reference and intracellular electrodes are set up. The program then polls the keyboard for one of the arrow cursor key codes. When received, this starts program control moving around the main electrode impalement loop (Fig. 10), as discussed in the next section. In each case, the input of a 'p' key code after the electrode impalements will allow the printout (or file output) of all the data values so far obtained, before control returns to the outer loop. This outer loop is repeated until enough solutions have been chosen. It may be terminated then, or at any earlier time, by pressing the ESC key.

The computer then displays a semilogarithmic graphics screen with either four log cycles for resting potential $[K]_o$ values or three for action potential $[Na]_o$ values. Again, a prompt allows one last opportunity for obtaining a printout of the data. The data are now entered with a simple re-entering and editing data entry loop using the arrow cursor keys. Each time the data point and its error information is entered, the point is plotted on the screen with its error bars and the values entered into a table in the right-hand panel (Fig. 5). Although the point number is automatically incremented for the next data value to be input, this value may be overridden, with a new (nonnegative) number to enter another data point out of sequence for editing purposes or with a negative number to delete a data point. Once all the points have been correctly entered, pressing the ESC key causes the data entry loop to be terminated.

The program then goes through a permeability-fitting loop, which prompts the input of trial values of P_{Na}/P_K to fit the data. The permeability value is used with the Goldman–Hodgkin–Katz equation to generate a theoretical curve, plotted as a solid line on the screen. In addition, the deviations of the data points from the theoretical curve are squared and, normally, each squared value is also divided by the square of the SEM to weight the curve to the most accurate data values (if required, this correction procedure can be avoided either by choosing the appropriate option with the MPSETUP program prior to program execution or at this stage of the program). The averaged summed value of the corrected (or uncorrected) deviations is then displayed in the top right-hand panel along with the chosen value of P_{Na}/P_K. Relative permeability values are then chosen to either find the curve that best

FIG. 9   Flow chart showing the program flow sequence of the main elements of the MEMPOT program. The double-sided MAIN ELECTRODE IMPALEMENT LOOP box is expanded in a separate flow chart in Fig. 10. Further details of the other elements are given in the text.

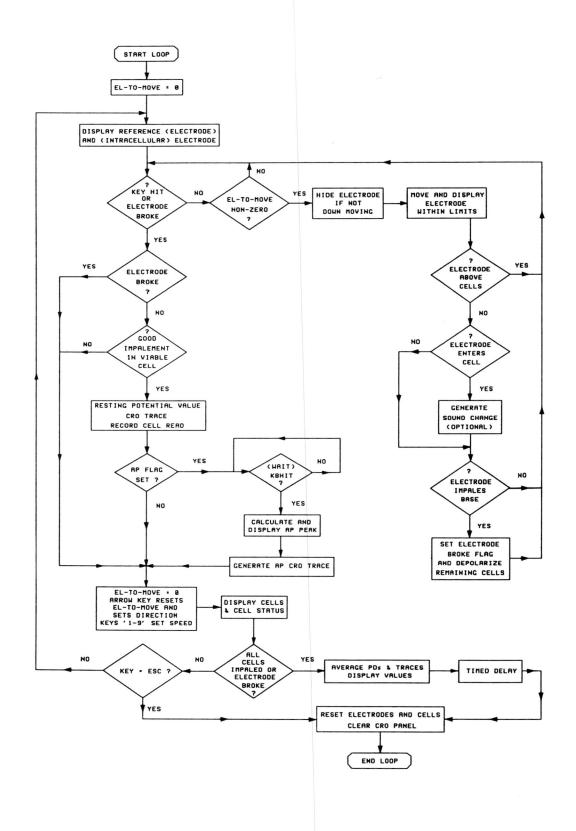

visually fits the data or to minimize the value of the sum of the deviations of the points. When an adequate fit has been found, pressing ESC terminates the loop. The program then determines which value of P_{Na}/P_K tried best fits the data and displays the curve corresponding to such a value. At this point, pressing 'p' will print out all the data values, together with that permeability ratio and the resultant predicted values for each of the solutions considered.

The whole procedure can then be repeated, with the alternative Action Potential Measurements or Resting Potential Measurements exercise, or else the program may be terminated by pressing ESC. This last alternative will also display a list of the program command line options, before stopping program execution.

Main Electrode Impalement Loop

On entering the main electrode impalement loop (depicted in Fig. 10), the incremental x and y electrode movement steps are both set to zero (i.e., el-to-move = 0 in Fig. 10). The reference electrode and intracellular electrodes are displayed. Then, until a key is hit, el-to-move = 0 and program execution remains within that immediate inner loop of the main right branch loop. If an arrow key is hit before any electrode movement is initiated, program operation first flows down the main left program branch, the appropriate x or y increment is reset as nonzero (i.e., el-to-move is nonzero in Fig. 10), and back. In the absence of any immediate further key hit, program control now flows around the right branch loop until a second key (e.g., the spacebar key) is finally hit. This program loop moves the intracellular electrode on the screen, by sequentially hiding (blanking) the electrode, moving it by the specified increment and then redisplaying it (using the C functions setfillstyle and fillpoly to color in and move the array defining the intracellular electrode). If the electrode tip enters any of the viable cells, there is a sound change (either a change in frequency or a beep of the appropriate frequency, unless the sound option has been completely turned off).

Program flow continues around the program loop, and the electrode continues to be moved on the screen until either a key is hit or the electrode breaks. If the key is now hit, after electrode movement is initiated, or the electrode

---

FIG. 10 Flow chart showing the program flow sequence of the double-sided MAIN ELECTRODE IMPALEMENT LOOP box shown in Fig. 9. Further details are given in the text.

has broken, the program flow again moves down the main program branch loop. If the key is hit while the electrode is in a viable cell, a resting potential value is generated and displayed, an oscilloscope trace is displayed in the top left-hand panel, and the cell read flag is set. In the event of the action potential flag (AP Flag) being set, the keyboard is polled waiting for a keyboard hit. When this occurs, the action potential peak is calculated and displayed, and the action potential waveform is scaled and displayed in the top left-hand panel. The electrode movement increments are again zeroed (el-to-move = 0 in Fig. 10).

If all the cells have been impaled or if the electrode has been broken, the potentials of the good impalements are averaged, standard errors of the mean calculated, and the averaged "oscilloscope trace" displayed together with dashed lines, indicating the positions of the potassium and sodium equilibrium potentials. This is followed by a timed delay during which any further keyboard entry is blocked, to ensure that the user has enough time to view this "oscilloscope" display without inadvertently pressing a key and leaving that screen. Pressing the spacebar then resets the position of the electrodes and clears the cell display (it should be noted that the ESC key will also terminate the main electrode impalement loop). Program control then leaves that loop and reverts to the main program loop illustrated in Fig. 9, ready for a printout of the data, the input of another solution value, or the beginning of the data reentering and permeability fitting section of the program.

Conclusions

This chapter describes the operation and principles underlying a graphical interaction program (MEMPOT)* to simulate an electrophysiological investigation of the properties of the membrane potential in excitable cells. MEMPOT simulates a microelectrode experiment devised to investigate the ionic dependence of both the resting potential and the peak of the action potential. It then enables the use of a least-squares routine to numerically and graphically fit the data obtained to the Goldman–Hodgkin–Katz Equation in order to determine the relative ion permeability ratio for each set of measurements.

* Program Availability: For further information about the full MEMPOT program (or a demonstration version) and its availability, within the United States, contact Dr. N. B. Datyner, P.O. Box 317, Stony Brook, NY 11790 [Telephone: 516-331-9623; Fax: (also) 516-331-9623]; within Western Europe (including the United Kingdom), contact M. de Courtenay, Commercial Service, BIO-LOGIC, 1, rue de l'Europe, Z. A. de Font Ratel, 38640 Claix, France [Telephone: 33-76.98.68.31; Fax: 33-76.98.69.09]; elsewhere, contact Dr. P. H. Barry, School of Physiology and Pharmacology, University of New South Wales, P.O. Box 1, Kensington, NSW 2033, Australia [FAX: 61-2-313 6043; Telephone: 61-2-697-2566].

Experiment–experiment and cell–cell variation are built into the program to increase realism and investigative interest.

Acknowledgments

I would like to thank Mr. Lumir Pek of the School of Physiology and Pharmacology, University of New South Wales, for helpful advice during the early development of this program, our own students, Dr. Alan Finkel of Axon Instruments Inc. (Foster City, CA), Professor Peter Gage of the Australian National University, and colleagues in the Departments of Physiology at the University of Sydney and Monash University for helpful suggestions with earlier versions of the program. In particular, discussions with Dr. Robert Bywater of Monash University on the contribution of electrogenic pumping to the membrane potential were much appreciated, as were other discussions with Associate Professor Donald Robertson of the University of Western Australia.

References

1. A. L. Hodgkin and P. Horowicz, *J. Physiol.* (*London*) **148,** 127 (1959).
2. A. L. Hodgkin and B. Katz, *J. Physiol.* (*London*) **108,** 37 (1949).
3. W. L. Nastuk and A. L. Hodgkin, *J. Cell Comp. Physiol.* **35,** 39–74 (1950).
4. R. H. Adrian, "Microelectrodes in Muscle." Rank Film Laboratories, Denham, Uxbridge, England, 1972.
5. P. H. Barry, *Am. J. Physiol.* **259,** (Adv. Physiol. Educ. 4), S15 (1990).
6. S. W. Jones, this volume [28].
7. D. Goldman, *J. Gen. Physiol.* **27,** 37 (1943).
8. P. H. Barry and P. W. Gage, *Curr. Top. Membr. Transp.* **21,** 1 (1984).
9. L. J. Mullins and K. Noda, *J. Gen. Physiol.* **47,** 117 (1963).
10. R. A. Sjodin, *in* "Electrogenic Transport: Fundamental Principles and Physiological Implications" (M. P. Blaustein and N. Lieberman, eds.), Raven, New York, 1983.
11. W. F. H. M. Mommaerts, *in* "Essentials of Human Physiology" (G. Ross, ed.), p. 35. Year Book Medical Publ., Chicago, 1978.

[30] Computer-Assisted Three-Dimensional Reconstruction and Dissection

Robert Nagele, Kevin Bush, Dale Huff, and Hsin-yi Lee

Background

Computer-assisted three-dimensional (3-D) imaging is a powerful tool for analyzing biological tissues that is now within the reach of many cell biologists, anatomists, and neuroscientists. The classic approach to 3-D reconstruction of both macroscopic and microscopic objects has been to obtain serial, two-dimensional (2-D) slices or images, make sheetlike models representing each section from wax, balsa wood, plastic, or other materials, and serially stack the 2-D models to recreate the 3-D object. The final quality and fidelity of such reconstructions can vary somewhat with the precision of the investigator, the specific technique used, and individual artistic skills. Obtaining high-quality 3-D reconstructions using this approach has proved to be an extremely labor-intensive task. The relatively recent availability of personal computers in the laboratory has led to the development of computer-assisted 3-D reconstruction systems (1, 2). Personal computers are ideally suited for this purpose, since they can be programmed to manipulate large volumes of spatial information and generate 3-D views from selected vantage points. The field is developing quickly and is yielding information on the 3-D structure of macroscopic and microscopic objects that can be gathered in no other way, except through artists' conceptions.

General Features of Three-Dimensional Reconstruction Systems

In this chapter we describe a system used in our laboratory for computer-assisted 3-D reconstruction of serial sections that can be obtained by any laboratory at a relatively low cost. We also provide technical suggestions regarding specimen handling, data entry, and 3-D image generation that are generally applicable to any 3-D reconstruction system and outline some basic features of 3-D reconstruction software that, it is hoped, will guide the reader toward obtaining hardware and software that are best suited to the task at hand. Three-dimensional reconstruction software packages have been developed that are useful for both macroscopic and microscopic objects. Depending on the specific features and calculation capabilities of the soft-

Methods in Neurosciences, Volume 10

ware, 3-D images can be produced that are derived from the data input of 2-D serial sections and include hidden-line removal, color video output, and stereoviewing capabilities (3–6). In some 3-D reconstruction programs, a series of 3-D images can be displayed in rapid succession on the video screen to "animate" the object of interest. Such animation sequences are commonly used to rotate the object about an axis for a better appreciation of true depth.

The 3-D reconstruction system employed in our laboratory is based on the use of PC3D, a 3-D reconstruction software package marketed by Jandel Scientific (Sausalito, CA). The system can be used across the entire range of magnification, from macroscopic down to the electron microscopic level. In our hands, PC3D is easy to learn and use, and it provides high-quality, camera-ready 3-D images that are suitable for publication. A set of eight subprograms runs on a standard personal microcomputer. Data are entered as trace points on a digitizing tablet from serial sections that have already been aligned. A 3-D view of the reconstructed object is generated in color and can be displayed from any perspective with hidden lines removed. Analysis of volume can be performed automatically. We have used this reconstruction system to produce publication-quality illustrations (7–9).

Hardware

The computer used for 3-D image processing is an IBM AT-compatible (12 MHz) with a math coprocessor, 2 megabyte RAM memory, a 40 megabyte hard disk, and a VEGA Video 7 Ultimate Enhanced Graphics Adaptor (Video 7, Inc., Freemont, CA). A digitizing tablet (GTCO Digipad 5) with a 16-button mouse-type cursor (Jandel Scientific) is used for input of 2-D data. Three-dimensional images are viewed and recorded directly from a NEC Multisync color monitor.

Software

The software system used in our laboratory for 3-D reconstruction of serial sections is PC3D (Jandel Scientific), which is composed of the following eight subprograms.

1. 3D-TRACE for data input from a digitizing tablet by tracing profiles of the selected features and creating trace files from individual sections
2. 3D-DESCRIBE for entering descriptions of selected structures, section thickness, magnification, and numerical sequence of the sections
3. 3D-EDIT for reviewing trace files of individual sections and editing struc-

ture descriptions while viewing a high-resolution color graphic display of each section

4. 3D-ALIGN for realigning misaligned sections with respect to any other related section
5. 3D-DISPLAY for generating 3-D images
6. 3D-PLOT for making color hardcopies of 3-D images using a Hewlett-Packard plotter
7. 3D-MOVIE for rapid playback of a series of 3-D images, image rotation, and animation (available as an option)
8. 3D-VOLUME for automatic calculation of the volume of any 3-D reconstructed structure

Tissue Processing

The basic steps involved in processing tissues for light and electron microscopy and the preparation of serial sections are well-established and need not be presented here. It should be stressed that the tissue sections must be free from obvious flaws. Prior microscopic evaluation of individual sections should reveal most section defects, such as fluctuations in section thickness, tissue breakage, and section compression due to variations in the consistency of the embedment or gradual dulling of the knife edge.

Data Input: Generation of Trace Files of Serial Sections

For 3-D reconstruction of serially sectioned tissue specimens, it should be obvious that the quality of the final 3-D output depends strictly on the quality of the 2-D input. Therefore, it should not be surprising that the data input step is the most time-consuming aspect of the 3-D reconstruction process. For this reason, before data input, one needs to clearly define the specific objectives of the 3-D reconstruction. If the goal is simply to get a general idea of the 3-D shape of one or more features within a specimen, it may not be necessary to input data from all available serial sections. For example, if the desired information can be gleaned from tracing 100 sections, instead of 500 sections, this will result in a considerable savings in time. On the other hand, if the purpose is to get a detailed 3-D view and to reveal the complex spatial interrelationships among several features within a specimen, it is most helpful to have all of the relevant information available. The 3-D reconstructed images of the mouse embryo shown in Figs. 1–4 were based on the input of 350 serial sections.

In our laboratory, data input is carried out on a GTCO Digipad 5 digitizing tablet (resolution 0.001 inch) using the 3D-TRACE subprogram of the PC3D 3-D reconstruction software and a 16-button mouse-type cursor. Each section can have up to 16 categories of structures, a maximum of 400 polygons, and up to 7000 data points. If the object is too large for the digitizing tablet, it can be photographically or xerographically reduced prior to data input. In the case of routine histological specimens embedded in wax or plastic, serial sections are viewed with a Nikon Optiphot microscope equipped with a camera lucida. Profiles of selected structures within each section are traced onto the digitizing tablet with the 16 button mouse cursor as a set of closed polygons. In this way, features of interest within sections are selected manually and are essentially ''extracted'' from the section, whereas those deemed by the investigator to be unnecessary are eliminated. To achieve a spatially accurate 3-D reconstruction, each element within a section should enclose a reasonable number of data points on the digitizing tablet. What is presented to the computer for 3-D reconstruction consists of a series of rather simplified 2-D pictures containing the features of interest that have been extracted manually from the tissue sections.

Data input by manual tracing is still the method of choice, since even with the most advanced 3-D reconstruction systems, the pattern recognition capabilities of humans cannot yet be accomplished by computers. While using 3D-TRACE, an image of the selected features within each section appears on the monitor while tracing for confirmation of the accuracy of 2-D input. The amount of time required to digitize a section is highly variable and, of course, depends on the complexity of the elements within it that are to be reconstructed. For example, an average of 3–4 min each was required to input the more complex sections in the mouse embryo shown in Fig. 1. Data representing each section can be stored as a separate trace file on disk. As many as 1000 sections can be used for a single 3-D reconstruction. Trace files representing each section are numbered consecutively in the filename extensions. The between section thickness is given from the keyboard. The entire set of trace files which has been entered can be viewed as a 3-D image at any step during data input using the 3D-DISPLAY program.

For 3-D reconstruction of cell and/or organelle shape from electron micrographs of serial sections, the most straightforward procedure is to trace the desired profiles on acetate sheets directly from the micrographs and enter them into the computer using the digitizing tablet and the 3D-TRACE subprogram. Care must be taken to ensure that all electron micrographs in the series are printed at exactly the same magnification. Colored pens can be used to facilitate visual identification of particular cells and/or organelles on the acetate sheets within the series.

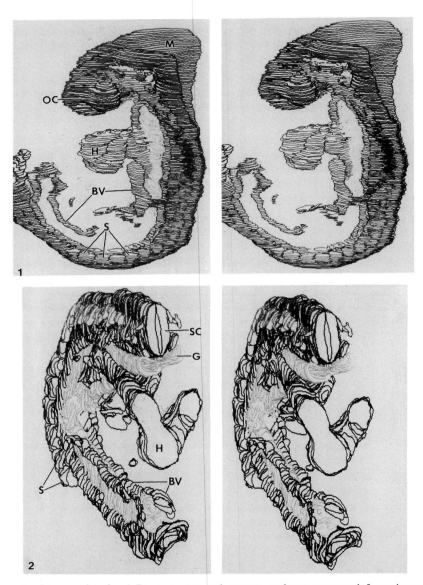

FIG. 1   Stereopair of a 3-D reconstructed mouse embryo prepared from input of 350 serial histological sections and photographically recorded directly from a video monitor using black-and-white film. Data from every other section were used to generate the image, and the skin (surface ectoderm) was "mathematically" dissected away to reveal underlying structures. M, Midbrain; OC, optic cups; H, heart; BV, blood vessels; S, somites.

Alignment of Serial Sections

To achieve a spatially accurate 3-D reconstruction from serial sections, one is required to align the sections or images strictly with respect to position and orientation. In histological specimens, this can be done by adding guide marks or fiducials to the tissue prior to sectioning by drilling small holes through the tissue margin or some other irrelevant region of the tissue at right angles to the plane of section (10, 11).

Alignment of object profiles in electron micrographs of serial sections is more difficult than it is for light microscopy (12). Currently, a reliable means to introduce fiduciary marks on thin sections used for electron microscopy is lacking. Consequently, the "best-fit" method is still the most practical approach. The contours of selected features from each consecutive section are traced directly from micrographs onto acetate sheets. The acetate sheets are then affixed to the surface of the digitizing tablet with adhesive tape for data input. After tracing the first section, a second acetate sheet containing the contours from the next section is then carefully placed over the first and aligned by a visual best-fit of several object profiles within the section. After securing the second acetate sheet to the digitizing tablet in the aligned position, the first sheet is then carefully removed. The acetate sheet containing object profiles from the third section is aligned with respect to the second in the same manner, and so on.

After the 3-D image is generated, some small misalignments may become apparent. In PC3D, the subprogram 3D-ALIGN enables fine-tuning of section alignment (including translation and rotation of individual section tracefiles) using another, superimposed section as a reference point.

Sectioning-Induced Deformations

The cutting and mounting of serial sections can introduce topological deformities into 3-D reconstructed images. Compression of sections during cutting, differential expansion while mounting sections on slides, variations in the

---

FIG. 2   Stereopair of the same 3-D reconstructed mouse embryo shown in Fig. 1. In this case the 3-D reconstructed image was generated from a different vantage point, giving the impression of image rotation. The block of sections representing the cephalic end of the embryo was intentionally deleted from the reconstructed image to give the embryo a "sectioned" appearance. Such views reveal morphological details that would otherwise be obscured by the brain region. S, Somite; SC, spinal cord; G, gut; BV, blood vessels; H, heart.

consistency of the embedding medium, and the quality and fidelity of tissue infiltration are all factors that impact negatively on the quality of the final 3-D reconstructed image. Care in handling tissues during processing can eliminate essentially all of these potential problems, except for section compression. When hundreds or thousands of sections are taken through a single specimen, the knife edge may gradually become dull, which causes compression (foreshortening) of the sections in the direction of cutting. The effects of compression are readily apparent as a change in the spatial arrangement of the reference markers that are incorporated into the tissue. If the knife is replaced while making serial sections, compression is usually relieved, but sudden changes in the dimensions of subsequent sections often introduces a serious error in the 3-D reconstructed image. In our work with relatively soft embryonic tissues, this has not been a problem, but section compression could prove to be particularly troublesome to those working with specimens that contain substantial amounts of tough connective tissue.

Three-Dimensional Image Display and Dissection

Multicolored 3-D images, based on data input from serial sections, are generated with PC3D using the 3D-DISPLAY subprogram by stacking the traced outlines and, after hidden line removal, displaying them on the video screen on a line-by-line basis. The final output on the monitor is a full-screen display (640 × 400 pixels) with 16 colors. Images can be rotated and redrawn from almost any angle by entering x, y, z rotation coordinates and saved on disk (Figs. 1–4). Selected portions of the specimen can be displayed by choosing to 3-D reconstruct only the block of sections that includes the area of interest (Fig. 2). Selecting the best orientation of the final 3-D image often involves choosing one from among a pool of several possible vantage points, each one of which has to be drawn separately. To facilitate this process, there is a provision for the 3-D display of every nth section, where n is typically a multiple of the trace file increment (Fig. 3). As expected, using fewer sections decreases the time required for both image calculation and display.

One of the most powerful features of some 3-D reconstruction programs is the fact that one or more structural features within a complex specimen can be selected for 3-D reconstruction and displayed with the exclusion of others. For microscopic specimens such as those presented in Figs. 3 and 4, this "dissection" capability provides microscopic views from almost any vantage point of tiny objects that are unobstructed by surrounding components. Such views would otherwise be impossible to obtain through a combination of manual dissection and conventional microscopy. Scaling of the image size can be done automatically to fill the monitor screen, or manually

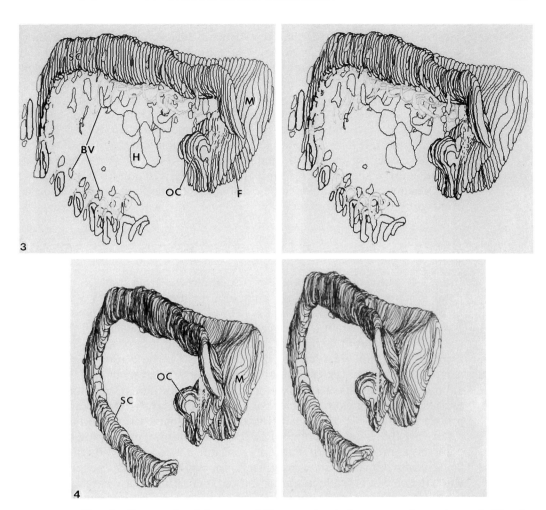

FIG. 3 Stereopair of the same 3-D reconstructed mouse embryo shown in Fig. 1. The image was generated to permit a view of the open brain region. The caudal half of the embryo was generated using every twentieth section to highlight major changes in the shape of the neuroepithelium. F, Forebrain; M, midbrain; SC, spinal cord; H, heart; BV, blood vessels; OC, optic cups.

FIG. 4 Stereopair of the same 3-D reconstructed mouse embryo shown in Fig. 1. In this image, the developing neural tube (brain and spinal cord) was 3-D reconstructed free from surrounding embryonic organ rudiments in order to reveal details of its 3-D shape without obstructions. SC, Spinal cord; OC, optic cups; M, midbrain.

relative to the entire object or some other component of the object. The more sections that are used for the reconstructed image, the smoother or more solid the image will appear. In fact, when the number of sections included in the 3-D reconstructed image begins to approach the number of pixels available on the monitor or the limits inherent within the software, the image appears to "solidify" with a consequent loss of topological information. Under these conditions, images should be rotated such that the long axis of the object corresponds to that of the color monitor (640 × 400 pixels). Another option is to incorporate every other section into the final 3-D reconstructed image for improved topological detail (Fig. 1). The image of the mouse embryo shown in Fig. 1 took approximately 10 min for our computer to calculate and draw. The calculation speed of the final image can be increased by using RAM disks.

Photographic Recording and Hardcopying Three-Dimensional Images

Final 3-D reconstructions can be recorded directly from the high-resolution color monitor with a camera as multicolor images (color prints or slides) or as black-and-white, halftone, images (Figs. 1–4). We use a Nikon F3 High Eyepoint 35-mm camera, equipped with a MicroNikkor lens and cable release, mounted onto a tripod. There is little or no loss of image resolution using this method. Hardcopies of 3-D images can be drawn in color with a Hewlett-Packard (or compatible) color plotter (driven by the 3D-PLOT subprogram) and subsequently recorded with color film. Three-dimensional images can also be plotted in black and white using a Hewlett-Packard laserjet printer and LASERPLOTTER (Insight Development Corp., Moraga, CA), a software that nicely interfaces PC3D with the laserjet printer and translates color assignments for individual polygons into shades of gray. Such halftone, black-and-white images are most useful for publication purposes, especially when the cost of reproducing color images is prohibitive or where the presentation of color images is not an option.

Stereopairs

The production of stereopair 3-D images allows one to best appreciate true depth and the spatial interrelationships among structures within the specimen (Figs. 1–4). Stereopair images are generated by having the computer draw two separate 3-D images that are tilted at an angle of 5°–7° on the y axis with respect to one another. Stereopair prints are published side-by-side for examination by the reader with an inexpensive, hand-held, stereoviewer. For presentation of 3-D images at conferences, red–blue stereopairs can be used, where one image of the pair is generated in red and the other in blue. Each 3-D image is recorded separately on the same negative using the "double-expose" option that is available on many 35-mm cameras. The superimposed images are viewed with red–blue glasses that can be purchased or manufactured in the laboratory from cardboard and sheets of red and blue acetate.

Rapid Three-Dimensional Image Display: Image Rotation and Animation

Once a series of separate 3-D images of a particular object are available on disk, the images can be recalled and viewed in rapid succession using the 3D-MOVIE subprogram. Developmental stages, live rotations, or movements can be simulated. Processing time between successive images is not a problem, since 3-D images can be stored in RAM memory and played at a rate of 1 image/sec. PC3D allows the rapid replay of as many as 20 separate 3-D images per "movie loop." Longer sequences can be made by recording successive movie loops on video tape.

Computer-Assisted Morphometry of Three-Dimensional Reconstructed Objects

Although the original focus for the development of 3-D reconstruction software was, of course, to generate 3-D images, the ability of the computer to produce accurate morphometric information derived from the same data is beginning to be exploited. Some programs now offer accurate quantitation of volumes and surface areas of rather complex 3-D objects. The outstanding advantage of interfacing 3-D reconstruction capabilities with morphometric measurements is that, unlike standard stereological methods, data are not estimated from a statistical sampling of random sections taken through a number of similar, but separate, objects. Instead, accurate measurements are made for each 3-D reconstructed object. In this way, individual variations between related objects can be clearly shown. Also, variations that occur only rarely still can be detected and are not lost in the statistics of the sampling method. The PC3D software package has 3D-VOLUME, a subprogram that automatically calculates the volume of any selected structural element after taking into account the magnification of the object, number of serial sections, and individual section thickness. Volumes of a selected portion of any structural element can also be determined by specifying the block of sections to be included in the volume calculation. As an example, we have calculated the volumes of neural tissue in the cephalic (future brain) ($5.13 \times 10^7 \ \mu m^3$) and caudal (future spinal cord) ($8.66 \times 10^6 \ \mu m^3$) regions of the mouse embryo shown in Figs. 1–4.

Conclusions

Computer-assisted 3-D reconstruction technology has now progressed to the point where accurate 3-D images can be generated from serial sections or images and used for research or presentation purposes. Input data can be derived from electron microscopy, light microscopy, macroscopic sectioning, CAT scans, or NMR imaging. Publication quality, color or black-and-white images can be presented from almost any vantage point with hidden lines removed. Images can be saved on disk, recalled rapidly, and strung together in a rapid-playback sequence to simulate rotation, animation, or a developmental sequence. Three-dimensional reconstruction systems are fairly easy to use after only a few days of training and inexpensive enough to be within the financial reach of most laboratories.

References

1. C. Levinthal and A. Ware, *Nature (London)* **236,** 207 (1972).
2. A. Veen and L. D. Peachy, *Comput. Graphics* **2,** 135 (1977).
3. E. R. Macagno, C. Levinthal, and I. Sobel, *Annu. Rev. Biophys. Bioeng.* **8,** 323 (1979).
4. R. G. Smith, *J. Neurosci. Methods* **21,** 55 (1987).
5. J. B. Upfold, M. S. Smith, and M. J. Edwards, *J. Neurosci. Methods* **20,** 131 (1987).
6. C. Keri and P. K. Ahnelt, *J. Neurosci. Methods* **37,** 241 (1991).
7. R. G. Nagele, K. T. Bush, M. C. Kosciuk, E. T. Hunter, and H. Lee, *Dev. Brain Res.* **50,** 101 (1989).
8. K. T. Bush, F. J. Lynch, A. S. DeNittis, A. B. Steinberg, H. Lee, and R. G. Nagele, *Anat. Embryol.* **181,** 49 (1990).
9. R. G. Nagele, K. T. Bush, F. J. Lynch, and H. Lee, *Anat. Rec.* **231,** 425 (1991).
10. K. Brandle, *Comput. Biomed. Res.* **22,** 52 (1989).
11. E. P. Meyer and V. J. Domanico, *J. Neurosci. Methods* **26,** 29 (1988).
12. C. P. Bron, D. Gremillet, D. Launay, M. Jourlin, H. Gautschi, T. Bachi, and J. Schupbach, *Eur. J. Cell Biol.* **49**(Suppl. 27), 15 (1989).

[31] Computer Simulations of Individual Neurons

William R. Holmes

Introduction

It is a common misconception that one has to be an expert in numerical analysis or computer science to do computer simulations of individual nerve cells. Although there may have been some truth to that idea a few years ago, it is not true today. A few years ago, modelers developed their own algorithms and their own code for modeling voltage changes in individual cells without being concerned (or willing to be bothered) to make the code available and sufficiently user-friendly for general distribution. Today, there are many software packages for doing single neuron simulations that are available for free or for reasonable prices. These programs can be run quickly on many different types of computers by individuals having very little mathematical background.

The major difficulty with doing computer simulations today is to make sure that the simulations provide biologically relevant results. Many neuroscientists harbor serious misgivings about mathematical and computational modeling, and too often these misgivings are justified. If computational models are to avoid being dismissed as "mindless modeling ventures that have become commonplace with the proliferation of computer hardware and software" (anonymous referee, personal communication), then the models must make biologically appropriate assumptions, address a particular biological question, and make specific predictions that can be tested experimentally. Models of neurons can have thousands of degrees of freedom (Rall, 1990). If these degrees of freedom are not constrained properly, then the old adage "give me four free parameters and I can make a dog, give me five and I can make its tail wag" (quoted by anonymous referee, personal communication) will apply, and the results will be meaningless.

This chapter discusses some of the issues that one must face when constructing models of individual nerve cells. These issues include choosing an appropriate level of morphological detail to use in a model, calibrating the passive electrotonic parameters of the cell, including dendritic spines, and incorporating voltage-dependent and synaptic conductances (their types, distribution, kinetics, and times of activation). We will describe how to approach these issues with examples from our own work. Once these issues have been addressed and one has selected the biologically reasonable as-

Methods in Neurosciences, Volume 10
Copyright © 1992 by Academic Press, Inc. All rights of reproduction in any form reserved.

sumptions that the model will make, one of the many available software packages can be used to pose particular questions to the model. This chapter describes the computational method that I use, and then briefly describes a few of the many software packages available for doing computer simulations of individual neurons.

Morphology

Reduced versus Full Morphology Models

The first issue one faces is to decide how much morphological detail to use in the model. The amount of detail that will be required by the model depends on the particular question being posed. If one is asking a general question, for example, about whether dendritic diameter can be made optimal for the effectiveness of a particular synaptic input, then a simple cylinder may be an appropriate starting point (Holmes, 1989). In fact many important insights have come from models using a simple cylinder (i.e., Rall, 1964). However, if one wants to study the function of a cell, more detail is required. Some workers have obtained considerable insight with models that use only 10–50 compartments to represent individual neurons (i.e., Traub *et al.*, 1991). If networks of neurons are to be modeled, computational limitations almost demand that simplified neuronal representations be used for the neural elements.

To study the function of individual neurons, my personal bias is to use the full morphological data from serial reconstructions. The reasons are 2-fold. First, if the data are available, why not use them? Detailed models may run more slowly on the computer, but hardware advances are making the differences in computer time between models using full morphology and models using reduced morphology much less of an issue. Second, we presently do not know the kinds of morphological simplifications that can be made that would preserve the essential computational properties of the neuron. Many morphological simplifications seem reasonable, such as the cortical pyramidal cartoon used by Stratford *et al.* (1989), but we will not know how reasonable such simplifications are without comparing results to those obtained with a model having the full morphology.

Essential Morphological Information

Anatomists have asked me to tell them what morphological information I need to construct models of individual cells. In models, neuronal segments are represented as cylinders. Consequently, a minimum requirement is to have the length and diameter of every dendritic process. Many years ago, I found that anatomists were careful to measure the lengths of dendrites, but, often, they did not measure the diameters. Measuring diameters is tedious, and the measurements can be distorted by tissue shrinkage during the fixation process. Computerized reconstruction techniques have made this less tedious, and details of such techniques are described in [19], [20], and [30] in this volume. One problem with diameter measurements is that the dendritic processes are not perfect cylinders (cf. Stevens and Trogadis, 1984). An average diameter or in some cases a minimum diameter (C. J. Wilson, personal communication) should be used in the model. Many reconstruction techniques give the three-dimensional coordinates of dendritic locations along with the diameter of the segment between consecutive pairs of coordinates. The three-dimensional coordinates are useful when the simulation software allows the user to visualize and rotate the neuron that is being modeled.

Dendritic Spines

Other morphological information needed is the number and dimensions of dendritic spines on each dendritic segment. To do this properly one would want a complete electron micrograph reconstruction of the cell, but this has been done only for one neuron to my knowledge (White and Rock, 1980). Thus spine data have to come from other studies done on the same neuron type. For example, in my models of dentate granule cells, I use spine data from Desmond and Levy (1985). Desmond and Levy categorize spines into long-thin, mushroom-shaped, and stubby and give the average and range of dimensions as well as the relative proportion and density of each spine type in proximal, mid-dendritic, and distal regions. These data allow one to make some assumptions about spine shape and density on dendrites in various dendritic regions.

Anatomists measure spine density in two ways. Spine densities are given either as the number of spines per unit length of dendrite or as the number of spines per unit dendritic membrane area. For dentate granule cells, whose dendrites in any one region of the molecular layer have similar diameters, both measurements will give similar information. However, for cortical pyramidal neurons, whose apical dendrite has a much larger diameter than any of its branches in a given dendritic region, these two measurements give very different information. In the latter case it might be necessary to give both measurements separately for the main apical dendrite and its branches in each dendritic region.

Calibrating Passive Electrotonic Parameters of Model

Ideally, one wants to have morphological and electrophysiological information from the same neuron. However, the model typically uses the morphology of one cell and electrotonic parameters estimated from electrophysiological data gathered from other cells of the same type. The electrotonic parameters we need to know are membrane resistivity, R_m (Ωcm^2), intracellular resistivity, R_i (Ωcm), and membrane capacity, C_m ($\mu F/cm^2$). In most models, R_m is 10–200 kΩcm^2, R_i is 50–300 Ωcm, and C_m is 1.0 $\mu F/cm^2$. Although it is convenient to select some arbitrary set of values and run the model, one should have some justification for the particular values chosen. This section discusses some of the uncertainties surrounding estimates of these parameters, and the next section describes methods to estimate these parameters from morphological and electrophysiological data.

Specific Capacitance

The traditional view has been that the capacity of biological membranes is about 1.0 $\mu F/cm^2$ (Cole, 1968). Some estimates are slightly higher (cf. Almers, 1979), but given the nonuniformity of membrane and membrane infolding a value of 1.0 $\mu F/cm^2$ seems reasonable. Although we like to think of C_m as being constant, the value of C_m in squid axon falls to about 0.5 $\mu F/cm^2$ at high frequencies and rises to a peak near 1.3 $\mu F/cm^2$ with depolarization halfway to the peak of the action potential (Almers, 1979; Takashima, 1976; Haydon et al., 1980). Larger estimates of C_m have been found in some cells (Durand and Carlen, 1985), but this may be due to membrane irregularities or the presence of dendritic spines (which can be considered to be a particular case of membrane infolding; also see below). Although modelers should keep these uncertainties in mind, it is usually reasonable to use a C_m value of 1.0 $\mu F/cm^2$ in the model.

Axial Resistivity

Intracellular or axial resistivity is difficult to measure, but values assumed in models have been fairly constant, until recently. Estimates of R_i in studies with invertebrates are close to 50 Ωcm (Foster et al., 1976; Gilbert, 1975). Estimates based on measurements in mammalian motoneurons placed R_i near 70 Ωcm (Barrett and Crill, 1974). However, the fact that the concentration of the dominant charge carriers in invertebrate axons is three times that in mammalian neurons leads one to believe that R_i in mammalian neurons should be closer to 150 Ωcm (W. Rall, personal communication). Some recent modeling studies have estimated R_i to be in the 200–400 Ωcm range (Shelton, 1985; Stratford *et al.*, 1989). The possibility exists that R_i may be nonuniform because of the presence of mitochondria, the spine apparatus, or other cellular components found in the cytoplasm. The truth is that we do not have good, consistent experimental estimates of R_i. The current wisdom places R_i in the 100–300 Ωcm range, and these are the values most often used in models.

Membrane Resistivity

Membrane resistivity estimates have risen through the years from about 1 to 100 kΩcm^2 in some cases today. Estimated values depend on the method of recording, the preparation used, and whether R_m is uniform or nonuniform. Estimates of R_m with intracellular electrodes are usually about 5–30 kΩcm^2. However, it is widely believed that there is a shunt conductance around the electrode caused by penetration of the electrode into the cell (Jack, 1979). If this is true then estimates of R_m with intracellular electrodes are too low depending on the size of the shunt. One way to avoid the problem of an artificial shunt is to estimate R_m from whole-cell patch electrode recordings. Estimates of R_m with whole-cell patch electrodes have been 3–10 times larger than estimates with intracellular electrodes (Mody and Otis, 1989; Storm, 1990a; Pongracz *et al.*, 1991). However, there are two factors that cast doubts on these extremely large resistivity values. First, many of these recordings are made on young cells, which are smaller than mature cells. Young cells have a larger input resistance than mature cells and may also have an intrinsically higher membrane resistivity. Second, whole-cell patch methods dialyze the cell. This process changes the internal environment of the cell and could affect the membrane resistivity. More recently, the perforated patch method has been used. This method has the advantages of not introducing a shunt and not changing the internal environment. Membrane resistivity estimates with this method seem to lie between those for intracellular recordings and whole-cell patch recordings.

The preparation used also affects the membrane resistivity estimates. One factor that will make membrane resistivity estimates lower *in vivo* is the presence of tonic afferent activity. Such tonic activity is drastically reduced by the slicing procedure used for tissue slice preparations and is virtually absent in tissue culture cells. Such tonic activity can affect membrane resistivity estimates significantly (Holmes and Woody, 1989; Bernander *et al.*, 1991).

Finally it is not known if membrane resistivity is uniform or not. Barrett and Crill (1974) proposed that membrane resistivity in motoneurons is higher in distal dendrites than in proximal dendrites. The exact nature of the nonuniformity is not known. Fleshman *et al.* (1988) could not distinguish between a step change in membrane resistivity or a sigmoidal change in membrane resistivity with their models. Borg-Graham (1989) was able to distinguish certain patterns of nonuniform resistivity by looking at the amplitude and phase of the frequency response, but the usefulness of this technique needs to be explored in cells whose full morphology is known.

Methods to Estimate Electrotonic Parameters

Given the uncertainties described above, how should one estimate the passive electrotonic parameters of a cell to use in a model? One method, given the morphology of the cell, is to assume a value for R_i and then choose different R_m values until one is found that gives an input resistance at the soma that matches the measured input resistance of the cell (Rall, 1959). Another method is to assume that the cell can be approximated as an equivalent cylinder (Rall, 1962), and then estimate R_m from the first time constant, τ_0, of a voltage transient. If R_m is uniform, then τ_0 is equal to the membrane time constant τ_m (Rall, 1969). The membrane time constant is equal to the product $R_m C_m$, and if C_m is assumed to be 1.0 μF/cm^2, then one can get R_m. Unfortunately, when both of these methods were applied to the same cell, they gave different R_m estimates unless C_m was assumed to be larger than 1.0 μF/cm^2 (Durand *et al.*, 1983; Fleshman *et al.*, 1988). It was results such as these that suggested that a soma shunt might be present and should be taken into account (Iansek and Redman, 1973; Durand, 1984).

A method that can be used to estimate the electrotonic parameters when there is soma shunt is to use what I call the inverse computation (Holmes and Rall, 1992a). In the inverse computation, parameters measured or estimated from experimental data are used as known parameters to estimate a set of unknown parameters. For example, suppose one can measure the input resistance, R_N, of the cell and estimate τ_0, the first time constant of a voltage transient following a current step, and τ_{vc1}, the first time constant of the voltage-clamp current transient for voltage clamp at the soma. The inverse computation can use this information to provide estimates of three electrotonic parameters, such as R_m, R_i, and C_m, given the morphology of the cell. Computational details are given in Holmes and Rall (1992a). Note that use of this method assumes that both morphology and electrophysiological measurements are available from the same cell.

With the inverse computation, electrotonic parameters can be estimated given different sets of constraints. For example, if there is compelling evidence that R_m is nonuniform, one can specify a function $f(x)$ describing this nonuniformity [and describe the membrane resistivity as the product $R_m f(x)$, for example] and let the inverse computation find R_m. Of course, different functions will give different R_m estimates, and one must decide which function is best for the particular cell being studied. The inverse computation will not tell which function is best, but, given a set of constraints, it will provide an estimate.

There are two problems with estimating electrotonic parameters with the inverse computation. One problem is that voltage-dependent conductances often are activated by the current pulse or step used to get estimates for R_N and τ_0. Pharmacological agents can be applied, however, to the preparation to block these voltage-gated conductances and allow one to get estimates of the passive electrotonic parameters. A second problem is that results of the inverse computation are sensitive to measurement error in certain regions of parameter space. Small errors in the values for R_N, τ_0, τ_{vc1}, and membrane area may in some cases produce large errors in the R_m, R_i, and C_m estimates. One should check the sensitivity of the results in each case to see how robust the estimates are. Despite these problems, the inverse computation provides perhaps the best means available for estimating the passive electrotonic parameters to be used in simulations.

Voltage-Dependent Conductances

Once the passive electrotonic parameters of the cell have been chosen, the next problem is to decide how to incorporate the voltage-dependent conductances. The number of different voltage-dependent conductances that have been described is close to 30 (e.g., Adams and Galvan, 1986; Brown *et al.*, 1990; Storm, 1990b). The modeler has to decide which voltage-dependent conductances to include, what kinetic descriptions to use, and how to distribute these channels in the neuron.

Although there usually are data describing the various types of voltage-dependent conductances in a particular cell, the kinetic descriptions are harder to find. Kinetic descriptions used in models usually follow those given by Hodgkin and Huxley (1952) for sodium and potassium in the squid axon. A generic description of the ionic current through a voltage-dependent channel following the Hodgkin–Huxley (HH) formalism is

$$I_i = \bar{g}_i m_i^x h_i^y (V - V_i) \tag{1}$$

where the subscript i refers to the ionic species i, $\bar{g}_i$ is the maximum conductance of this type in the compartment or segment (all channels open), and V_i is the reversal potential for ion i. In the HH formalism, the voltage-dependent channel is gated by a certain number of particles that can be in the open or closed position, and which act independently. In this example there are x activation particles and y inactivation particles. The variable m is the probability of an activation particle being in the open state, and h is the probability of an inactivation particle being in the open state. Only when all particles are in the open position is the channel open. Hence the conductance equals the probability that all particles are in the open state times the maximum possible conductance, or $\overline{g}_i m_i{}^x h_i{}^y$.

The probability of a particle moving from the open state to the closed state or vice versa is a function of voltage. For example, if m is the probability of a particle being in the open state and $(1 - m)$ is the probability of being in the closed state, then the transition from one state to the other can be described as

$$m \underset{\alpha}{\overset{\beta}{\rightleftharpoons}} (1 - m) \tag{2}$$

where α and β are functions of voltage, or by the differential equation

$$dm/dt = -\beta m + \alpha(1 - m) \tag{3}$$

For a constant voltage, the steady-state probability of being in the open state, m_∞, equals $\alpha/(\alpha + \beta)$. For a step from one voltage to another, m can be obtained from Eq. (3) as

$$m = m_\infty + (m_0 - m_\infty) \exp[-t(\alpha + \beta)] \tag{4}$$

where m_0 is the initial value of m at the time of the voltage change and m_∞ is the steady-state value of m for the new voltage. The quantity $1/(\alpha + \beta)$ is often called τ.

The rate functions α and β have to be determined from voltage clamp experiments as in Hodgkin and Huxley (1952). Functions that fit the data are of the form

$$\alpha = \frac{p_1(p_2 + V)}{p_3 + \exp[(p_2 + V)/p_4]}$$

$$\beta = \frac{q_1(q_2 + V)}{q_3 + \exp[(q_2 + V)/q_4]} \tag{5}$$

where the p_i and q_i are determined by curve-fitting. The modeler can take one of three approaches here. First, p_i and q_i and the exponents x and y can be obtained from experimental data if available. For example, Traub *et al.* (1991) used the data of Sah *et al.* (1988a,b) to get kinetic descriptions of the fast sodium conductance and the delayed rectifier potassium conductance for hippocampal cells. A second approach, taken by Yuen and Durand (1991), is to vary the p_i and q_i for fixed x and y for each conductance until the action potential generated by the model matches that seen experimentally. Although this method will not produce a unique solution, it will provide a spike-generating mechanism that gives the proper results for a wide range of problems. A third approach is to plug the HH kinetics into the model. This assumes that the spike-generating mechanism of the squid axon applies to all neurons. Although this approach is useful for general questions that are not applied to a particular cell type, it should not be used for specific cells except as a first approximation.

The discussion up to this point applies to those voltage-dependent conductances that are involved in spike generation at the soma or in the axon hillock. Kinetic descriptions of these conductances are possible because the regions in which these conductances are being studied can be voltage clamped with an electrode in the soma. However there is evidence that voltage-gated conductances exist in the dendrites of some cells. Determining the activation and inactivation kinetics of voltage-dependent channels in dendrites in the usual way (HH) will not yield good results because of inadequate space clamp (Johnston and Brown, 1983). In these cases it may be possible to determine the kinetics from patch electrodes (i.e., Fisher *et al.*, 1990). Even if channel kinetics descriptions could be obtained, there is little information at present about the distribution and densities of voltage-gated channels in the dendrites. The development of antibody stains specific for particular classes of channels may alleviate this problem in the future (Westenbroek *et al.*, 1990; Wollner and Catterall, 1986), but for now the modeler has the problem of distributing these conductances in a manner that can be justified by experimental results.

A special problem exists for those conductances that are both voltage dependent and ion concentration dependent as with the calcium-dependent potassium conductance. With this conductance the model must keep track of both voltage and calcium concentration. This is a difficult thing to do given what little is known about calcium dynamics in a cell. Once again, whatever assumptions the modeler makes about the calcium dynamics must be based on available experimental results.

Synaptic Conductances

The current entering a dendrite due to a synaptic conductance change is described simply as

$$I_{syn} = g_{syn}(V - V_{syn}) \tag{6}$$

where g_{syn} is the synaptic conductance change and V_{syn} is the synaptic reversal potential. Sometimes the synaptic current is represented as an α function or a function of the form $t \exp(-\alpha t)$ (Rall, 1967; Jack and Redman, 1971). Although this approximation may be valid for single inputs of small magnitude, it ignores the nonlinear effect imposed by the reduction in driving force as the potential changes. A better way to model synaptic input is to assume that the synaptic conductance g_{syn} has a time course given by an α function or a double exponential function. I will describe cases when these are appropriate choices.

Let us assume that the process of synaptic transmission can be represented simply by the following binding reaction:

$$A + R \underset{k_{-1}}{\overset{k_1}{\rightleftharpoons}} AR \underset{\alpha}{\overset{\beta}{\rightleftharpoons}} AR_{open} \tag{7}$$

where A is the neurotransmitter concentration in the synaptic cleft and R is the receptor concentration. The transmitter combines with the receptor to form the complex AR, which then undergoes an allosteric transformation to the open channel state AR_{open}. This is the general reaction scheme studied by Magleby and Stevens (1972). The k_1, k_{-1}, α, and β represent the rate constants for the indicated reactions. If one assumes that the transmitter disappears from the cleft very quickly, then the reaction scheme can be simplified:

$$A + R \underset{k_{-1}}{\longleftarrow} AR \underset{\alpha}{\overset{\beta}{\rightleftharpoons}} AR_{open} \tag{8}$$

From here one can write differential equations for AR and AR_{open} and solve them. The solution for AR_{open} assuming that AR_{open} at time $t = 0$ is zero is

$$AR_{open} = AR(0) \, \beta(r_1 - r_2)^{-1}[\exp(r_1 t) - \exp(r_2 t)] \tag{9}$$

where $AR(0)$ is the transmitter–receptor complex formed at $t = 0$ due to a bolus of transmitter that disappears after $t = 0$ and $r_1 = -a + (a^2 - b)^{1/2}$,

$r_2 = -a - (a^2 - b)^{1/2}$, $a = (k_{-1} + \alpha + \beta)/2$, and $b = (k_{-1}\alpha)$. For repetitive inputs, $AR_{open}(0)$ may not be zero and the expressions are slightly more complicated (Holmes and Levy, 1990). Equation (9) is a double exponential. With appropriate choices of α, β, k_{-1}, and $AR(0)$, one can use AR_{open} to calculate the number of open channels at the synapse. The number of open channels at time t multiplied by the single-channel conductance gives the synaptic conductance.

Perkel *et al.* (1981) give a derivation of the α function and the double exponential function for a slightly different reaction equation. Other possible kinetic equations are discussed by Gage (1976).

To make use of these descriptions, one needs values for r_1 and r_2. These can be selected at random to produce a desired EPSP wave form, but much greater insight is possible if one determines r_1 and r_2 from the parameters α, β, k_{-1}, and $AR(0)$. The value of α can be estimated from the mean channel open time, but values for the other parameters are more difficult to determine. We do not know the number of receptors at a synapse, and we do not know how many of these receptors are bound with transmitter following transmitter release. However, given a number that seems reasonable for $AR(0)$, values of the other parameters can be estimated (although nonuniquely) by fitting values to give an EPSP at the soma similar to that observed experimentally. These values may imply something about the process being studied. For example, when I tried to find values for these parameters for the NMDA receptor-mediated synaptic conductance in dentate granule cells, I was driven to the conclusion that, during a synaptic event, only one or two NMDA receptor-channels are ever open at a given moment in time (Holmes and Levy, 1990). I was led to postulate that the transition from the transmitter–receptor complex to the open channel state must be a very difficult one. This observation was important for explaining the asynchronous activation results of Levy and Steward (1983) and Gustafsson and Wigstrom (1986).

A full description of a synaptic conductance change should include the diffusion and reuptake of the neurotransmitter, and the dissociation of the transmitter from the receptor. However, at present, such a model would have far too many unknown parameters to provide much insight. Consequently, it seems most reasonable at present to use the double exponential function described above and to try to estimate the four parameters it requires.

Modeling Methods

All simulation methods are based on the one-dimensional cable equation. For the derivation and assumptions of this equation one should see Rall (1977, 1989), Jack (1979), or Jack *et al.* (1975) among the many references available. In a general form the voltage in a dendritic segment can be described by the cable equation:

$$\tau \, \partial V/dt = \lambda^2 \, \partial^2 V/dx^2 - V + \Sigma \, I_i + \Sigma \, I_{syn} + I_{app} \qquad (10)$$

where τ is the membrane time constant, λ is the space constant, the I_i are the ionic currents specified by Eq. (1), the I_{syn} are the synaptic currents specified by Eq. (6), and I_{app} is the applied current, if any. There is one equation of the form given by Eq. (10) for each dendritic segment. At the junction of two or more dendritic segments, the solution of Equation (10) for each segment must satisfy the boundary conditions of continuity of voltage and conservation of current. There are several methods that can be used to solve the system of equations and their boundary conditions. One is to convert the partial differential equations into the ordinary differential equations of a compartmental model and then solve this simpler system in any of a number of ways (Rall, 1964; Perkel *et al.*, 1981). Other methods are to solve the system of equations with explicit (i.e., Hodgkin and Huxley, 1952) or implicit (i.e., Joyner *et al.*, 1978; Moore *et al.*, 1978) numerical schemes, or else with transform methods (Turner, 1984; Koch and Poggio, 1985; Holmes, 1986). For a good discussion of various numerical methods, see Mascagni (1989).

Inverse Laplace Transform Method

Although implicit methods are favored today, I will briefly describe the method I use which is based on Laplace transforms, noting the new features that have been added since the original description (Holmes, 1986). In this method each dendritic segment is modeled with the cable equation. By taking the Laplace transform of the cable equation (assuming conductances are constant), the partial differential equation is converted into an ordinary differential equation. This ordinary differential equation can be solved analytically. By applying the boundary conditions (conservation of current and continuity of voltage) the arbitrary constants in the analytic solution can be expressed in terms of the voltage at the start and end of the dendritic segments. One then has to solve a linear system of algebraic equations for the voltage at each node in the Laplace transform domain. This is done with a sparse matrix routine as by Sherman (1977). The voltage in the Laplace domain is inverted back to the time domain with a fast and accurate method developed by Stehfest (1970a,b).

Since the time this method was first described, the general procedure of going to the Laplace domain, solving the equations and inverting back to the time domain with the Stehfest algorithm has been proposed as an effective method for solving stiff systems of differential equations (Luxon, 1987). The method is stable and accurate although it is slower than a good Crank–Nicholson method (Hines, 1984) for problems with rapidly varying conductances. Time points need to be computed only when there is a change in conductance or whenever a time point is needed to make a graph look nicer. There is no Δt with which to be concerned. Thus this method is much faster than other methods when changes in conductances or current inputs are infrequent. The stability properties are evident when one models the HH axon with large time steps. However, one does not want to assume that the conductances are constant for longer than 0.1 msec during the generation of an action potential regardless of the stability of the method.

Although the implementation of the algorithm, as presented in Holmes (1986) but with subsequent additions, can be done in many ways, I use slight variations of the following procedure:

Initialization of parameters

GETDM   Read dimensions file and connectivity file.

Set up default R_m, R_i, and C_m and other values for each segment.

Set up bookkeeping arrays for the sparse matrix routine.

Long loop

CHGSPC   Allow changes to be made to any dimension, parameter, spine density, or replication factor.

CONDIS   Enter or change conductances on a segment, that is, constant conductances or HH-like conductances.

CONDIP   Enter or change point conductances at a node, that is, synaptic conductances.

Compute resistivity for each segment given the various conductances.

PRTEL   Print the electrotonic length of the segments and the electrotonic distance from the soma if desired.

CURIP   Enter or change current input or voltage clamp.

LINV   Initialization routine for the inverse Laplace transform (do once only).

Enter time point at which to evaluate the voltage.

Main Computational Loop

UPSCON   Update segment conductances that have changed since the last time point.

UPPCON   Update point (synaptic) conductances that have changed since the last time point.

GETPQR   Compute the matrix elements in the Laplace do-

main (Holmes, 1986); solve the sparse system in
the Laplace transform domain. Do this as often
as needed by the inverse Laplace transform pro-
cedure.
Finish the inverse Laplace transform calculation
Save results
Should the voltages just calculated be used as initial conditions
for the next time point?
Answer YES if the conductances will change.
Answer NO if the conductances will remain constant.
Should any specs be changed for the next time point?
If YES then go back to the beginning of the long loop.
If NO enter next time point and do main computational
loop unless done.
End of main computational loop
End of long loop

I will describe two features present in this modeling method that are not
readily apparent in the above description, but which are extremely useful.
The first of these is how the model includes dendritic spines, and the second
is how large models can be built with few segments by taking advantage of
replication.

Including Dendritic Spines

The dimensions file read by the routine GETDM, besides specifying the
length and diameter of each segment, contains fields giving the spine density
and the average spine size for each dendritic segment. The model uses this
information to "include" spines in the model without specifying compartments for every spine. A segment with many spines is modeled as an equivalent segment without spines by reducing R_m and increasing C_m for the segment
according to the increment in membrane area caused by the spines. For
example, if the membrane area of spines on a dendritic segment equals the
membrane area of the dendritic segment alone, then the model halves R_m
and doubles C_m for that particular segment (Fig. 1). The validity of this
approximation has been shown previously (cf., Holmes and Rall, 1992b). Of
course, if a spine is to receive synaptic input during the simulation, then it
must be specified in the model with separate segments defined for the spine
neck and head. However, models rarely activate more than a few hundred
synapses on spines, even when the neuron has tens of thousands of spines.

Structure   Equivalent modeled structure

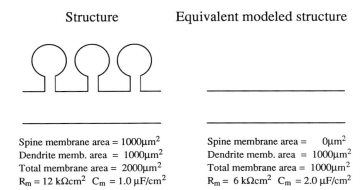

Spine membrane area = 1000μm²   Spine membrane area =   0μm²
Dendrite memb. area = 1000μm²   Dendrite memb. area = 1000μm²
Total membrane area = 2000μm²   Total membrane area = 1000μm²
R_m = 12 kΩcm² C_m = 1.0 μF/cm² R_m = 6 kΩcm² C_m = 2.0 μF/cm²

FIG. 1   Including spines in a model. A dendrite with spines can be modeled as a dendrite without spines if R_m is reduced and C_m is increased according to the contribution of the spines to the total membrane area of the dendrite.

Not having to code for every spine in the neuron reduces the computation time significantly.

Replication

The dimensions file also contains a field giving the number of times a particular segment (or the portion of the dendritic tree starting with that segment) should be replicated in the model. This is very useful if one is constructing a hypothetical neuron with many branches or trees that are symmetrical. The model only has to code the segment or tree once, and the replication factor makes it seem that the segment or tree appears multiple times without any additional computational time (Fig. 2A). I also have found replication useful when modeling synaptic inputs on spines. I can increase the number of spines at a given location by increasing the replication factor (Fig. 2B). If the model defines one spine in each of 10 different locations, I can activate synapses on 50 different spines by making the replication factor equal to five for each of the 10 defined spines. Of course, the synaptic input to each of the five spines at a given location must be identical, but in general this is not a significant restriction given the computational savings and the ease with which replication can be coded.

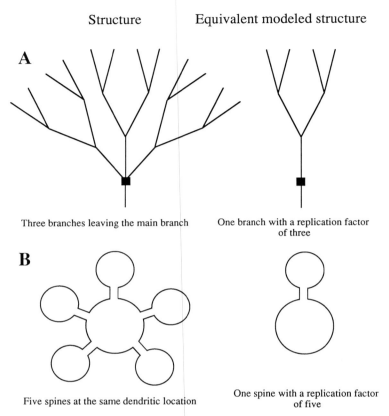

Structure   Equivalent modeled structure

A

Three branches leaving the main branch

One branch with a replication factor
of three

B

Five spines at the same dendritic location

One spine with a replication factor
of five

FIG. 2   Advantages of replication factors. (A) Branching structures. The branching structure at left, with three symmetrical branches leaving the main branch, can be modeled as the structure given at right, having one branch with a replication factor of three. (B) Replication of spines. A dendrite with spines is shown in cross-section. The dendrite with five spines at a given distance along the dendrite can be modeled as a dendrite with one spine replicated five times. The use of replication factors allows one to increase the complexity of the dendritic structure without increasing computation time.

Software for Neuronal Modeling

Programs for doing simulations were initially written by individuals for particular problems, for example, Cooley and Dodge (1966), Traub (1977), and Moore and Joyner (1974). Because the cable equation describes the voltage in a nerve cell in terms of resistances and capacitors, it was natural that some investigators began doing modeling studies with circuit analysis programs or general purpose simulators such as NET2 (Lev-Tov *et al.*, 1983) or SPICE (Segev *et al.*, 1985; Bunow *et al.*, 1985) and more recently SABER (Carnevale, 1990). These programs, not being written for neurons and their voltage-dependent conductances, are somewhat unwieldy and carry excess baggage not needed in neural simulations [although this is much less true of SABER than the other methods; in fact, SABER does have some definite advantages (see Carnevale, 1990)]. Today there are several simulation software packages designed specifically for doing simulations of individual neurons or networks of neurons. The potential modeler does not have to be concerned about knowing the particular details of the numerical algorithms to make use of these packages. We will give brief descriptions of five of the available software packages: NODUS, GENESIS, NEURON, NEURONC, and NEMO-SYS. One should be cautioned that these packages are constantly being improved and ported to new machines, and the remarks noted here may become quickly outdated.

NODUS (DeSchutter, 1989) is a simulator that runs on the Apple Macintosh line of computers. Version 3.1.4 runs on the Macintosh II or SE30 models while an older version (2.3) runs on some of the older Macs. It has the typical Macintosh interface, and, like most software for the Mac, one can figure out how to use it (at least on an elementary level) without reading the manual. Still, the manual is a good one, and several tutorial examples are described in detail. Single neurons and networks of neurons can be modeled. The user can define the kinetics of voltage-dependent conductances as long as they are in HH form. The one drawback of NODUS is really not a problem with NODUS itself, and that is speed. The Macintosh computers do not possess the computing speed of present UNIX workstations, and this might frustrate those who want to develop large, detailed models. However, for those used to doing models on the old PDP computers or small VAX computers, NODUS on the Mac is a good choice. It is available at reasonable cost from Erik DeSchutter at the Department of Neurology, University of Antwerp (UIA), Universiteitsplein 1, B2610 Antwerp, Belgium. A bitnet address is gen2@ccv.uia.ac.be. (At present, DeSchutter can be reached at erik@smaug.cns.caltech.edu.)

GENESIS (Wilson *et al.*, 1989) is a simulator developed at Cal Tech (Pasadena) that runs on SUN and DEC UNIX workstations. It has been ported successfully to other systems, including the IBM RS/6000. GENESIS has been the simulator used in the course in Computational Neuroscience taught at Woods Hole (MA) during the past few summers, and so has gained a considerable following. GENESIS can be used to simulate single neurons or large networks of neurons. There is a 200-page (but still incomplete) manual and several tutorials to help the user learn how to use GENESIS.

There is a free version of GENESIS available via ftp, but to get the current version (1.3) and the manual, one must become a member of the users group called BABEL. Besides the current program and the documentation, BABEL membership allows access to an electrotonic bulletin board where BABEL users can post their special additions to GENESIS (voltage-dependent conductance descriptions) and share information with other users. For information on GENESIS send e-mail to genesis@caltech.bitnet, genesis@caltech.edu, or genesis@smaug.cns.caltech.edu.

NEURON (Hines, 1984, 1989) is a simulator developed by Michael Hines at Duke University (Durham, NC) that has gained a considerable following judging by the number of abstracts at the last Society for Neuroscience meeting (USA) that mention using it. It succeeds a previous version called CABLE. The program uses an interpreter (based on HOC, an extension of the high order calculator described by Kernighan and Pike, 1984) and the emacs screen editor for the specification of nerve properties. Given this specification, the program sets up the equations and solves them. The particular algorithm used (Hines, 1984) has proved to be an excellent one, and a variation of this algorithm is now available in GENESIS. Versions are available for both DOS and UNIX systems. Although early UNIX versions ran only on SUN workstations, it has been ported to a variety of UNIX systems, including the IBM RS/6000 and the Cray. Although the manual is inadequate, the program comes with a tutorial and several examples, and the best way to learn NEURON is to study these. NEURON is available via ftp. For information send e-mail to hines@neuro.duke.edu.

NEURONC (Smith, 1992) is a simulator developed by Robert G. Smith at the University of Pennsylvania. NEURONC is a language and simulator program for simulating individual neurons or large neural circuits. Complicated neural circuits can be described in physiological terms with the "nc" language, which is an interpreter language based on HOC. The program forms the equations from the circuit description and solves them. Although the simulator is a general one, it also has some unique features that pertain specifically to the visual system and has been used to model large retinal circuits. It has been ported to AT-compatible UNIX systems, SUN systems, Cray, Convex, IBM RS/6000, Stardent 3000, and the IBM PC with DOS. There is an extensive manual with a series of tutorials and exercises. As with the above two programs, the best way to learn NEURONC is to do the tutorials and exercises. NEURONC is available via ftp. For information send e-mail to rob@retina.anatomy.upenn.edu or write to Robert G. Smith at the Department of Anatomy, University of Pennsylvania, Philadelphia, PA 19104-6058.

NEMOSYS or NEural MOdeling SYStem was developed at the University of California, Berkeley, by J. Tromp, F. Eeckman, and F. Theunissen. This simulator has a highly graphical front-end with extensive menus to allow one to specify or change parameters and control runs. The program uses a combination of explicit and implicit methods to solve the equations. At the present time the documentation is very much at a preliminary stage of development, although one can comb through the menus to figure out how to use the many available features. It runs on the IBM RS/6000, SUNs, Silicon Graphics machines, and other UNIX systems. NEMOSYS is available via ftp. For information, send e-mail to John Miller at jpm@acheta. berkeley.edu.

The choice of which simulator to use is a matter of personal taste, much the same way the word processor one uses is a matter of personal taste. The choice may be limited by the type of machine that one uses, the interface one prefers, or raw speed. There will be frustrations when a simulator will not do what a user may want it to do. For example, the simple tricks I use in my method to include spines or replicate parts of the dendritic tree may not be easy to do in these simulators at present. Nevertheless, many of the simulators allow the user to build in special functions, although learning how to do this may not be something everyone wants to do.

One should be aware of the fact that different simulators may give different results without one being right and the other wrong. For example, I have used NODUS, NEURON, and the inverse Laplace transform method to model the action potential produced in a HH axon by a 0.5-msec current pulse of 3 μA into one end of the axon begining at $t = 1$ msec. The different action potentials are shown in Fig. 3. The action potential shapes are similar, but the peaks are slightly different, and the times of the peaks are different. If one looks at the algorithms closely it is possible to understand why these differences occur. NODUS is a compartmental model, and the current is injected into the middle of the compartment rather than into the end. Furthermore, NODUS stopped the current 0.025 msec (one major time step) later than the other simulators, and consequently its action potential occurred first. NEURON uses the potential at the end of a segment (Δx) to update the voltage-dependent conductance values over the length of the segment, whereas the inverse Laplace transform method averages the potential at the start and end of the segment to update the conductances. Hence, the action potential in NEURON occurred later. Nevertheless the results with the three methods are closer to each other than any one is to the action potential observed experimentally by Hodgkin and Huxley.

Concluding Remarks

Although the proliferation of simulation software and fast computing machines has increased the use of computer modeling, it does not guarantee that good models will be developed or that the development of good models

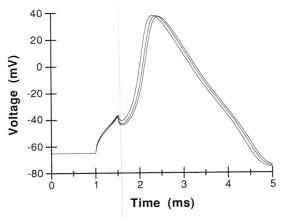

FIG. 3   Action potentials computed with three different computational methods. Action potentials in a HH axon were computed at the point of current injection with NODUS, NEURON, and the inverse Laplace transform method discussed in the text. A current of 3 μA was applied for 0.5 msec begining at $t = 1.0$ msec. For NODUS 20 compartments were used. Each had a diameter of 476 μm and a length of 500 μm. The major time step was 0.025 msec with the program determining the minor time steps. For NEURON 20 segments with a total length of 20,000 μm were used. NEURON divided the segments into subsegments 500 μm long. Values every 0.025 msec are plotted. For the inverse Laplace transform method there were 20 segments with a diameter of 476 μm and a length of 500 μm. The time step used was 0.05 msec. The action potential with NODUS peaked first, followed by that of the inverse Laplace method and then NEURON. Conductance descriptions and other parameters were standard HH.

will be any easier. This chapter has described methods that can enable one to calibrate and make appropriate assumptions about parameters included in a model. With the powerful simulation packages that are available today, there is a tremendous potential for abuse if one does not justify the assumptions made in a model. However, these powerful techniques also have a tremendous potential to expand significantly our knowledge of how nerve cells function.

References

Adams, P. R., and Galvan, M. (1986). *Adv. Neurol.* **44**, 137.
Almers, W. (1979). *Rev. Physiol. Biochem. Pharmacol.* **82**, 97.
Barrett, J. N., and Crill, W. E. (1974). *J. Physiol. (London)* **239**, 301.

Bernander, O., Douglas, R., Martin, K., Koch, C., and Niebur, E. (1991). *Soc. Neurosci. Abstr.* **21,** 1321.

Borg-Graham, L. (1989). M.S. Thesis, MIT, Cambridge, Massachusetts.

Brown, D. A., Gahwiler, B. H., Griffith, W. H., and Halliwell, J. V. (1990). *Prog. Brain Res.* **83,** 141.

Bunow, B., Segev, I., and Fleshman, J. W. (1985). *Biol. Cybernet.* **53,** 41.

Carnevale, N. T., Woolf, T. B., and Shepherd, G. M. (1990). *J. Neurosci. Methods* **33,** 135.

Cole, K. S. (1968). "Membranes, Ions and Impulses." Univ. of California Pres, Berkeley.

Cooley, J. W., and Dodge, F. A. (1966). *Biophys. J.* **6,** 583.

DeSchutter, E. (1989). *Comput. Biol. Med.* **19,** 71.

Desmond, N. L., and Levy, W. B. (1985). *Neurosci. Lett.* **54,** 219.

Durand, D. (1984). *Biophys. J.* **46,** 645.

Durand, D., and Carlen, P. L. (1985). *J. Neurophysiol.* **54,** 807.

Durand, D., Carlen, P. L., Gurevich, N., Ho, A., and Kunov, H. (1983). *J. Neurophysiol.* **50,** 1080.

Fisher, R. E., Gray, R., and Johnston, D. (1990). *J. Neurophysiol.* **64,** 91.

Fleshman, J. W., Segev, I., and Burke, R. E. (1988). *J. Neurophysiol.* **60,** 60.

Foster, K. R., Bidinger, J. M., and Carpenter, D. D. (1976). *Biophys. J.* **16,** 991.

Gage, P. W. (1976). *Physiol. Rev.* **56,** 177.

Gilbert, D. S. (1975). *J. Physiol.* (*London*) **253,** 257.

Gustafsson, B., and Wigstrom, H. (1986). *J. Neurosci.* **6,** 1575.

Haydon, D. A., Requena, J., and Urban, B. W. (1980). *J. Physiol.* (*London*) **309,** 229.

Hines, M. (1984). *Int. J. Biomed. Comput.* **15,** 69.

Hines, M. (1989). *Int. J. Biomed. Comput.* **24,** 55.

Hodgkin, A. L., and Huxley, A. F. (1952). *J. Physiol.* (*London*) **117,** 500.

Holmes, W. R. (1986). *Biol. Cybernet.* **55,** 115.

Holmes, W. R. (1989). *Brain Res.* **478,** 127.

Holmes, W. R., and Levy, W. B. (1990). *J. Neurophysiol.* **63,** 1148.

Holmes, W. R., and Rall, W. (1992a). *J. Neurophysiol.* (in press).

Holmes, W. R., and Rall, W. (1992b). *In* "Single Neuron Computation" (T. McKenna, J. Davis, and S. K. Zornetzer, eds.), pp. 7–25. Academic Press, Boston.

Holmes, W. R., and Woody, C. D. (1989). *Brain Res.* **505,** 12.

Iansek, R., and Redman, S. J. (1973). *J. Physiol.* (*London*) **234,** 613.

Jack, J. J. B. (1979). *In* "The Neurosciences: Fourth Study Program" (F. O. Schmitt, and F. G. Worden, eds.), pp. 423–438. MIT Press, Cambridge, Massachusetts.

Jack, J. J. B., and Redman, S. J. (1971). *J. Physiol.* (*London*) **215,** 283.

Jack, J. J. B., Noble, D., and Tsien, R. W. (1975). "Electric Current Flow in Excitable Cells." Oxford Univ. Press (Clarendon), Oxford.

Johnston, D., and Brown, T. H. (1983). *J. Neurophysiol.* **50,** 464.

Joyner, R. W., Westerfield, M., Moore, J. W., and Stockbridge, N. (1978). *Biophys. J.* **22,** 155.

Kernighan, B. W., and Pike, R. (1984). "The UNIX Programming Environment." Prentice-Hall, Englewood Cliffs, New Jersey.

Koch, C., and Poggio, T. (1985). *J. Neurosci. Methods* **12,** 303.

Lev-Tov, A., Miller, J. P., Burke, R. E., and Rall, W. (1983). *J. Neurophysiol.* **50,** 399.

Levy, W. B., and Steward, O. (1983). *Neuroscience* **8,** 791.

Luxon, B. A. (1987). *Bull. Math. Biol.* **49,** 395.

Magleby, K. L., and Stevens, C. F. (1972). *J. Physiol. (London)* **223,** 173.

Mascagni, M. V. (1989). *In* "Methods in Neuronal Modeling" (C. Koch and I. Segev, eds.), pp. 439–484. MIT Press, Cambridge, Massachusetts.

Mody, I., and Otis, T. S. (1989). *Soc. Neurosci. Abstr.* **19,** 1309.

Moore, J. W., and Joyner, R. W. (1974). *J. Theor. Biol.* **45,** 249.

Moore, J. W., Joyner, R. W., Brill, M. H., and Waxman, S. G. (1978). *Biophys. J.* **21,** 147.

Perkel, D. H., Mulloney, B., and Budelli, R. W. (1981). *Neuroscience* **6,** 823.

Pongracz, E., Firestein, S., and Shepherd, G. M. (1991). *J. Neurophysiol.* **65,** 747.

Rall, W. (1959). *Exp. Neurol.* **1,** 491.

Rall, W. (1962). *Ann. N.Y. Acad. Sci.* **96,** 1071.

Rall, W. (1964). *In* "Neural Theory and Modeling" (R. Reiss, ed.), pp. 73–97. Stanford Univ. Press, Stanford, California.

Rall, W. (1967). *J. Neurophysiol.* **30,** 1138.

Rall, W. (1969). *Biophys. J.* **9,** 1483.

Rall, W. (1977). *In* "Handbook of Physiology (Section 1): The Nervous System I. Cellular Biology of Neurons" (E. R. Kandel, ed.), pp. 39–97. American Physiological Society, Bethesda, Maryland.

Rall, W. (1989). *In* "Methods in Neuronal Modeling" (C. Koch and I. Segev, eds.), pp. 9–62. MIT Press, Cambridge, Massachusetts.

Rall, W. (1990). *In* "The Segmental Motor System" (M. D. Binder and L. M. Mendell, eds.), pp. 129–149. Oxford Univ. Press, Oxford.

Sah, P., Gibb, A. J., and Gage, P. W. (1988a). *J. Gen. Physiol.* **91,** 373.

Sah, P., Gibb, A. J., and Gage, P. W. (1988b). *J. Gen. Physiol.* **92,** 263.

Segev, I., Fleshman, J. W., Miller, J. P., and Bunow, B. (1985). *Biol. Cybernet.* **53,** 27.

Shelton, D. P. (1985). *Neuroscience* **14,** 111.

Sherman, A. H. (1977). *ACM Trans. Math. Software* **4,** 391.

Smith, R. G. (1992). *J. Neurosci. Methods* (submitted).

Stehfest, H. (1970a). *Commun. ACM* **13,** 47.

Stehfest, H. (1970b). *Commun. ACM* **13,** 624.

Stevens, J. K., and Trogadis, J. (1984). *Annu. Rev. Neurobiol.* **5,** 341.

Storm, J. (1990a). *Soc. Neurosci. Abstr.* **16,** 506.

Storm, J. F. (1990b). *Prog. Brain Res.* **83,** 161.

Stratford, K., Mason, A., Larkman, A., Major, G., Jack, J. J. B. (1989). *In* "The Computing Neuron" (R. Durbin, C. Miall, and G. Mitchison, eds.), pp. 296–321. Addison-Wesley, Workingham, U.K.

Takashima, S. (1976). *J. Membr. Biol.* **27,** 21.

Traub, R. D. (1977). *Biol. Cybernet.* **25,** 163.

Traub, R. D., Wong, R. K. S., Miles, R., and Michelson, H. (1991). *J. Neurophysiol.* **66,** 635.

Turner, D. A. (1984). *Biophys. J.* **46,** 73.

Westenbroek, R. E., Ahlijanian, M. K., and Catterall, W. A. (1990). *Nature (London)* **347,** 281.

White, E. L., and Rock, M. P. (1980). *J. Neurocytol.* **9,** 615.

Wilson, M. A., Bhalla, U. S., Uhley, J. D., and Bower, J. M. (1989). *Adv. Neural Inf. Process. Syst.* **1,** 485.

Wollner, D. A., and Catterall, W. A. (1986). *Proc. Natl. Acad. Sci. U.S.A.* **83,** 8424.

Yuen, G. L. F., and Durand, D. (1991). *Neuroscience* **41,** 411.

[32] Factors Influencing Neuronal Density on Sections: Quantitative Data Obtained by Computer Simulation

C. Duyckaerts, P. Delaère, C. Costa, and J.-J. Hauw

I. Introduction

The evaluation of neuronal density is often necessary in neuropathology and neuroanatomy. Neuronal density, in this chapter, refers to numerical density of neurons, that is, the number of neurons per unit area or per unit volume. The density of neurons per unit volume cannot be obtained directly from counts performed on usual histological sections. The overestimation made in evaluating cell density is obvious when one considers a series of contiguous sections: some cells are cut several times and are thus counted more than once. This effect is sometimes colloquially referred to as the "tomato slices" effect: the number of tomato slices in a salad is obviously much higher than the number of tomatoes. The naive and erroneous way to avoid recuts is to perform the counts on only one section. To take only one section, however, is actually equivalent to sampling the total number of sections; the sampling does not avoid the bias. The probability of picking up a slice coming from a large tomato is higher because the large tomatoes have been cut into more slices than the small ones. In mathematical terms, the counting of slices is biased in favor of the large tomatoes.

The inference of density in a volume [three-dimensional (3-D) density] from density in a section (2-D density) is only one of the difficulties which are met with in the evaluation of neuronal densities. In the 2-D space of the section, the location of each neuron in the plane can be represented by one single point defined by a pair of X and Y coordinates. The distribution of these points, when a large area is considered, is probably never random, since it conforms to the organization of the brain in layers, columns, or nuclei. Commonly used statistics that have been devised for random, normally distributed populations cannot be considered valid in these settings. Moreover, theoretical distributions which would correctly account for the observed data are presently lacking or mathematically intractable. Computer simulation is often the only way of obtaining the basic parameters (mean, variance, confidence interval) of the theoretical distributions which have been devised to describe the spatial distribution of points.

Methods in Neurosciences, Volume 10

Finally, the significance of a given 3-D density value can be questioned: highly significant large- and small-scale variations in neuronal density are present in the organization pattern of nervous tissue, even between structures located only a few microns apart or within a single given nucleus. In these heterogeneous structures, a global value of neuronal density wipes out much of the interest of the quantitative assessment. In this chapter, we consider two main topics: (1) the magnitude of the error made in inferring 3-D from 2-D density values and (2) the patterns of neuronal distribution in the 2-D space of the histological section.

II. From Three to Two Dimensions

A. Definitions

We used the terminology recommended by Weibel (1). A 2-D section of a given 3-D structure such as a neuronal perikaryon is called a profile. The 3-D structure is called a particle. We will use indiscriminately the terms neurons, cells, and particles. The number of particles per unit volume is N_v. The number of profiles per area is N_a. The letter t stands for the thickness of the section. A section is said to be ultrathin when its thickness is so small that it may be considered equal to 0 in comparison with the dimensions of the particles under study. The dimension of a particle, perpendicular to the plane of the microtome knife, is named the caliper diameter (C) (Fig. 1); it is the distance between two planes which are parallel to the plane of the section and touch the object tangentially (1). The probability of cutting a particle is linearly proportional to its caliper diameter. The caliper diameter of a sphere is the same whatever the orientation of the knife. For all other shapes, the caliper diameter depends on the relative orientation of both the knife and the particle. In some instances, the caliper diameter may change dramatically with the orientation of the particle. This is particularly evident with nerve processes. An axon oriented perpendicularly to the plane of the knife is necessarily involved by the section. By contrast the caliper diameter of the axon is small when it runs parallel to the plane of the section; it has only a low probability of being cut and of being visible on the side.

B. Overestimation of Numerical Density

1. Illustrative Data Obtained by Computer Simulation

The bias in favor of large particles can be illustrated by computer simulation. To perform it more easily, we made the simplifying assumption that neurons, although of different sizes, were all spherically shaped. (Theoretically, any

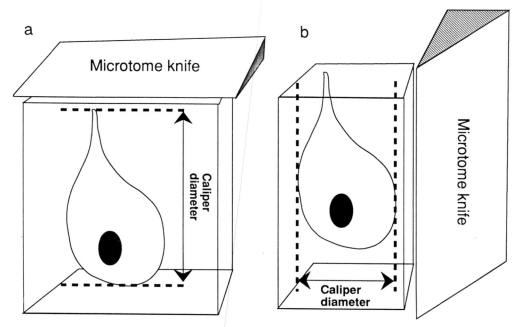

FIG. 1   The caliper diameter (*C*) is the distance between two planes (dotted lines) which are parallel to the plane of section (here represented by the microtome knife) and tangent to the particle (here a neuron). The caliper diameter depends on the orientation of the section: it is larger in (a) than in (b) even though the neurons have the same shape and size.

shape can be imagined as long as all particles affect the same shape.) The following experiment is then simulated: a fragment of cerebral cortex is cut at a known thickness. The neurons are drawn on a digitizing tablet and their diameter is measured. A histogram of diameters is obtained; increasing diameters are ranked by classes on the *X* axis. The number of particles in each class is on the *Y* axis. Once cut, each neuron yields two types of profiles, defined by different diameters: one type has a diameter which is equal to the maximal 3-D diameter of the whole neuron; the other type includes all the profiles that have not been cut at the equatorial plane and thus have a smaller diameter than the whole neuron. The diameters which belong to the first type will be called primary, the others, secondary.

   The histogram class at the extreme right includes only the primary diameters of the largest neurons (Fig. 2b,c). All the other classes may include both primary and secondary diameters. This observation has lead to the "unfolding" procedure, first proposed by Wicksell (2). This procedure is aimed at subtracting all the secondary diameters from the histogram columns (Fig. 2). The histogram class on the extreme right contains only primary diameters; the number of secondary diameters which are due to the recut of the largest particles, located at the extreme right of the histogram, can be calculated and subtracted from all the other histogram classes. After this subtraction has been performed, the last two columns located on the right of the histogram contain only primary diameters. The number of secondary diameters due to the recut of the particles whose primary diameter falls in the next to last column may then be subtracted from the remaining columns, leaving only primary diameters in the last three columns on the right; the procedure is repeated iteratively with all the size classes of the histogram, from right to left (Fig. 2). This calculation necessitates an *a priori* knowledge concerning the contribution of the large particles to the small profiles: in case of spherical shapes, this contribution has a simple mathematical form (1). We used (3) the simplest algorithm devised by Cruz-Orive (mentioned in Ref. 1). It is straightforward and can be implemented on microcomputers. Its mathematical expression is simple when written in matrix algebra. The transformation matrix from 2-D to 3-D data can be readily inverted to obtain 2-D from 3-D data. Cellular atrophy or loss may then be simulated in 3-D and converted into a 2-D histogram. Data coming from a size histogram of cortical neurons were used in the simulation in order to closely mimic a real situation (3).

2. Influence of Section Thickness

The influence of section thickness on the density of profiles is depicted in the simulation shown in Fig. 2. The first histogram (Fig. 2a) shows the "real distribution" of diameters in 3-D space. The two lower histograms (Fig. 2b,c) show the data which can be collected in a section.

a. Relative Number of Profiles

The histogram in Fig. 2b expresses the data as a percentage per unit area; the same simulation has been repeated for three different section thicknesses. The mode (= peak) of the distribution is similar for the three thickness values. On the left side of the mode, the 2-D histogram is richer than the initial 3-D histogram. This is due to the fact that small profiles ("secondary diameters" of the preceding section) have been produced by the recutting of the large neurons. The percentage of small profiles is highest for the thinnest section (1 μm) and lowest for the thickest one (100 μm). The thicker the section, the larger the fragment of the neuron that it may include. The diameter which is attributed to the fragment of neuron by an observer is the largest diameter

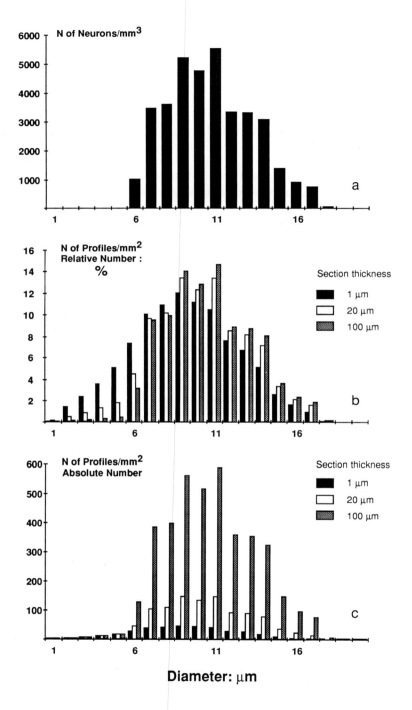

of the included fragment ("Holmes effect," Ref. 1). In terms of sampling, "a large fragment" of neuron can be regarded as a set of contiguous thin profiles, piled together into one single volume. Systematically choosing the largest diameter obviously reduces the proportion of small diameters in the histogram. On the contrary, in thin sections, the Holmes effect is minimal: small discrete profiles cut from large neurons are identified as such. Each profile, once cut, is visible as a discrete shape, because it not masked by the projection of the enclosed fragment of the neuron having a larger diameter.

b. Absolute Number of Profiles

The third histogram (Fig. 2c) presents the same data in absolute number of profiles per square millimeter. The total number of neurons included in one section is obviously larger with a thicker section; this is demonstrated by the larger surface area covered by the histogram obtained with a section thickness of 100 μm. The histogram classes situated on the left-hand side of the 6 μm diameter class include small profiles produced by cutting the neurons close to their poles. The number of small profiles included in these classes is identical whatever the section thickness (yet their relative number, as we have just seen, is much smaller in thick sections, see Fig. 2b). This is explained by the fact that a thick section is, so to speak, made of a volume sandwiched between two planes of section, one above and one below, which together constitute one and only one "ultrathin section": the upper plane of the section acts as ultrathin for neurons which are cut by the inferior pole (when the upper pole is sectioned, the rest of the neuron remains embedded

FIG. 2 Influence of the section thickness on the measured density of neuronal profiles. (a) Histogram of the number of neurons per cubic millimeter. The histogram depicts the density in 3-D space and was obtained by the deconvolution method in Brodmann's area 22 of one case (aged 33), with no known neurologic or psychiatric illness, who died from cardiac insufficiency. X axis, diameter of the neurons in μm; Y axis, number of neurons per mm^3. Neurons were supposed to be spherical. (b) Histogram of the number of neuronal profiles per square millimeter, expressed as a percentage, with three different section thicknesses, obtained by calculation using the 3-D data shown in (a). The number of recuts yielding small profiles is highest with the thinnest section (1 μm). X axis, same as in (a); Y axis, number of neuronal profiles per mm^2. (c) Histogram of the number of neuronal profiles per square millimeter, expressed as n absolute numbers, with three different section thicknesses. The calculation was performed as in (b). The profiles having a diameter smaller than 6 μm are always produced by the recut of neurons at a distance from the equatorial plane. Their absolute number is the same whatever the section thickness (see text). (Modified slightly from Ref. 3, with permission.)

in the section and the profile appears larger: no small profile is produced); the lower plane of the section acts as ultrathin for neurons which are cut by the superior pole for the same reason. This explains why both the upper and the lower planes of the section together make one—and not two—ultrathin sections. The 100 μm thick section of the histogram in Fig. 2c is made of an ultrathin section (left part of the histogram) plus a volume of 100 μm height; in a similar way, the 1 μm thick section is made of an ultrathin section plus a volume of 1 μm height. The ultrathin part of each section is the same whatever the thickness and is always represented on the left part of the histogram. The subtraction method which we consider below makes use of this phenomenon (see Section IIC,2, subtraction method).

c. Pathological Changes of Caliper Diameter and Pseudo-Loss

In the course of a pathological process, the caliper diameter may change, either chronically or acutely. Perikaryal atrophy, for instance, has been known for a long time under various names: Nissl's chronic cell disease, shrinkage of nerve cells (4), cell sclerosis (5), and simple atrophy (6). Ischemic cell changes, a frequent consequence of heart failure in the agonic state, can also modify perikaryal volume. These changes reduce the caliper diameter and thus the 2-D density of neuronal profiles. A count of the number of profiles per unit area may lead the observer to conclude that the neuronal 3-D density has decreased when only the volume of the neurons has changed. Size histograms, based on the measurement of profile diameters, might provide a reasonable solution to this problem. Neuronal atrophy would then be defined as the shift of a given neuron from one size class of the histogram to another class on its left, the total number of neurons remaining the same. Such a shift without loss of profiles would take place if the atrophy of particles could be examined in a volume.

On a tissue slice, however, the histogram itself is biased: imagine a neuron of caliper diameter C cut into ultrathin sections. After atrophy, its caliper diameter is $C/2$. The probability of cutting this neuron has also been divided by 2. The number of cells classified in the cell class C was N; after atrophy the number of cells recorded in class $C/2$ is only $N/2$: the total number of cells has decreased, and this could be falsely interpreted as a consequence of cell loss whereas it is due to cell atrophy. This is illustrated in Fig. 3. The upper histogram (Fig. 3a) represents the atrophy in 3-D space. The black columns show the number of neurons before atrophy. The white columns depict the situation after atrophy: the total surface (i.e., the total number of neurons) of the two histograms (before and after atrophy) is by definition the same, since atrophy increased the height of some (small diameter) columns in the exact proportion of the decrease of the height of some (large diameter) columns. By contrast, the total areas covered by the 2-D histograms of the

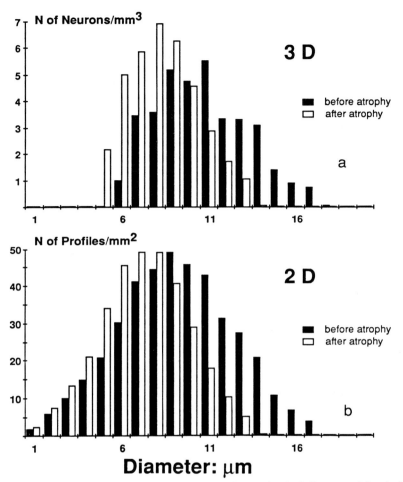

FIG. 3 Pseudo-loss. (a) Illustration of neuronal atrophy in 3-D space (simulation). An atrophy of neurons above 10 μm in diameter, by 50% of their diameter, was simulated. Solid columns, before atrophy; empty columns, after atrophy. The total area of the histogram before and after atrophy is, by definition, the same, since the decrease in the number of large neurons (before atrophy) is equal to the increase in the number of small neurons (after atrophy). (b) Two-dimensional representation of the histogram in (a), which refers to 3-D space. Section thickness is 1 μm. The total area of the histogram before atrophy (solid columns) is larger than the total area after atrophy (empty columns): the atrophy of neurons decreases their probability of being cut. (Modified slightly from Ref. 3, with permission.)

same populations of neurons (histogram of Fig. 2c) are not equal: the total number of profiles after atrophy is smaller than before atrophy. The decrease in caliper diameter has decreased the probability of cutting the atrophied neurons. Their profiles are thus less numerous.

In conditions involving thin sections and severe atrophy the loss of profiles can become significant and lead to the wrong conclusion of neuronal loss. This phenomenon has been called pseudo-loss (3): when neuronal atrophy occurs, the number of profiles on sections decreases, whereas the number of neurons per unit volume remains the same. The loss of profiles on section is proportional to the neuronal atrophy:

$$L = MA$$

where L stands for profile loss on section, A for atrophy, and M for the proportionality factor between these two values. Factor M tends toward 1 when the ratio of section thickness to neuronal caliper diameter tends toward zero; under these conditions, an atrophy of 50% leads to a pseudo-loss of 50% of the profiles. It tends toward zero when section thickness is high compared to the caliper diameter of the neuron (Fig. 4).

C. Techniques to Avoid Overestimation

Several techniques have been devised to avoid the overestimation of cellular density on sections and the bias that it introduces in statistical work. Three methods have been identified by Clarke (7): calculation, subtraction, and reconstruction. The reconstruction method belongs to a family of techniques such as 3-D counting (8) and disectors (9), which will be considered in the same section because they were developed to obtain directly an unbiased count of the particles within a sampling volume. The advent of confocal microscopy will probably add a fourth, even more straightforward technique which we shall consider briefly.

1. Calculation: Abercrombie Formula and Unfolding Procedure (Deconvolution)

a. Characteristic Point

In order to count each neuron once and only once, it is necessary to imagine that each neuron is reduced to one single point, which was called a nuclear point by Abercrombie (10). It is called a characteristic point here because it might not be nuclear: the characteristic point could indeed be any point in the particle volume, such as the center of gravity, a point on its surface, or

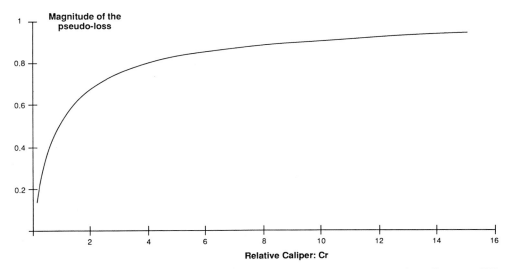

FIG. 4   Magnitude of the pseudo-loss as a function of relative caliper diameter. When a neuron undergoes atrophy, the density of its profiles on sections decreases (pseudo-loss) in such a way that the drop in neuronal density on the section is equal to M times the atrophy of the neuron. The factor of proportionality M is called magnitude of the pseudo-loss and is represented on the Y axis. M varies with the relative caliper diameter (C_r, which is the caliper diameter of the neuron/section thickness). At the most, when the relative caliper diameter is high (e.g., above 10 as is the case with an ultrathin section of large neurons), the drop in profile density tends to be equal to the neuronal atrophy: a 50% atrophy of the caliper diameter may cause a 50% loss of profiles on sections. The magnitude of pseudo-loss, M, is then equal to 1. At the opposite end of the curve, that is, when the relative caliper diameter is low (e.g., below 1 as when counting nucleoli in 100 μm sections), the magnitude of the pseudo-loss becomes small, for example, 0.33 for a relative caliper diameter of 0.5: a 50% atrophy of the neuronal diameter would then cause a 0.33 × 50% drop in profile density on sections or 16.5%. The derivation of M is given in Duyckaerts *et al.* (3) by $M = C_r/(1 + C_r)$.

a pole, with the restriction that a characteristic point cannot overlap two adjacent sections. Let us admit, for the time being, that each neuron is reduced to its center of gravity. Theoretically, a neuronal profile on the slice should now be counted only if it includes the characteristic point, that is, the center of gravity of the neuron. If only those profiles were counted, 3-D density could be immediately calculated:

$$N_v = N_a/t \qquad (1)$$

In other words, to obtain the density per cubic micron, one should divide the density per square micron (N_a) by the thickness of the section (t). No punctiform marker exists to indicate the center of gravity of the cell; numerous profiles seen on the slice should be discarded because their center of gravity lies outside the section. Yet, the profiles which should not be counted cannot be distinguished from those including the characteristic point.

b. Thickness of Slice

Among the profiles which are visible on the slide but which should not be counted are some cut through their inferior pole (the center of gravity being above the section), the others being through their superior pole (their center of gravity lying below the plane of the section). To include these profiles with their characteristic point, one would have to increase the volume where the count is performed. To t, the thickness of the slice, should be added two layers, one above and one under the original slide. Figure 5 shows that each layer (the one above and the one below the section) is 0.5 C thick, C being the caliper diameter. This transforms Eq. (1) to the Abercrombie formula (10):

$$N_v = N_a/(t + 0.5C + 0.5C) \rightarrow N_v = N_a/(t + C) \qquad (2)$$

In other words, the number of neurons per unit volume (N_v) is proportional to the number of neurons per unit area, that is, examined on the section (N_a), and inversely proportional to the size of the particle (C, caliper diameter) and to the thickness (t) of the section. Actually, very small profiles may not have been detected, either because they fall off the section or because they are not recognized as significant by the observer. This "lost cap" effect necessitates a corrective factor (h_0) in Abercrombie's formula. h_0 is the depth that a particle has to reach within the slice before it is recognized as a profile. $N_a = N_v(C + t - 2h_0)$, i.e., the Floderus formula (11). To simplify the formulas, h_0 will not be taken into account in the following sections. The reader may add this corrective factor afterward, when necessary.

Equation (2) makes it clear that neither the size of the particle nor the thickness of the section are significant by themselves: when the section thickness is very large compared to the diameter of the cells, cell size does not significantly influence the numerical density: $t \gg C \rightarrow t + C \approx t \rightarrow N_v \approx N_a/t$ [i.e., Eq. (1)]. On the other hand, the use of thin sections leads to a greater sensitivity of numerical density toward particle size. For thin sections:

$$C \gg t \rightarrow t + C \approx C \rightarrow N_v \approx N_a/C \qquad (3)$$

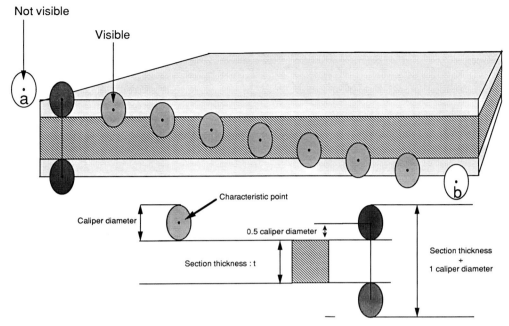

FIG. 5   Illustration of Abercrombie formula. A histological section of thickness t is shown (hatched). Neurons are depicted as ovoids. The characteristic point of each neuron is considered here to be its center of gravity. A neuron is counted only when its profile is included in the section: neurons a and b are not cut and therefore are not counted. All the light gray neurons are cut and counted. The characteristic point of all these neurons lies in a volume that is larger than the volume of the slice: this volume includes the section (of thickness t) sandwiched between two fictitious layers with a thickness equal to 0.5 times the caliper diameter. The volume where the counting is performed is therefore larger than the volume of the section; the volume of the section is equal to the area of the section times the section thickness. The actual volume in which the counting is performed is equal to the area of the section times the thickness plus the caliper diameter.

At the most, with an ultrathin section, a drop of 50% of the neuronal density could be a consequence either of the loss of half the neurons per unit volume or of neuronal atrophy with a 50% reduction in caliper diameter. In statistical terms, the evaluation of numerical density is then severely biased in favor of the large neurons.

Equations (1)–(3) show that the most significant parameter influencing the evaluation of numerical density is actually the relative caliper (C_r) of the particles, that is, the ratio of the caliper diameter to the section thickness, C/t (3). The overestimation of neuronal density can be easily calculated: the raw estimate of the numerical density per unit volume N_v is $N_v' = N_a/t$. The actual value is $N_v = N_a/(t + C)$. The overestimate is

$$N_v'/N_v = (N_a/t)[(t + C)/N_a] = (t + C)/t = 1 + (C/t) = 1 + C_r \quad (4)$$

In other words, the factor of overestimation equals 1 + relative caliper [Eq. (4)].

Some examples may help to grasp the significance of Eq. (4). When, for instance, the density of medium-sized pyramidal neurons 20 μm in diameter is evaluated with a 1 μm thick slice, the overestimation is $1 + 20/1 = 21$, that is, the real density of the neuron is 21 times less than the uncorrected estimate. When nucleoli (diameter 3 μm) are counted in 50 μm thick sections, the overestimation is $1 + 3/50 = 1.06$, that is, the uncorrected estimate is 6% too high in comparison with real data, a bias that most morphologist would probably accept in a large number of situations.

c. Mean Caliper Diameter

Various methods have been described to assess the caliper diameter (C) of a particle. If the particle shape can be considered spherical, the caliper diameter is equal to the largest diameter measured on the section. For all the other shapes, Abercrombie (10) proposed to recut the slice at a right angle to the plane of the section and to measure directly the caliper diameter. In practical situations, the cells under study do not all have the same caliper diameters. If the population is rather homogeneous, one may assume that the mean caliper diameter is representative of value C in Eq. (2). The mean value of the diameter measured on sections is an underestimation of the true mean diameter, since both complete particles and fragments of them (which lower the mean) are taken into account. According to Abercrombie (10), the error is small, however, and can be neglected in most practical applications. Woody *et al.* (12) have suggested that a computer-aided 3-D reconstruction of the particle might be performed. Mean value of C can then be calculated by simulating all the possible orientations of the particle in space. The problem becomes more complex when cells are heterogeneous in size as is usually the case, for example, in the cerebral cortex. If so, some unfolding algorithm (similar to the one described earlier in this chapter) has to be used.

Finally, the caliper diameters themselves may change during a pathological process; this modifies the bias, leading ultimately to pseudo-loss. The magnitude of this pseudo-loss can be assessed with the Abercrombie formula:

$$N_a = N_v(C + t)$$

Atrophy reduces C to $C' = aC$ with a being a factor of atrophy < 1.

$$N'_a = N_v(aC + t)$$

$$N_a/N'_a = (C + t)/(aC + t)$$

$$= (C/t + 1)/(aC/t + 1)$$

$$= (1 + C_r)/(1 + aC_r) \tag{5}$$

As above, $C_r = C/t$ is the relative caliper diameter.

Equation (5) again demonstrates that the significant value is actually the relative caliper. This has some practical implications. Imagine that only nucleoli are counted to assess neuronal density. Nucleoli in the temporal neocortex have a diameter of 3.58 μm in normal aged controls and 2.84 μm in Alzheimer's disease [data calculated from Mann *et al.* (13)]. With a section thickness of 20 μm, C_r is 0.179 in normal cases and 0.142 in Alzheimer's patients. The pseudo-loss with such an atrophy remains below 4%, a low enough value, worth considering before planning more complex methods of counting. For the same values, the overestimation of neuronal density in normal cases would be 1.179 (i.e., around 18%) and 1.143 (i.e., around 14%) in Alzheimer's cases. These error values are, in our opinion, not sufficiently high to lead the reader to definitely reject, as irremediably biased, data previously obtained in the literature with these classic methods. The simplicity of the classic techniques of counting also advocates for their current use in some specific situations, after the extent of the bias has been assessed and considered acceptable. Moreover, the calculation method is useful in evaluating the weight of the bias; unfolding algorithms are easily implemented on a microcomputer and can furnish illustrative data through simulation.

2. Subtraction Method

The subtraction method (method 2) was first described by Abercrombie (10). We previously noticed that the profiles consisted of two populations: one includes the characteristic points and provides an unbiased estimate of the density [group A according to Abercrombie (10)]; the other includes profiles whose characteristic point lies outside the slice, the counting of which leads to an overestimation of the numerical density (group B). In the simulation study (see Section II,B,2,b, Absolute number), the same data were expressed in a slightly different way; we considered that a thick section was made of a volume (actually including group A profiles) and of one ultrathin section (including group B profiles). When the section thickness is increased, the volume is also increased but the ultrathin section remains the same; the number of group A profiles increases but the number of group B profiles does not change.

$$N(\text{profiles}) = N(\text{A profiles}) + N(\text{B profiles})$$

After increasing the section thickness:

$$N(\text{profiles})' = N(\text{A profiles})' + N(\text{B profiles})$$

Subtraction of the two equations gives:

$$N(\text{profiles})' - N(\text{profiles}) = N(\text{A profiles})' - N(\text{A profiles})$$

The difference between the number of profiles after and before an increase in section thickness cancels the B profiles, that is, those which do not include the characteristic point and lead to an overestimation of cell density. The subtraction technique consists in counting profiles using two different section thicknesses. The difference between these two counts gives an unbiased estimate of the number of cells included in a volume, which is equal to the difference in volume of the two sections.

The elegance of the subtraction method is counterbalanced by a practical difficulty: the accuracy of the estimate relies heavily on the evaluation of section thickness. It has been shown by Clarke (7) that the actual thickness of a section does not correspond to the nominal thickness setting on the microtome and may change even in the same paraffin ribbon. Sensor technology now makes it possible to obtain sensitive length gauges which can be attached to the microscope stage and permit repeated measurements of section thickness. The efficiency of the subtraction method has been demonstrated by Clarke (7). Its aplication would probably be simplified by this new way of measuring section thickness but, to our knowledge, the pros and cons of the subtraction method have not yet been compared to those of the disector method.

3. Disector

Up to now in this chapter, the center of gravity was considered the characteristic point of the particle. Several authors have noted that it would be more efficient to select one of the poles of the particle (9). A new method may then be used to count particles: take two consecutive and parallel sections (hence the term disector), separated by a known distance h that is less than the height of the particles; some cells will be caught by the first section and escape the second one. This occurs only if the first section involves a pole of the particle. Counting the number of cells present on one section and absent on the next is thus equivalent to counting the inferior pole of the cell. This count is not biased and gives a reliable and efficient way of evaluating numerical density per unit volume. The theoretical perfection of the technique might mask some practical problems, one of them being to determine which cells are present in only one of the two slices. The optical disector was developed to solve this problem: one focuses down through a known thickness h in one continuous movement and counts how many new neuronal nuclei come into focus within the counting frame (14).

4. Confocal Microscopy

Confocal microscopy now allows faithful 3-D reconstructions of cells which could permit a cellular count within a volume. Unfortunately, only fluorescent stains can be used in practice; together with the cost, this will probably limit its use in morphometry.

III. Characteristics of Neuronal Distribution in Two-Dimensional Space

Avoiding the bias arising from the transition from 3-D to 2-D space is only one of the several problems faced when studying the distribution of neurons in a given region. The next step is to study the spatial distribution of the neurons within the plane of section itself. At least four difficulties can be mentioned.

1. Heterogeneity of the structure: A global value of density does not reflect possible regional variations (e.g., those observed in the multilayered cerebral cortex, or within nuclei).

2. Borders: Anatomical structures only exceptionally exhibit straight borders or clear limits. However, the graticules which are used to evaluate the numbers of cells or discrete lesions have geometrical shapes, such as squares, rectangles, and circles, which do not fit the anatomical shapes.

3. Confidence interval of the numerical density: The usual way of obtaining the confidence interval of a numerical density assessed by manual counting is to calculate the mean and standard deviation of the number of neurons, N, per sampling area. The standard error is then calculated as being the standard

deviation of $N/$(number of sampling areas)$^{1/2}$. This way of calculating the standard error is purely empirical and requires a sufficiently large number of samples.

4. Distribution of neurons: Up to now, there has been no way to describe in precise, mathematical terms the distribution of neurons in the plane of the section. A given neuronal 2-D distribution can be viewed as a "spatial point pattern," that is, according to Diggle (15), "a set of locations, irregularly distributed within a region of interest, which have been generated by some unknown random mechanism." The mechanism that generates these locations is called a spatial point process. The main problem met in studying these spatial distribution is their intrinsic complexity. Even apparently simple distributions cannot always be described in closed form. Their parameters can only be obtained by repeated simulation on a computer (Monte Carlo testing). Remarkably little is known about the statistical properties of the random, or rather partially random, mechanisms that govern the distribution of different groups of cells. The understanding of such mechanisms would make it possible to simulate a specific cell distribution with the computer. Our present inability to simulate neuronal distributions partly explains why quantification of neuronal loss and recognition of pathological cell distributions are still unsatisfactory.

A practical way to start the analysis of a given spatial point pattern is to test it against known theoretical distributions (15, 16). The first question which arises is whether the points (cells, neurons) are distributed in a completely random manner.

A. Complete Spatial Randomness: Poisson Distribution

In case of complete spatial randomness (CSR), the mean density of points per unit area is constant, whatever the size or the location of the sampling areas: a spatial point process with that property tends to follow a Poisson distribution which is described by the following equation (understanding of which is not compulsory for the rest of the chapter):

$$P(r) = \frac{(ls)^r}{r!} e^{-ls} \tag{6}$$

$P(r)$ is the probability of finding r points in a surface area S where the density of points is l; e is 2.71828, the natural logarithm base.

By contrast, in an aggregated point process, the density is lower than the mean in areas located between the aggregates and higher than the mean in the aggregates. In a regular point process, the density value is dependent on the sampling procedure: regularly spaced samples could lead to a systematic over- or underestimation of the mean density.

When the number of cells, n, encompassed by a microscopic graticule is counted over several fields used as sampling areas, the variance of n is equal to mean n in case of CSR. This is a distinctive property of a Poisson distribution. Actually, mean and variance of the number of cells per graticule seldom exhibit equal values in neuronal counts, suggesting that the CSR hypothesis should not be *a priori* taken for granted in practical situations.

The Poisson point process can be simulated on any microcomputer: two numbers, >0 and <1, are taken at random. The first number is the X coordinate of a point and the second is its Y coordinate. Iterating the process n times leads to a Poisson distribution with density n, over a plane of unit area.

Complete spatial randomness (CSR) represents a minimal standard that has to be tested first for any unknown spatial distribution. Further studies are useless most of the time if CSR is not rejected; moreover, CSR is used as a dividing line between aggregated patterns on the one hand and regular patterns on the other (15).

·r Distributions

⸱ CSR hypothesis is likely to be rejected in most neuronal counts. Other of point processes could be found which might be applicable to some ·l distributions. It is intuitively clear that in some regions, the neurons re regularly spaced than in the case of CSR and in other, probably ⸱n locations, be more aggregated. Some examples of point pro- ·ng such distributions (inhomogeneous Poisson, Cox, Markov can be found in Diggle (15): to our knowledge, neuronal ·ot yet been tested against these theoretical distributions.

·⸱s the often underestimated problem of edge ⸱e of the greatest difficulties encountered ⸱s. A Poisson distribution has precise, ·t are valid only when the points are ⸱r the area is limited, either by the ⸱erformed or by the boundaries of ⸱nger applies, at least to the fields adjoining the edges. For example, regardless of the distribution, neurons situated along the borders of the area have fewer neighbors than those located near the center, and density values vary accordingly.

This leads to the as yet unsolved problem of determining the minimal area for which a spatial distribution remains valid, that is, the minimal area (or the minimal number of cells) for which edge effects are negligible. On the other hand, edge effects are particularly significant in regions which have a large perimeter/area ratio, such as the cerebral cortex. The shape of the structure under study may also modify the magnitude of the edge effects.

D. Parameters

The formulas which describe the above-mentioned spatial point processes are actually no more than mathematical recipes; most of their properties cannot be directly apprehended in algebraic form. These recipes can, however, be translated into a computer language in order to obtain iterative representations of the model. The next step consists in accumulating experimental data, for example, coordinates of neurons, in order to test them against one of the aforementioned models, starting with CSR. The subjective judgment concerning the match between a theoretical and an observed point pattern is obviously insufficient, and some parameters describing the data (e.g., nearest-neighbor distances) should be used to evaluate the goodness of fit between the experimental data and the computer-simulated model.

Several parameters have been used to characterize a spatial point pattern interpoint distances, nearest-neighbor distances (distance between each poi and its nearest neighbor), distances between each point and some referer points, and quadrat count, that is, counting the number of points per samp area (15).

E. Monte Carlo Testing

Most of the parameters that were enumerated cannot be obtain theoretical equation, either because the equation itself is not kn cause it is too complex to be solved. The chosen random point it a Poisson, or any other process which can be simulated by t is then realized n times using the random number generator of For each realization of the point process, any statistic, here (e.g., mean nearest-neighbor distance), is calculated. The m mal values, Y_{max} and Y_{min}, of S after n random realization The experimental statistic S has the probability $1/n$ of fal and Y_{min} (e.g., 1/100 if the random process has been realized 100 times to obtain the values Y_{max} and Y_{min}, 1/1000 if 1000 iterations have been performed). The advantages of this so-called Monte Carlo technique are numerous: any random process, as long as it can be simulated, may be treated by this type of statistics; moreover, up to a certain extent, it has freed the morphometrist of the requirement of handling complex mathematical formulas. This pragmatic and descriptive approach is often used because the clear mathematical understanding of the model is lacking. Monte Carlo testng might well appear obsolete once full comprehensive of the model is reached.

F. Density and Dirichlet Tessellation

Most of the essential characteristics of a neuronal distribution are probably dependent on short-range interactions. A neurohistologist can recognize a specific area of the brain even at a high magnification since the relative position of no more than a few neurons is significant. Moreover, the neuropathologist almost always deals with variations of cellular density in very restricted areas of the brain. One global value of density calculated over a large area A does not have much significance in such a heterogeneous organ as the brain. A more precise picture would be achieved if not just one mean density was available, but several values, each corresponding to a different subregion of A. Ideally, one could increase the number of subregions, to the point where each neuron itself becomes a subregion; at that point, a density value would be affected to each neuron.

This description of very small-scale density alterations has been made possible through a new way of evaluating numerical density (17). Instead of counting the number of neurons in a sample of known area, the "area left free" around each cell is measured and the statistics performed on this value. The area left free around each cell is inversely proportional to the numerical density: it is small when cells are tightly packed together and large otherwise. When all the points are considered, the various free areas (each one depending on a cell) perform a tessellation of the plane, that is, cover the plane as a pavement. Each area is a polygon called the Dirichlet polygon. An individual density [= 1/(Dirichlet polygon area)] may be attributed to each individual neuron. Monte Carlo testing showed that the coefficient of variation for the Dirichlet polygon areas ranged from 30 to 81% in a Poisson point process (these values are conservative). Below 30%, the distribution of neurons may be considered too regular, as compared to a Poisson distribution; above 81%, the distribution is aggregated or clustered. Giving different colors to the various polygons according to their individual surface area generates color density maps (Fig. 6).

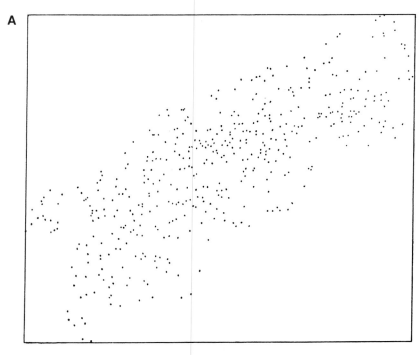

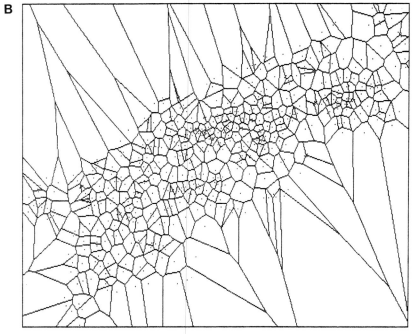

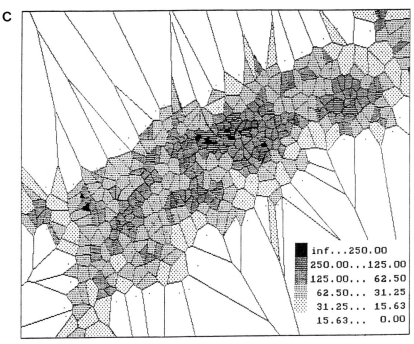

FIG. 6 Dirichlet tessellation to evaluate neuronal density. Nucleus basalis of Meynert. The coordinates of the neurons have been plotted with an image-analyzer (Biocom 200). (A) Each point represents a neuron. (B) Dirichlet tessellation. Polygons were drawn around each neuron. Any point in a polygon is closer to the included neuron than to any other. (C) Densities are attributed to each polygon according to its area. See figure for relationship between densities and polygon areas. In parentheses are colors of the same areas shown in the color photo of this panel on the cover of the comb-bound edition of this volume.

Color	Area	Density
(Brown)	0–4000 μm^2	250/mm^2 or higher
(Red)	4000–8000 μm^2	126/mm^2 $\leq$ density < 250/mm^2
(Orange)	8000–16,000 μm^2	62.5/mm^2 $\leq$ density < 125/mm^2
(Green)	16,000–32,000 μm^2	31.25/mm^2 $\leq$ density < 62.5/mm^2
(Blue)	32,000–64,000 μm^2	15.625/mm^2 $\leq$ density < 31.25/mm^2
(Dark blue)	64,000 or higher	Lower than 15.625/mm^2

IV. Conclusions

Two main difficulties are encountered in cell density assessment. (1) A histo-logical slide is a 2-D projection of a 3-D volume; quantification is distorted by the transition from 2-D to 3-D space. This causes an overestimation of cellular density and may lead to the erroneous conclusion of cell loss in cases of cell atrophy (pseudo-loss). (2) The difficulties encountered in describing the 2-D distribution of cells are still incompletely solved. Although numerous techniques have improved the understanding of the spatial distribution of points, their application to neurohistology has so far been limited. These methods resort to modeling of spatial point patterns and Monte Carlo testing. They should permit a better understanding of how cells, and more specifically neurons, are distributed within a tissue section, in normal and in pathological cases.

References

1. E. R. Weibel, "Stereological Methods, Volume 1: Practical Methods for Biological Morphometry." Academic Press, London, 1979.
2. S. D .Wicksell, *Biometrika* **17,** 84 (1925).
3. C. Duyckaerts, E. Llamas, P. Delaère, P. Miele, and J.-J. Hauw, *Brain Res.* **504,** 94 (1989).
4. M. Bielschowsky, *In* "Cytology and Cellular Pathology of the Nervous System," pp. 147–188. P. B. Hoeber, New York, 1932.
5. W. Spielmeyer, "Histopatholgie des Nervensystems," pp. 63–67. Springer-Verlag, Berlin, 1922.
6. W. Blackwood and J. A. N. Corsellis, "Greenfield's Neuropathology," 3rd Ed., p. 16. Edward Arnold, London, 1976.
7. R. Clarke, *J. R. Microsc. Soc.* **88,** 189 (1967).
8. R. Williams and P. Rakic, *J. Comp. Neurol.* **278,** 344 (1988).
9. D. C. Sterio, *J. Microsc.* **134,** 127 (1984).
10. M. Abercrombie, *Anat. Rec.* **94,** 239 (1946).
11. S. Floderus, *Acta Pathol. Microbiol. Scand.* **53**(Suppl.), 1 (1944).
12. D. Woody, E. Woody, and J. D. Crapo, *J. Microsc.* **118,** 421 (1980).
13. D. M. A. Mann, P. O. Yates, and B. Marcyniuk, *J. Neurol. Sci.* **69,** 139 (1985).
14. M. J. West and H. J. G. Gundersen, *J. Comp. Neurol.* **296,** 1 (1990).
15. P. J. Diggle, "Statistical Analysis of Spatial Point Patterns." Academic Press, London, 1983.
16. B. D. Ripley, "Statistical Inference for Spatial Processes." Cambridge Univ. Press, Cambridge, 1991.
17. C. Duyckaerts, C. Costa, D. Seilhean, P. Delaère, and J.-J. Hauw, submitted 1991.

[33] Physical Modeling of Neural Networks

Lipo Wang and John Ross

Introduction

There has been extensive research in the area of artificial neural network (NN) models (1–17). All models are radically simplified representations of experiments or intuition. The aim of such models is the attainment of a balance between incorporation of a semblance to observations and tractability for systematic studies and analytic predictions. We present studies on a general class of neural network models, that is, a randomly diluted network of McColluch–Pitts neurons which interact via Hebbian-type connections. We derive and solve exact dynamic equations that describe how the system evolves from its initial state under various conditions. The motivation for these studies is the incorporation of neurobiological features into theoretical models and studying their effects on the emergent network performance of associative memory. Specifically, we study interactions of neural networks, a variable threshold, hysteresis at the level of a single neuron, and higher order synaptic interactions in the presence of noise. Our models simulate a wide variety of phenomena such as distraction, concentration, selective attention, (de)sensitization, anesthesia, noise resistance, oscillations, chaos, crisis, and multiplicity.

Noise in Neurons and Little–Hopfield Models

In this section we first present a general discussion on noise in neurons and then give a brief review on neural network models discussed by Little (2) and Hopfield (4), together with a version studied by Derrida, Gardner, and Zippelius (10). Consider N spinlike neurons that have two states, namely, active and quiescent, or, $S_i = \pm 1$, where $i = 1, \ldots, N$. Each neuron receives signals from its neighboring neurons, and the signals are affected by synaptic weights T_{ij}. The neuron then either fires if the total input h_i exceeds a threshold θ_i, or remains quiescent otherwise, that is,

$$S_i(t + \Delta t) = \text{sign}[h_i(t) - \theta_i] \tag{1}$$

Here the input $h_i(t)$ is equal to

$$h_i^\circ(t) = \sum_{j=1}^{N} T_{ij} S_j(t) \tag{2}$$

for deterministic (zero-noise) dynamics, such as in the original model by Hopfield (4); alternatively, $h_i(t)$ may fluctuate randomly around $h_i^\circ(t)$ in the presence of noise, which models probabilistic releases of synaptic vesicles of neurotransmitters (18). These spinlike systems, though crude compared to biological NNs, already display intriguing features, such as a form of learning and recalling of associative memory. Suppose the probability distribution that the total input takes the value h_i is $D(h_i - h_i^\circ)$, where we assume that $D(x)$ is an even function and peaks at $x = 0$ so that h_i° is the average and the most probable value of the total input h_i. With this distribution the probability that neuron i will fire if updated according to Eq. (1), that is,

$$P(+1) = \int_{\theta_i}^{+\infty} D(h_i - h_i^\circ)\, dh_i = \int_{\theta_i - h_i^\circ}^{+\infty} D(x)\, dx \tag{3}$$

is a ''sigmoid function'' of h_i°: for $h_i^\circ \gg \theta_i$, $P(+1) = 1$, for $h_i^\circ \ll \theta_i$, $P(+1) = 0$, for $h_i^\circ = \theta_i$, $P(+1) = \frac{1}{2}$.

Noise may therefore be introduced by specifying either the probability distribution $D(x)$ or the probability function $P(+1)$. Little (2) first introduced a stochastic feature in the model and chose the Fermi probability function

$$P_{\text{little}}(+1) = \frac{1}{e^{-\beta(h_i^\circ - \theta_i)} + 1} \tag{4}$$

which is a sigmoid function. In neurobiological systems, the noise is Gaussian distributed (18):

$$D(h_i - h_i^\circ) = \frac{1}{(2\pi)^{1/2}\sigma_0}\, e^{-(h_i - h_i^\circ)^2/2\sigma_0^2} \tag{5}$$

Using Eq. (3), one can calculate the probability that neuron i will fire (3):

$$P(+1) = \frac{1}{(2\pi)^{1/2}} \int_{(\theta_i - h_i^\circ)/\sigma_0}^{+\infty} e^{-x^2/2}\, dx = \frac{1}{2}\left[1 + \text{erf}\left(\frac{h_i^\circ - \theta_i}{(2\sigma_0)^{1/2}}\right)\right], \tag{6}$$

where

$$\text{erf}(y) = \frac{2}{\pi^{1/2}} \int_0^y e^{-x^2}\, dx \tag{7}$$

is the standard error function. The probability given in Eq. (6) is also a sigmoid function. In fact, it differs from the Little probability function, which is normalized to unity, by 0.01 at most at any value of the variable, if one chooses (3, 6) the inverse temperature in Eq. (4) to be $\beta = 2(2/\pi)^{1/2}\, 1/\sigma_0 = 1.6/\sigma_0$.

Once the probability function $P(+1)$ is known, the noise distribution $D(x)$ can be calculated by a simple differentiation, that is, $D(x) = -\partial/\partial x$ $[P(+1)|_{\theta_i - h_i^\circ = x}]$, according to Eq. (3). For instance, the noise defined by Little, Eq. (4), has a distribution function $D(x) = \beta \exp(-\beta x)[\exp(-\beta x) + 1]^{-2}$, which is very close to a Gaussian distribution and is also symmetric with respect to $x = 0$. Further, it is convenient to introduce the variable (11) $\eta_i \equiv h_i - h_i^\circ$, which has an average of zero and the same standard deviation and distribution function as h_i.

Hopfield (4) studied a network in which neurons are updated sequentially and synaptic connections are chosen to be

$$T_{ij}^{\text{H}} = \sum_{\mu=1}^{p} S_i^\mu S_j^\mu \tag{8}$$

according to the Hebbian rule (19); $\vec{S}^\mu$ is the μth stored pattern, and p is the number of patterns stored. An energy function can be defined and is shown never to increase in the absence of noise: neurons update until an energy minimum (a stored pattern) is reached. The Hopfield model has been solved exactly (5) with powerful tools of statistical physics, and the system makes a phase transition to a disordered (no-memory) state if the temperature (noise) and the number of store patterns exceed certain critical values. The Little model (2), in which all neurons are updated in parallel, cannot be solved exactly; however, a diluted version does have exact solutions (10). In this model, the synaptic connections are

$$T_{ij} = C_{ij} T_{ij}^{\text{H}} \tag{9}$$

where the random number C_{ij} assumes 1 with probability C/N and 0 with probability $1 - C/N$. Hence only C of N synapses remain for each neuron after dilution. If $C \ll \ln N$, the dilution process eliminates correlations among neurons and facilitates exact solution of the network dynamics. Because C_{ij}

and C_{ji} are independently chosen, one has $T_{ij} \neq T_{ji}$ and hence the network becomes asymmetric after dilution, which simulates the sparse connectivity and asymmetry in biological neural networks (1). The model shows a phase transition at a critical noise level quite different from the fully connected Hopfield model (5, 10). We shall incorporate this random synaptic dilution in our studies throughout this chapter.

Neural network models may function without noise, but the introduction of noise into such networks brings additional features. Noise reduces stabilities of memories in the models discussed above and is hence undesirable for pure pattern storage/recall. In optimization problems, however, the existence of noise helps the network to search a larger portion of the phase space and find a deeper energy minimum, which corresponds to a better solution to the problem, after the noise is gradually reduced to zero (20). This simulates the annealing process in metal. It has been found (21) that a substance called norepinephrine, secreted by the cells of the locus coeruleus (LC), preferentially inhibits spontaneous neural activity (noise) over stimulated activity, and it has be conjectured (12) that the LC may be involved in a type of simulated annealing (20).

There are interesting similarities, and some essential differences, between neural networks models and the evolution reactor model of Eigen (22) for prebiotic evolution. A detailed comparison is given in Ref. 23 in regard to the role of noise and other characteristics. In evolutionary models noise is an essential component; without noise no errors in replication, that is, no evolution, can occur.

Interactions of Neural Networks: Models for Distraction and Concentration

In this section, we model distraction and concentration by studying interactions among groups of neurons (14) and relate the results with experiments on infant rats. For simplicity, we consider a diluted NN A and a fully connected Hopfield network B, whose dynamics and synapses are defined in the previous section, with $\theta_i = 0$. Suppose that the initial state of the network A is set in the neighborhood of pattern $\vec{S}^1$, that is, $m^1(0) = \max\{m^\mu(0)|\mu = 1, 2, \ldots, p\}$, where $m^\mu(t) = (1/N)\vec{S}^\mu \cdot \vec{S}(t)$ is the overlap between the state of the system at time t and the μth pattern. $m^\mu(t)$ yields a measure of the state of the system at a given time. Let network B settle down to another stored pattern, $\vec{S}^2$. Suppose each neuron in network B projects a field proportional to its state to the corresponding neuron in network A; thus, we consider a local field of neuron i in network A, $h_i(t) = \sum_{j=1}^{N} T_{ij}S_j(t) + \lambda S_i^2 + \eta_i$, where λ is the strength of the projection. Exact equations that describe the dynamical evolution of network A from its initial state can then be derived (14) in a similar fashion as in Ref. 10, which we will not repeat here. The basic idea is to separate the total input to a neuron into two parts (4): a signal term that "pulls" the state of the network toward stored pattern(s), and a noise term that models the combined effects from the probabilistic nature of neural signaling and interferences from other stored patterns. Let all neurons update their states simultaneously at a given time (synchronous updating). The time development of the overlap between the state of the network and stored pattern(s) can then be computed.

We first consider the case where the projecting network B is in a pattern different from the initial attracting state of network A, that is, $\vec{S}^1 \neq \vec{S}^2$. The results (14) are

$$
\begin{aligned}
m^1(t + 1) &= \left(\frac{d}{N}\right) \text{erf}\left[\frac{m^1(t) - m^2(t) - (\lambda/C)}{(2)^{1/2}\sigma_0}\right] \\
&+ \left(1 - \frac{d}{N}\right) \text{erf}\left[\frac{m^1(t) + m^2(t) + (\lambda/C)}{(2)^{1/2}\sigma_0}\right] \\
m^2(t + 1) &= \left(\frac{d}{N}\right) \text{erf}\left[\frac{m^2(t) - m^1(t) + (\lambda/C)}{(2)^{1/2}\sigma_0}\right] \\
&+ \left(1 - \frac{d}{N}\right) \text{erf}\left[\frac{m^2(t) + m^1(t) + (\lambda/C)}{(2)^{1/2}\sigma_0}\right]
\end{aligned}
\tag{10}
$$

where $\sigma \equiv [(p - 2)/C + (\sigma_0/C)^2]^{1/2}$ and d is the Hamming distance between $\vec{S}^1$ and $\vec{S}^2$, that is, they are d bits different.

By analyzing Eq. (10), we find a critical value $\lambda_c \equiv C[m^1(0) - m^2(0)]$ such that, for $\lambda > \lambda_c$, network A is pulled out of its initial basin of attraction and into that of the projecting pattern. In the case of λ below λ_c, the final state of network A can be calculated by iterating Eq. (10), and the results are given in Fig. 1. There exists a noise threshold σ_λ corresponding to each λ. For the case where $\lambda < \lambda_c$ and the noise level $\sigma < \sigma_\lambda$, there are two possible retrievals, with different probabilities: the initial attracting state of network A and the projecting pattern. If $\lambda < \lambda_c$ and $\sigma > \sigma_\lambda$, stable states of network A disappear. This analysis provides a model for distraction.

When the projecting network B is in the same pattern as the initial attracting state of network A, the projection acts as an external reinforcement, which enables the network A to retrieve in highly noisy conditions. The dynamics of network A is described by

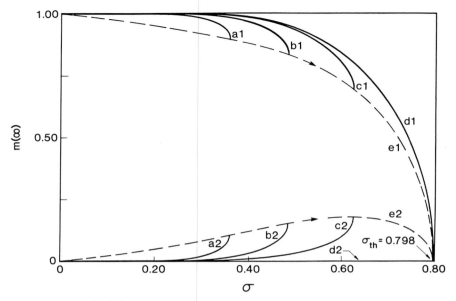

FIG. 1   A model of distraction: two possible retrievals when the projecting pattern is not the same as the initial attracting pattern. Solid lines labeled with (a), (b), (c), and (d) are the average overlaps between the final state of network A and the two patterns $\vec{S}^1$ and $\vec{S}^2$ according to Eq. (10), for projection strengths $\lambda/C = 0.34$, 0.20, 0.08, and 0, respectively, for case (i) and $\lambda < \lambda_c$ as a function of σ, the noise level. The numerals 1 and 2 represent the corresponding overlaps with patterns $\vec{S}^1$ and $\vec{S}^2$, respectively. The dashed lines labeled (e) are the envelops of the overlap curves at noise thresholds for different projection strengths λ. The arrows on these dashed lines indicate the directions in which λ decreases ($0 < \lambda/C < 1$). The noise thresholds for curves (a) through (d) are $\sigma_\lambda = 0.360$, 0.483, 0.624, and 0.798, respectively.

$$m(t + 1) = \text{erf}\left[\frac{m(t) + (\lambda/C)}{(2)^{1/2}\sigma_0}\right] \qquad (11)$$

Sharp noise thresholds for nonzero retrievals are shown to be eliminated by the projection (see Fig. 2). The second case serves as a model of concentration.

In the remainder of this section we present some discussions of this model. By definition (24), distraction and concentration are opposite ways of directing one's attention. Attention is the act of focusing one's mind on a subject: to concentrate is to direct one's thoughts or efforts, or to fix one's attention, whereas to distract is to draw one's attention away in another direction.

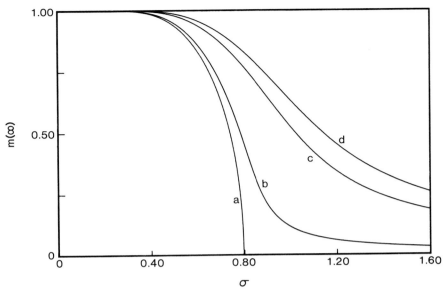

FIG. 2 A model of concentration: one enhanced retrieval when the projecting pattern is the same as the initial attracting pattern. Average overlap between the final state of network A and the projecting pattern as a function of noise level σ at different projection strength: (a) $\lambda/C, = 0$, (b) $\lambda/C = 0.03$, (c) $\lambda/C = 0.19$, and (d) $\lambda/C = 0.27$, in the case where the projection reinforces the processes of evolution in network A, according to Eq. (11).

The basic mechanisms by which the brain accomplishes its diverse acts of attention remain largely unknown, despite much research done on attention mechanism (see, e.g., Ref. 25), in particular on its physiology. It has been suggested that a subportion of the brain called the posterior parietal cortex, which is generally referred to as area 5 and 7, plays a major role in directing attention (25), whereas other parts of the brain, for instance, the basal ganglia and the cerebellum, are more important for execution of an action. "Anticipatory sets" represent a physiological organization in the brain that facilitates processes of expectancy, especially in the field of conditioned response (26). The neural counterparts and derivatives of the psychological expectancies are thought to lie at the interface between the sensory and motor systems of the brain, exert their influences on these domains, and conceivably play a role in such mental processes as selective attention, anticipation, and motivation (26).

In the models presented in this section, we considered a neural system consisting of networks A and B. Suppose that network A is the decision-making network, for example, it is connected to a motor system, and network B directs A's attention, through a projection, to pattern $\vec{S}^*$, or in a broader sense, a decision, of a particular interest according to past experiences. We also assume that B evolves much faster than A so that when A and B are exposed to the same (sensory) input, B immediately reaches a stable state $\vec{S}^*$. We assume that $\vec{S}$ is the pattern to be recognized, or the correct decision to be made, and $\vec{S}$ may be either different from or the same as $\vec{S}^*$. Hence networks A and B, in a highly simplified model, assume roles analogous to the cerebellum and the posterior parietal cortex, respectively. Pattern $\vec{S}^*$ plays the role of the expectation of the system.

In the first case, the projecting pattern is different from the pattern to be recognized, that is, $\vec{S}^* \neq \vec{S}$; the influence from B to A is incorrect. The case may be used to model distraction: the attention of the system is distracted toward pattern $\vec{S}^*$, which is different from the pattern $\vec{S}$ to be recognized. Our analysis shows that the chance of making the correct pattern recognition is decreased by the distraction modeled by a projection from network B. If the projection strength λ is below a critical value λ_c and the noise level σ is below a threshold σ_λ, our results predict that there exists a certain probability for network A to retrieve the projecting pattern $\vec{S}^*$, though the probability to retrieve $\vec{S}$ is greater. When $\lambda < \lambda_c$ and $\sigma > \sigma_\lambda$, there are no retrievals. The noise threshold σ_λ decreases as the strength of the projection λ increases, which indicates that the retrieval ability of the system decreases with increasing distraction. If the projection strength is equal or above the critical value λ_c, the retrieval of network A is most likely to be the distracting pattern $\vec{S}^*$, whereas the probability of retrieving pattern $\vec{S}$ vanishes. For a task of pattern recognition, if the input is too "fuzzy," which is represented by a small initial overlap between the state of network A and the pattern to be recognized, we know that λ_c is also small, and hence a small λ ($>\lambda_c$) suffices for network A to retrieve $\vec{S}^*$.

In the second case, where $\vec{S}^* = \vec{S}$, the influence from B to A is the correct one for recognition of $\vec{S}$. Our results show that the probability of making the correct recognition is enhanced by the influence of B, and this is our model for concentration.

There are interesting experiments (27) on how expectancies of infant rats can influence their performances in connection with our models. In these studies (27) infant rats are trained to reach the correct goal box in a simple T-maze after deprivation. The reward for the rats after successfully approaching the correct goal box is either the opportunity to suckle or milk in the absence of the mother. The performances of the rats are found to be in the following decreasing order: (i) preweanling rats with reward to suckle, or milk in the presence of maternal odors (the so-called shavings effect, cf.

Ref. 27), (ii) weaning rats with milk alone, and (iii) preweanling rats with milk alone. These results suggest that (27) preweanling rats deprived of nutrients, having never obtained nutrients from any source other than the mother, have an expectancy that their nutritional needs will be met by locating and approaching the mother. The experimental results may be explained with our theory as follows. The correct "pattern" $\vec{S}$ to be recognized is "milk is available in the goal box," no matter whether the mother is present or not. The expectancies of the rats vary in these three cases. In experiment (i), either the mother or maternal odors are present and the preweanling rats think that the mother is in the goal box. Hence the expectancy, which is represented by pattern $\vec{S}*$, is also "milk is available in the goal box," that is, $\vec{S} = \vec{S}*$ (concentration). In this case the probability that the rats find the correct goal box is therefore enhanced by the expectancy and the performance is the best. However, in experiment (iii), the absence of the mother and maternal odors implies to the preweanling rats the absence of milk, therefore the expectancy in this case is "milk is not available in the goal box," that is, $\vec{S} \neq \vec{S}*$ (distraction). Because the rats are in the preweanling stage, we anticipate the expectancy, a distraction, to be large (λ_3 large). The distraction reduces the chance of reaching the correct goal box, and the performance is the worst. After the infant rats have acquired early weaning experience, as in experiment (ii), the association between mother and nutrients becomes weakened. In this stage the expectancy is still $\vec{S}* = $ "milk is not available in the goal box" $\neq \vec{S}$, but the distraction is less in (ii) than in (iii), and hence $\lambda_2 < \lambda_3$. The performance in (ii) is therefore intermediate between (i) and (iii). We note that λ_2 and λ_3 must be below the critical value, since the rats are not yet completely incapable of finding the correct goal box.

Variable Threshold as Model for Selective Attention, (De)sensitization, and Anesthesia

In conventional associative NN models (4, 10), the threshold for a neuron to fire an action potential is fixed ($\theta_i = 0$), and stored patterns are created randomly. As a result, there are equal amounts of "on" bits and "off" bits in each pattern stored. This type of pattern is called unbiased and has an activity level, which is defined as the fraction of "on" bits, $r = 50\%$ (or 0.5).

In this section we present a brief review on our studies (16) on the effects of a variable threshold on convergence rates and fixed points of an associative neural network model in the presence of noise. We allow a random distribution in the activity levels of the patterns stored. A modification to the standard Hebbian learning rule given by Eq. (8) is proposed (16) for storage of patterns with an activity distribution. Dynamical formulation with arbitrary threshold values can be derived (16). For simplicity, we concentrate on the case where thresholds do not depend on individual neuron, that is, $\theta_i = \Theta$, $i = 1, ..., N$. The result (16) is

$$m(t + 1) = r\, \mathrm{erf}\left[\frac{m(t) - \theta}{(2)^{1/2}\sigma}\right] + (1 - r)\, \mathrm{erf}\left[\frac{m(t) + \theta}{(2)^{1/2}\sigma}\right] \qquad (12)$$

where $\theta \equiv \Theta/C$.

For patterns with a given activity level r, we find an optimal threshold as a function of the noise level and the time,

$$\theta_{\max} \equiv \frac{\sigma^2}{2m(t)}\, \ln\left(\frac{1}{r} - 1\right) \qquad (13)$$

at which the retrieval ability, for example, the average final overlap and the convergence rate, of the network for these patterns are the greatest (see Figs. 3 and 4). While the neuronal threshold is set to its optimal value corresponding to patterns with one activity level, we show that performances of the network with respect to patterns with different activity levels are significantly reduced. Hence the network can achieve selective attention by choosing threshold values. The effects of a constant (noise, time, and pattern independent) threshold can also be discussed through Eq. (12). For high- (low-) activity patterns, the average final overlap is shown to be increased at high noise levels and decreased at low noise levels by a negative (positive) threshold, whereas a positive (negative) threshold always reduces the final average overlap. When the magnitude of the threshold exceeds the certain critical value, there is no retrieval.

These results are related to sensitization, desensitization, and anesthesia; to establish the relation we discuss the evidence of a variable neuronal threshold in neurophysiological systems and interesting phenomena caused by variations in thresholds. We begin by first relating the present model to neurophysiological systems (28).

In a quiescent real neuron, differences in the permeabilities and the initial concentrations of sodium and potassium ions (both extra- and intracellular) cause an electrical potential difference across the membrane, which is called the resting potential V_r and ranges normally from -95 to -75 mV, the negative sign meaning that the potential inside the membrane is lower than that outside the membrane. If this potential difference is made less negative (depolarized), for example, by applying an external electric field, to a firing potential V_f, a voltage impulse (or action potential) is generated. Hence the

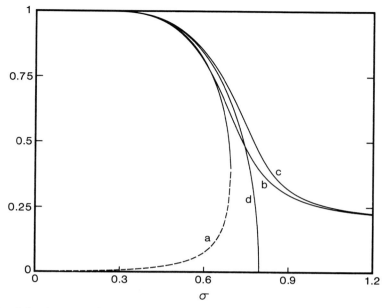

FIG. 3   Selective attention and optimal threshold: final overlap. Plot of final overlap $m(t = \infty)$ between the state of the network and a stored pattern, according to Eq. (12), with activity level $r = 0.6$, versus the standard deviation of the Gaussian noise σ, while the network focuses its attention on stored patterns with activity level r' by setting the neuronal threshold to $\theta = \theta_{max}(r')$ [Eq. (13)]: (a) $\theta = \theta_{max}(r' = 0.4)$, (b) $\theta = \theta_{max}(r' = 0.8)$, (c) $\theta = \theta_{max}(r' = r = 0.6)$. For comparison, we have also included curve (d), in which the neuronal threshold $\theta = 0$ and activity level r is arbitrary.

threshold signal for a neuron to fire is $\Theta = V_f - V_r$. We emphasize the difference between the definitions given in the present chapter for the firing potential V_f and the threshold θ, since V_f is sometimes referred to as "firing threshold." Hence the neuronal threshold can be varied by changing either the resting potential or the firing potential, or both, which can in turn be controlled by a number of variables, for example, permeabilities and concentrations of key ions such as K^+, Na^+, and Ca^{2+}, temperature, and other chemicals. In fact, any condition that favors the potassium (sodium) transport across the membrane, rather than the sodium (potassium) transport, tends to raise (reduce) the neuronal threshold (28).

For example, the drug veratrine increases the membrane permeability to sodium and can significantly increase neuronal excitability. A lower-than-average calcium ion concentration in the extracellular fluids has the same

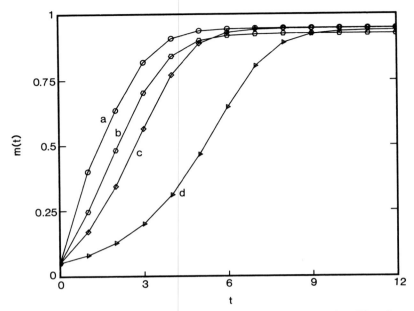

FIG. 4   Selective attention and optimal threshold: temporal dynamics. Plot of overlap $m(t)$ given by Eq. (12) versus time t for various choices of the neuronal threshold: (a) $\theta = \theta_{max}(r = 0.7)$, (b) $\theta = -0.3$, (c) $\theta = -0.15$, and (d) $\theta = 0$. The stored pattern to which the network converges has an activity level $r = 0.7$. The standard deviation of the Gaussian noise is $\sigma = 0.5$, and the initial condition is $m(t = 0) = 0.05$. We compare the average final overlaps as follows: $m(t = \infty; \theta = -0.3) < m(t = \infty; \theta = 0) < m(t = \infty; \theta = -0.15) < m(t = \infty; \theta = \theta_{max})$.

effects on a neuron. For patients who have lost their parathyroid glands and who therefore cannot maintain normal calcium ion concentrations, this condition may cause spontaneous respiratory muscle spasm and can be fatal. On the other hand, a high calcium concentration increases the neuronal threshold. Other "stabilizers," for example, procaine, cocaine, and tetracaine, may be used as local anesthetics. When the concentrations of these drugs are high enough, the neuronal threshold can be raised so high that the maximum signal received by a neuron is always below its firing threshold and the neuron thus never responds to signals of action potentials from other neurons, which stops other neurons from affecting the anesthetized neurons and vice versa. Changes in neural excitability can also be casued by damage to sensory fibers, which produces reduced (increased) sensation and is also called hype(re)sthesia (29). An overall change (either increase or decrease) in neuronal threshold, for example, caused by special chemicals and physical injuries discussed above, is related to the constant threshold discussed in this section.

Cognitive factors, such as the attentional state of a higher animal or the significance of an event, can alter the excitability of sensory neurons. For instance, athletes and soldiers frequently fail to notice painful injuries until after the game or battle. Human subjects asked to focus their attention elsewhere while receiving a painful stimulus rate the pain as less than that experienced when they are allowed to attend to the pain (30). Thermally sensitive neurons in a monkey (31) are much more excitable when the monkey is encouraged to attend to the thermal stimulus in comparison with situations in which the monkey is encouraged to attend elsewhere, such as a visual cue. This "task-related" modulation in neuronal excitability is closely analogous to our optimal neuronal threshold, which is pattern dependent.

We have incorporated the concepts of (de)sensitization and anesthesia, which are usually confined to sensory neurons (28), into associative neural systems. We have shown that an associative network can achieve such higher function as selective attention by purposefully changing its neuronal threshold in a prescribed manner; (de)sensitization in an associative neural network causes a variety of interesting effects on stationary points and dynamical properties (such as relaxation time) of the system.

The excitability of a neuron may also change according to its dynamics. For example, a neuron becomes considerably less excitable if the neuron is repetitively stimulated below its firing potential V_f. A slow depolarization of the membrane has the same effect. This phenomenon is called accommodation. The opposite phenomenon is potentiation (28): a neuron becomes more excitable if the neuron is repetitively stimulated above the firing potential V_f. This type of threshold change is in the spirit of hysteresis in a single neuron proposed recently by Hoffmann (8) and studied in detail in associative neural network models by the authors (13); this is the subject of the next section.

Effects of Hysteresis at Single Neuron Level

Many observations suggest that some neurons have higher excitabilities when they have just actively fired [see references cited by Wang and Ross (13)]. This translates to a hysteresis in the neural response function [compare Fig. 5A, which is the conventional response function given by Eq. (1), with Fig. 5B, one with hysteresis]. Other evidence has shown nonlinear (multiplicative) neuronal interactions (7, 28). We discuss effects of these two additional features of hysteresis and nonlinear interaction in an associative NN.

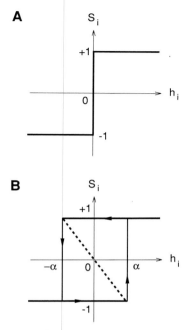

FIG. 5   Input–output response function for (A) the conventional model without hysteresis and (B) the present model with hysteresis.

Instead of the first-order neural interaction described by Eq. (2), we introduce second-order interaction:

$$h_i^{\circ}(t) = \gamma_1 \sum_{j=1}^{N} T_{ij} S_j(t) + \gamma_2 \sum_{j,k=1}^{N} T_{ijk} S_j(t) S_k(t) \tag{14}$$

where γ_1 and γ_2 measure the relative strengths of the first-order and second-order interactions and

$$T_{ijk} = C_{ijk} \sum_{\mu=1}^{p} S_i^{\mu} S_j^{\mu} S_k^{\mu} \tag{15}$$

and C_{ijk} is the dilution parameter for the second-order interaction (15), which takes the value 1 with probability $2/N^2$ and the value 0 with probability $1 - 2/N^2$. Without dilution ($C_{ijk} = 1$ for all ijk), Eq. (15) was first introduced by Peretto (7). Only C out of $N^2/2$ second-order synapses remain after dilution. Using the response function shown in Fig. 5B, we find the dynamical equation (13):

$$
\begin{aligned}
m(t + 1) = & \\
& \left[\frac{1 + m(t)}{2}\right] \text{erf}\left\{\frac{\gamma_1 m(t) + \gamma_2[m(t)]^2 + (\alpha/C)}{(2)^{1/2}\sigma}\right\} \qquad (16) \\
& + \left[\frac{1 - m(t)}{2}\right] \text{erf}\left\{\frac{\gamma_1 m(t) + \gamma_2[m(t)]^2 + (\alpha/C)}{(2)^{1/2}\sigma}\right\}
\end{aligned}
$$

where α is the half-width of the bistable region in the hysteresis and the noise level now has to include the second-order effects, that is, $\sigma \equiv [\gamma_1^2 + \gamma_2^2)(p - 1)/C + (\sigma_0/C)^2]^{1/2}$.

By analyzing Eq. (16) for various widths of the bistable region at various noise levels, we find that the overall retrieval ability in the presence of noise and the memory capacity of the network in the present model are better than in conventional models without hysteresis (see Fig. 6). The convergence rate is increased by the hysteresis at high noise levels but is reduced by the hysteresis at low noise levels (13). Explicit formulas are derived for calculations of average final convergence and noise threshold as functions of the width of the bistable region (13).

Oscillations, Chaos, Crisis, and Multiplicity in Neural Networks

Oscillations and chaos have been the subject of extensive studies in many chemical, physical, and biological systems. The Belousov–Zhabotinsky reaction (32), Rayleigh–Benard convection (33), and glycolysis (34) are well-known examples. Oscillatory phenomena frequently occur in living systems; some of these are a result of rhythmic excitations of the corresponding neural systems, such as locomotion, respiration, and heartbeat. Oscillatory and chaotic dynamics have been reported in electroencephalography (EEG) and have been discussed in relation to brain functions (35). It is interesting to see that oscillations and chaos are also found in the present class of NN models (15).

In the previous sections, we see that the network dynamics are nonoscillatory at the macroscopic level, that is, as far as the overlap is concerned, this is independent of the updating algorithms, such as parallel or sequential. In some particular regions of the parameter space, the dynamics can be oscillatory if neurons are updated synchronously (15). For instance, for $\alpha = 0$ (no hysteresis), $\gamma_1 = 1$, and $\gamma_2 = -1$, Figure 7 shows the fixed points of Eq.

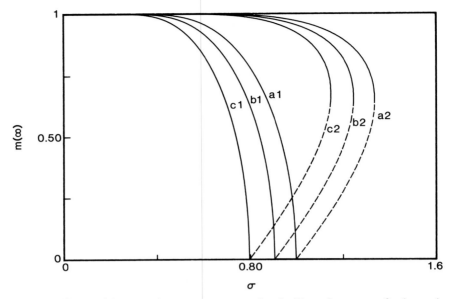

FIG. 6   Effects of hysteresis on memory retrieval. Plot of average final overlap $m(t = \infty)$ given by Eq. (19) versus standard deviation of the Gaussian noise σ for various widths α of the bistable region shown in Fig. 5B. (a) $\alpha/C = 0.3$, (b) $\alpha/C = 0.15$, and (c) $\alpha/C = 0$ (the conventional curve). The numerals denote (1) the first-order ($\gamma_1 = 1, \gamma_2 = 0$) and (2) the second-order ($\gamma_1 = \gamma_2 = 1$) curves, respectively. The dashed portions in the second-order curves are unstable.

(16) [results after many iterations from an initial condition $m(0) = 0.1$]. A larger variety of dynamical behaviors, including crisis and multiplicity, may be found in Refs. 15 and 17. We have shown that oscillatory amplitudes decrease when the synchronicity (number of neurons updated together at a given time step) decreases and vanish when updating is entirely asynchronous (17).

Our results may be applied to temporal sequence storage, which has been studied by means of delays (9). Cyclic and chaotic sequences may be stored in and recalled from the present system by specifying the bifurcation parameters and the initial conditions. Chaos has been discussed in connection with perception processes of the brain by a number of authors, and it has been claimed that the brain makes chaos in order to make sense of the world (36). The present analytical model may add some intuition to these discussions. Questions have also been raised as to whether the brain uses stable fixed points or oscillatory/chaotic attractors to store memory and to achieve per-

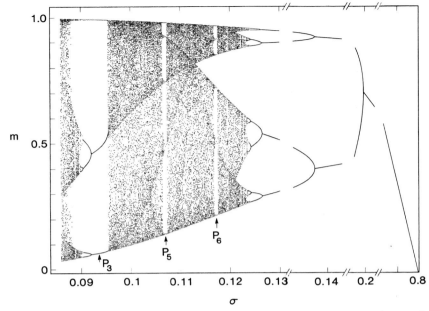

FIG. 7   Oscillations and chaos: fixed points according to Eq. (19) versus the standard deviation of the Gaussian noise σ, with $\gamma_1 = 1$, $\gamma_2 = -1$, and $\alpha = 0$.

ception and information processing. Recent discovery of the so-called 40-Hz oscillations in the cat (37) seem to support the latter.

Summary and Outlook

We have discussed the incorporation of several important neurobiological aspects such as interaction of neural networks, variable firing thresholds, hysteresis at a single neuron level, and diluted higher order synaptic interactions in a series of studies on associative artificial neural network models in the presence of noise. In light of these models, we are able to discuss a variety of interesting phenomena such as distraction, concentration, selective attention, (de)sensitization, anesthesia, noise resistance, oscillation, and chaos. Because of the extreme complexity of real biological neurons and their network systems, there are many more features, both at a single neuron level and a network level, yet to be explored. With systematic studies of physical models such as ones presented in this chapter coupled closely with experiments, we may hope to determine the pertinent biological features responsible for essential information transmission and storage.

Acknowledgments

This work was supported in part by the National Science Foundation.

References

1. For a recent review on NNs, see, for example, H. Sompolinsky, *Phys. Today* **41,** 70 (1988); for a recent book, see J. Hertz, A. Krogh, and R. Palmer, "Introduction to the Theory of Neural Computation." Addison-Wesley, Reading, Massachusetts, 1991.
2. W. A. Little, *Math. Biosci.* **19,** 101 (1974).
3. G. L. Shaw and R. Vasudevan, *Math. Biosci.* **21,** 207 (1974).
4. J. J. Hopfield, *Proc. Natl. Acad. Sci. U.S.A.* **79,** 2554 (1982).
5. D. J. Amit, H. Gutfreund, and H. Sompolinsky, *Phys. Rev. Lett.* **55,** 1530 (1985); D. J. Amit, H. Gutfreund, and H. Sompolinsky, *Phys. Rev. A* **32,** 1007 (1985); D. J. Amit, H. Gutfreund, and H. Sompolinsky, *Ann. Phys.* **173,** 30 (1987).
6. P. Peretto, *Biol. Cybernet.* **50,** 51 (1984).
7. P. Peretto and J. J. Niez, *Biol. Cybernet.* **54,** 53 (1986).
8. G. W. Hoffmann, *J. Theor. Biol.* **122,** 33 (1986).
9. H. Sompolinsky and I. Kanter, *Phys. Rev. Lett.* **57,** 2861 (1986); D. Kleinfeld, *Proc. Natl. Acad. Sci. U.S.A.* **83,** 9469 (1986).
10. B. Derrida, E. Gardner, and A. Zippelius, *Europhys. Lett.* **4,** 167 (1987).
11. A. J. Noest, *in* "Neural Information Processing Systems" (D. Z. Anderson, ed.), p. 584. American Institute of Physics, New York, 1988.
12. J. D. Keeler, E. E. Elgar, and J. Ross, *Proc. Natl. Acad. Sci. U.S.A.* **86,** 1712 (1989).
13. L. Wang and J. Ross, *Proc. Natl. Acad. Sci. U.S.A.* **87,** 988 (1990).
14. L. Wang and J. Ross, *Proc. Natl. Acad. Sci. U.S.A.* **87,** 7110 (1990).
15. L. Wang, E. E. Elgar, and J. Ross, *Proc. Natl. Acad. Sci. U.S.A.* **87,** 9467 (1990).
16. L. Wang and J. Ross, *Biol. Cybernet.* **64,** 231 (1991).
17. L. Wang and J. Ross, *Phys. Rev. A* **44,** R 2259 (1991).
18. B. Katz and R. Miledi, *J. Physiol. (London)* **192,** 407 (1967).
19. D. O. Hebb, "The Organization of Behavior," p. 44. Wiley, New York, 1949.
20. S. Kirkpatrick, C. D. Gelatt, Jr., and M. P. Vecchi, *Science* **220,** 671 (1983).
21. S. L. Foote, F. E. Bloom, and G. Aston-Jones, *Physiol. Rev.* **63,** 844 (1983).
22. M. Eigen, *Naturwissenschaften* **58,** 465 (1971); I. Leuthäusser, *J. Chem. Phys.* **84,** 1884 (1964).
23. E. E. Elgar, J. D. Keeler, and J. Ross, *Complex Syst.* **4,** 75 (1990).
24. V. Neufeldt and D. B. Guralnik (ed.), "Webster's New World Dictionary." Webster's New World, New York, 1988.

25. R. A. Anderson, *in* "Higher Functions of the Brain" (V. B. Mountcastle, ed.), Part 2, p. 483. American Physiological Society, Bethesda, Maryland, 1987.
26. R. W. Sperry, *Br. J. Anim. Behav.* **3,** 41 (1955).
27. G. J. Smith and N. E. Spear, *Science* **202,** 327 (1978); G. J. Smith and E. R. Croft, *Behav. Neural. Biol.* **43,** 250 (1985).
28. R. H. S. Carpenter, "Neurophysiology." Univ. Park Press, Baltimore, Maryland, 1984.
29. W. D. Willis, Jr., and R. G. Grossman, "Medical Neurobiology," 2nd Ed., p. 410. Mosby, St. Louis, Missouri, 1977.
30. H. Leventhal, D. Brown, S. Shacham, and G. Engquist, *J. Personality Social Psychol.* **37,** 688 (1979).
31. M. C. Bushnell, G. H. Duncan, R. Dubner, and L. F. He, *J. Neurophysiol.* **52,** 170 (1984).
32. R. J. Field and M. Burger, "Oscillations and Traveling Waves in Chemical Systems." Wiley, New York, 1984.
33. R. P. Behringer, *Rev. Mod. Phys.* **57,** 657 (1985).
34. B. Hess and A. Boiteux, *Annu. Rev. Biochem.* **40,** 237 (1971).
35. E. Basar (ed.), "Chaos in Brain Function." Springer-Verlag, New York, 1990.
36. C. A. Skarda and W. J. Freeman, *Behav. Brain Sci.* **10,** 161 (1987).
37. R. Eckhorn, R. Buaer, W. Jordan, M. Brosch, W. Kruse, M. Munk, and R. J. Reitboeck, *Biol. Cybernet.* **600,** 121 (1988); C. M. Gray and W. Singer, *Proc. Natl. Acad. Sci. U.S.A.* **86,** 1698 (1989).

Index